T0296967

Pitt Press Series

HERODOTOS

VI

ERATO

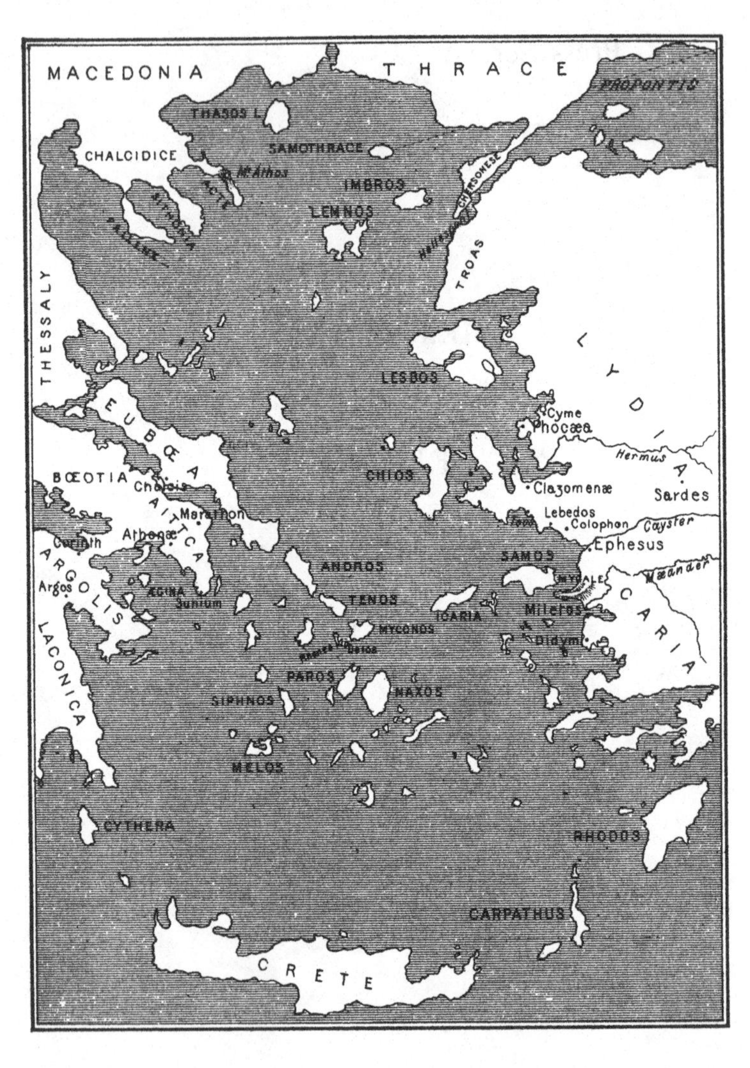

THE IONIAN CITIES AND THE ISLANDS OF THE AEGEAN

HERODOTOS

VI

ERATO

EDITED BY

E. S. SHUCKBURGH

LITT.D.

CAMBRIDGE UNIVERSITY PRESS

CAMBRIDGE

LONDON · NEW YORK · MELBOURNE

CAMBRIDGE UNIVERSITY PRESS
Cambridge, New York, Melbourne, Madrid, Cape Town, Singapore,
São Paulo, Delhi, Dubai, Tokyo

Cambridge University Press
The Edinburgh Building, Cambridge CB2 8RU, UK

Published in the United States of America by
Cambridge University Press, New York

www.cambridge.org
Information on this title: www.cambridge.org/9780521141161

First published 1889
Reprinted 1891, 1896, 1908, 1950, 1959, 1965, 1976
This digitally printed version 2010

A catalogue record for this publication is available from the British Library

ISBN 978-0-521-05248-1 Hardback
ISBN 978-0-521-14116-1 Paperback

PREFACE.

THIS Edition of the Sixth Book of Herodotos has been drawn up at the request of the Syndics of the University Press on the same lines as those of the Eighth and Ninth Books already published. All information which I thought the reader would require as to persons and places mentioned has been put together in the Historical and Geographical Index. The Index to the explanatory Notes has been formed with the special design of enabling the reader to find easily all examples of exceptional usage, either in regard to words or construction, which occur in the book, as well as those which are noteworthy without being exceptional or irregular. The Introduction contains a sketch of the previous history down to the point at which the Sixth Book begins; and an attempt to separate the various subjects treated of in the Book itself and to assign them to their right place in the general plan of the work of Herodotos. The text has been carefully revised by the help of the conspectus of MS. readings given by Stein in his earlier critical edition, and a few of the more important variations are appended to the Introduction.

The books consulted have been many; but as before my chief obligations are to Bähr, Rawlinson, Stein and Abicht.

CAMBRIDGE,
January, 1889.

TABLE OF CONTENTS.

INTRODUCTION.

THE main object of the History of Herodotos is to picture the struggle of East and West. Among endless digressions this theme is never out of sight. It was necessary to describe the growth and character of the nations between whom the struggle was to take place; the difficulties they had had to encounter; the enemies they had had to subdue; the customs and characters of the ancestors from whom they sprang, and which helped to account for their existing peculiarities. Herodotos was, besides, a traveller and a man of insatiable curiosity, delighting in stories which were either picturesque in themselves, or served to illustrate the character of a nation, or of the individuals who had to play parts in the great drama which he had undertaken to compose; and of which the decisive incidents were to be enacted at Marathon, Artemisium, Thermopylae, Salamis, Plataea and Mykale. He had no scruple in interrupting the course of his narrative for the sake of a story, however remotely connected with his subject; and we must try therefore to trace the thread of his history through these divergencies, and see how he is always leading to his main point.

The main object of the History of Herodotos.

Remembering, then, that the object of the previous books had been to bring the combatants face to face before an audience made fully acquainted with their characters and previous history, we may first ascertain at what point in his narrative the Sixth Book opens, and how far it carries us towards the final catastrophe.

Progress made in the story in previous books.

In the previous books he has told us of the rise of the Persian Monarchy under Kyros, and how his conquest of the Lydian Monarchy brought the Persians a supremacy over the Asiatic Hellenes. His successor Cambyses was mostly employed in subduing Egypt. But Darius (522—485) was the great organiser of the vast empire of which he obtained possession. He divided it out into twenty Satrapies, each with a Governor or Satrap, and each paying a fixed tribute to the Royal Exchequer, amounting altogether to a sum equivalent to nearly £4,000,000 of our money yearly [3, 90—4]. The Satrapy which concerns us principally was the first, including 'the Ionians, the Magnesians of Asia, the Aeolians, the Karians, the Milyans, and the Pamphylians' [3, 90]. It was this element in the Persian empire which was destined to bring the forces of the East and West into decisive contest. For the Ionians, though living in Asia, were Hellenes, and as such had the sympathy, though not it appears the high respect[1], of their European kinsfolk. And Darius, early in his reign, seems to have turned his attention to the policy of extending his supremacy to the European Hellenes; and had gone so far as to send commissioners to investigate Greek affairs and report to him [3, 134—6]. His attention, however, had been mean-

[1] 1, 143 οἱ μὲν νυν ἄλλοι Ἴωνες καὶ οἱ Ἀθηναῖοι ἔφυγον τὸ οὔνομα οὐ βουλόμενοι Ἴωνες κεκλῆσθαι· ἀλλὰ καὶ νῦν φαίνονταί μοι οἱ πολλοὶ αὐτῶν ἐπαισχύνεσθαι τῷ οὐνόματι. Cp. 5, 66, 69.

while turned to another quarter. The Skythians had about B.C. 625 invaded Media, and remained in the country for nearly 30 years. The tradition of this invasion of northern Barbarians over the Caucasus seems to have suggested to Darius—anxious for military prestige and incited, it was said, by his wife Atossa—the desirability of subduing these wild tribes and securing his frontiers from any farther incursion. His plan was to cross the Bosporus into Thrace near Byzantium and advance to the Danube; thence to march through the land to the Caucasus, and then to pass back to his own country. But the dangers were great and unknown, and it was important for him to have the bridge of boats constructed over the Danube safely guarded, that he might keep a way of retreat open in case of need. The tyrants of the various Greek cities, Ionian and others[1], who depended for their position at home on the support of the Persian Court, and had accompanied him on his expedition, were left in charge of the bridge, with orders to wait sixty days, and then break it up and retire. The sixty days passed. No news of Darius reached the Greeks who were guarding the bridge. Still they waited without breaking it up; when suddenly a band of Skythian horsemen appeared on the north bank, and announced to them that Darius was in full retreat, and that they could easily cut him off, if the Greeks would only break the bridge. The Greek tyrants held anxious consultation. It

Darius invades Skythia.

The Greek tyrants at the bridge over the Danube.

[1] For a list of them see Herod. 4, 138, but there were probably more than he mentions there. For instance, Koes, though not a tyrannus at the time, received as a reward afterwards the rule of Mytilene for his services at the bridge, and therefore his position as general of the Mytileneans must have been sufficiently absolute to give him a voice among the other tyranni [5, 11].

was a great opportunity of striking a blow at the oppressor of the Asiatic Greeks; and perhaps of freeing their cities altogether from the yoke of the barbarian. Miltiades, the future hero of Marathon, urged that it should be done. But other counsels prevailed. The tyrants were reminded that if the cities no longer felt the restraint of the Persian Court, their first step would be to expel their absolute rulers and establish democracy. This argument sufficed to deter them from following the advice of Miltiades. They indeed broke a part of the bridge near the north bank, in order to delude the Skythians; but, as soon as they had departed to intercept Darius, this was easily repaired. And when the king, having baffled the intercepting force, arrived on the Danube with his discomfited army[1], the boats were swung round into their place again, and he crossed in safety [4, 141].

It is necessary to note this expedition [circ. B.C. 513—9[2]], because the action of these Greek tyrants, and the division of opinion on what it was right to do, which was doubtless immediately communicated to Darius, increased his desire to extend his supremacy over the European Hellenes[3]; and

[1] His adventures in Skythia are detailed by Herodotos in 4, 98—141. Grote believes them to be 'a great illustrative fiction.' The strong points on the other side are put by Rawlinson, vol. 3, p. 95. Particulars of such a campaign doubtless could not be satisfactorily obtained; but the great Historian of Greece was perhaps less able to judge of the probabilities of such a history than those who have been personally acquainted with the East and its ways.

[2] Grote puts it as early as B.C. 515: others as late as 508 B.C. The truth is hard to arrive at, and depends much on the interpretation of μετὰ οὐ πολλὸν χρόνον in 5, 28.

[3] When Darius returned to Sardis, he left Megabazos with 60,000 men to conquer the Greek towns in Thrace. This he easily did, and was then ordered to enter Makedonia, when he transported the Paeonians to Phrygia, and received earth and water from king Amyntas [5, 1—10].

also led in an unexpected way to the Ionian revolt, which finally brought him into direct collision with Greece.

The Sixth Book of Herodotos opens with the account of the suppression of the Ionian revolt [cc. 1—40]. We must therefore try to learn something about the Ionians and see who they were, how they came to be subjected to Darius, and why they revolted.

The Sixth book.

'Now the Ionians of Asia, who meet at the Panionium, have built cities in a region where the air and climate are the most beautiful in the whole world: for no other region is equally blessed with Ionia, either above it or below it, or east or west of it[1].' Thus Herodotos speaks of a country which he knew well, and near which he was himself born. The traveller Pausanias also speaks of the unrivalled climate of Ionia and the splendour of its temples[2]. But in spite of these advantages, 'Ionia,' we are told, 'was always in a dangerous state, and property was continually changing hands[3].' This was caused partly by the inherent defect in Greek politics, the inability to combine for national existence, and the exaggerated love of local autonomy; and partly by the fact that the Greeks living in this country were constantly the prey of powerful neighbours. The Ionians formed a community of twelve states, which signalized their ties of blood by a yearly meeting at the Panionium, a temple of Poseidon on the promontory of Mykale. This, like the meetings of other Amphiktyonies, was religious and had nothing in common with a political league. Yet, like other Amphiktyonies also, at times of great danger threatening the community, we

The Ionians.

[1] Herod. 1, 142. [2] Pausan. 7, 5, 2. [3] Herod. 6, 86.

find the members acting for a time in concert[1]. But such concert was short-lived and precarious, and liable to dissolve at the first touch of difficulty or hardship.

The twelve states which formed this Union were *Miletos*, *Myus*, *Priene*, all in Karia and using the same dialect: *Ephesos*, *Kolophon*, *Lebedos*, *Teos*, *Klazomenae*, *Phokaea* in Lydia, and using a dialect differing from the former group: two islands *Samos* and *Chios*, and one town in Lydia, *Erythrae*, using again a different dialect. The difference in dialect probably arose from the mixture in a different degree with the Karian and Lydian inhabitants when the Ionian settlers arrived. In some cases these people would be driven out or killed; in others large numbers would be admitted to live side by side with the new comers. Ephesos and Miletos were the richest and most important of these towns, the latter especially showing its power and populousness by the great number of colonies which it sent out: and Ephesos was so important an element in the Union, that the Panionian festival some time before the time of Thucydides (4, 104) came to be called the Ephesia. Though never powerful, in the ordinary sense of the word, these Ionians had been an adventurous, busy, and thriving mercantile people. Their colonists had fringed the coasts of the Propontis and Euxine with Hellenic towns; their seamen had made their way to Italy and Spain. To the Asiatic or Island Greeks, Ionian and Aeolian, belong nearly all the great names in literature between the age of Homer, and the outburst of Attic literature in the 5th century B.C. Thus Kallinos (690), probably Tyr-

Early importance of Asiatic Greeks:

literary,

[1] Thus in 5, 109 τὸ κοινὸν τῶν Ἰώνων is spoken of as a political body capable of corporate action.

taios (605), Archilochos (670), Simonides of Amorgos (660), Mimnermos (620), Phokylides (540), Xenophanes the poetical philosopher (510), Simonides of Keos (480), Terpander (670), Alkaios (611), Anakreon (530), may all be claimed by various sections of this branch of the Hellenes. Their influence and dialect embraced Dorian towns, such as Halicarnassus, the birthplace of Herodotos, and he could compose a history intended for the whole Hellenic world in the Ionic dialect, as the literary language of the day [Cauer *Gr. Inscrip.* p. 322].

political.

The Asiatic Greeks had also led the way in a political movement which was reproduced in other parts of Greece. Some time after B.C. 750 monarchy began to give way to oligarchy in the Greek towns of Asia and the Islands; that again to tyrannies; and on the expulsion of the tyrants democracies were established in most of the states. It seemed as though the centre of Hellenic life was to be Asiatic and Insular rather than European. But these prospects were overclouded by the rise of the great Lydian kingdom under Alyattes (625—560) and Kroesos (560—546), who gradually reduced the Greek towns in Asia. The Ionian towns like the others were made tributary,—the first-fruits of the jealous and separatist policy which caused them afterwards to reject the advice of Thales to found on the Panionian assembly a common council and government. The fall of Kroesos before the victorious arms of Kyros in B.C. 546 did not set the Greeks free. It simply transferred their allegiance to a new master in the person of the Great King, although they offered a feeble and spasmodic resistance for a time. The chief political effect it had upon them, besides probably an extension of tribute, was from the Persian policy, which was to depress democracy and insist upon the government

Their subjugation by the Lydians

and by the Persians.

of a tyrannus dependent for his position on the support of the Satrap.

One member of the Ionic body indeed, if it had not wholly retained its independence, had yet risen to a high pitch of power. During a reign of over twenty years Polykrates of Samos had collected a larger navy than any other state in the Aegean; had made an alliance with Amasis of Egypt; had conquered neighbouring islands and towns on the mainland, and had conceived the idea of a Panionian empire. Before Cambyses' invasion of Egypt (B.C. 525) Polykrates had become detached from his alliance with Amasis and furnished, like other Ionian states, a contingent to the king's army. But his power was a standing menace to the Persian ascendency over the Greeks, and the Satrap Oroetes entrapped him into his dominions, by a feigned offer of vast treasures which would enable him to make himself master of all Greece, and there put him to death (B.C. 522). The disputes as to the succession to the throne of Samos, which followed the death of Polykrates, resulted in its more complete subjection to the Persian Government, and in the establishment of Syloson, the banished brother of the late tyrant, in the tyranny, which he was content to hold as an acknowledged tributary of the Great King[1]. The treatment of Samos is instructive as to the line of policy followed by the Persian Court towards Hellenic states, which were allowed their own dissensions and revolutions with the certainty that eventually a leader of one or the other party would appeal to the Persians for help, and give the Satrap an excuse for interfering. The subjection of Samos in B.C. 521 may be said to have marked the final submission of Ionia to Persia. The authority of the Great

Samos under Polykrates. 535—522.

Grievances of the states.

[1] Herodot. 3, 39—47; 54—6.

King continued to be exercised over it for the next 20 years without any open outbreak. The towns were left with ostensible independence except in two respects; (1) they had to pay their quota of tribute, and (2) in most of them the constitution was placed in abeyance by the usurpation of some tyrannus, nominated or connived at by the Persian Court, and supported by its influence and arms when he had gained power. But these two exceptions really made the ostensible autonomy almost valueless. The tribute might be borne; but the tyrant was a standing grievance, more especially as he was the creature of a foreign Satrap, whose behests he was obliged to obey. The amount of personal inconvenience caused varied no doubt with the character of the particular tyrant, and of the Satrap. But at best it was subjection to a power other than that of the law of the state; and under such subjection the cities were always restive. We must see how a chain of accidental circumstances provoked this smouldering discontent into a blaze.

Origin of the Ionian revolt. B.C. 510—501.

Among the consequences of the Skythian expedition of Darius was the promotion of Histiaios of Miletos. When Darius got safe back to Sardis, he sent for Histiaios and Koes of Mytilene, and bade them name their reward for their services at the bridge over the Danube. Koes asked to be made despot of Mytilene. Histiaios asked for the grant of Myrkinos with a surrounding territory in the country of the Thrakian Edonians, near the river Strymon. Both requests were granted [5, 11]. But before Megabazos left Thrace, he began to suspect the loyalty of Histiaios. He was fortifying Myrkinos, which was in an admirable position, in the midst of an extensive plain, well supplied with timber for ship-building, and rich with silver-

Histiaios.

mines. Near it, or perhaps on its very site, afterwards rose the town of Amphipolis, which was long an object of contention between Athenians, Spartans, and Makedonians. Informed of this, Darius, ever jealous of his vassals rising to independent power, summoned Histiaios to Sardis, under the complimentary pretext of requiring his advice. Then he requested him to give up both Miletos and Myrkinos to deputy-governors and accompany the court to Susa. Histiaios who at first had been charmed with the compliment, soon discovered that he was practically a state prisoner[1]. But he was obliged to conceal his chagrin, and submit to the will of the king.

Darius returned to Susa about B.C. 506, leaving his half-brother Artaphernes in command at Sardis: and for a few years the Greeks of Asia seem to have been unmolested, and to have enjoyed a brief period of quiet and prosperity[2].

Seeds of mischief. Hippias. But there were not wanting indications of coming trouble. The Athenian despot Hippias was expelled in B.C. 510, and retiring to Asia, began after a time intriguing with Artaphernes to secure his restoration. To counteract these intrigues the Athenians sent envoys to Sardis; who, however, could get no other answer than a peremptory command, 'if they valued their safety to restore Hippias' [5, 96]. Herodotos affirms that after the expulsion of the Peisistratidae the Athenian power rapidly rose [5, 66]; the people were in no mood to brook such

[1] Herod. 5, 24—25. It was a common device of eastern Sovereigns in regard to subjects of whom they had reason to stand in fear. See 3, 132. Xenoph. *An.* 1, 8, 25. 2 Sam. 9, 7; 19, 33. 1 Kings 2, 7.

[2] See Herod. 5, 28 and Mr Grote's note at the beginning of c. xxxv. If the Skythian expedition was as early as B.C. 515 this date would have to be pushed back also. But I cannot accept Grote's reading of μετὰ δὲ οὐ πολλὸν χρόνον in 5, 28.

dictation, and determined that henceforward they would be at open enmity with the Persians. In these circumstances only an opportunity was needed to make the breaking out of hostilities certain. This was not many years in coming.

Naxos. B.C. 502.

The island of Naxos was the largest and most powerful of the Cyclades. It had not long before been under the tyrant Lygdamis, the friend of Peisistratos [1, 64]: it had then become oligarchical [5, 30]; and about this time a popular rising had driven out the oligarchs, some of whom took refuge in Miletos. The government of Miletos had been entrusted by Histiaios to his son-in-law Aristagoras. The exiles were guest-friends of Histiaios; and Aristagoras was willing to undertake their cause, not indeed with his own forces, but by obtaining aid from Artaphernes.

The Persian Satrap was induced by the promises of Aristagoras to furnish a fleet of 200 ships under the command of Megabates. Not only was the wealth of the island represented to him as great; but the restoration of rulers in Naxos under Persian influence would be a step towards extending the supremacy over the Cyclades, which as yet were free [5, 30]. But the two commanders Aristagoras and Megabates soon quarrelled. And in his anger Megabates gave the Naxians secret information of the purpose for which the fleet was assembled. Hitherto it had been given out that it was bound for the Hellespont; and the idea had been to take the Naxians by surprise. But now, when the fleet arrived at Naxos, every preparation had been made to stand a siege; and after an ineffectual blockade of four months the Persian fleet retired baffled to Asia [5, 33—4]. Aristagoras had now to fear the vengeance of the Persian Satrap for having induced him to risk the ships and men of the king on an

Perplexity of Aristagoras.

expedition which had proved a failure. He had also promised to pay the expenses; but he had no means of doing so. In his perplexity and alarm his thoughts turned to the idea of a revolt. While he was revolving this plan, a slave arrived from his father-in-law Histiaios bearing only the message that he was to shave his head and look what was on it. On doing so he found branded upon the scalp the words 'raise Ionia' ['Ιωνίαν ἀναστῆσον]. This message, the result of Histiaios' weariness of his gilded captivity at Susa, chimed in well with the necessities of Aristagoras. He at once held council with his friends, and they were unanimous for the revolt, with the exception of the historian Hekataios: who first urged the impossibility of resisting the king; and then, when he failed to convince them, urged that they should at least seize the treasures of the temple of Branchidae, which would supply them with means, and which in any case would be pillaged by the Persians. But he was not listened to in either respect. The conspirators were determined on the revolt; but they feared to shock Greek feeling by rifling a temple.

A message from Histiaios.

The immediate steps taken were, first, to seize the fleet which had been lately employed at Naxos; and next, by its help, to arrest the tyrants of the several towns, who had been under Persian protection. This done, and the several tyrants being allowed to go away by their subjects, except in the case of Koes at Mytilene who was stoned, democracy was proclaimed all through Ionia, Aristagoras setting the example by ostensibly resigning his despotic power in Miletos. Each state however chose a strategus to command their forces, who was also apparently a civil magistrate as well. At any rate Aristagoras seems under another name to have really retained his power at Miletos. So far the states had not technically

First step. The tyrants deposed.

revolted from Persia or refused tribute. All they had done was to alter their internal constitutions in a way which they knew would be unacceptable to the Persian Court. It was however well understood that they were in revolt, and both sides made preparations. Aristagoras went in the autumn of 501 B.C. to Sparta, as the head of Hellas, to apply for help, but was repulsed [5, 50]. He then went to Athens. The Athenians, as we have seen, were in a high state of prosperity, eager to play a conspicuous part in Hellenic politics, and already incensed with the Persians. Miletos moreover was believed to have special ties of blood and friendship with Athens; and the political movement in Ionia against tyrannies accorded with the interests of the Athenians, on whom the Persians had endeavoured to force back their own tyrant Hippias. In Athens therefore Aristagoras had a greater success. The people voted that 20 ships under the command of Melanthios should be sent to aid the Ionians [5, 96—7].

Both sides prepare for war.

Relying on this and other aids Aristagoras, on his return to Miletos, pushed on the revolt. He began by sending word to the Paeonians, whom Darius had settled in Phrygia, that they might return to their native land; which they managed to do in spite of being pursued by some Persian cavalry [5, 98]. This was an overt act of rebellion; and Artaphernes, though he still remained without moving at Sardis, sent for reinforcements from the main encampment of the Cis-Halysian army, and summoned the Phoenikian fleet to sail up the coast. His land forces apparently at once commenced besieging Miletos[1];

B.C. 500. *Active measures taken by Aristagoras.*

[1] This is the statement of Plutarch *de Malign. Herod.* c. 24, which helps to account for what seems strange, that Artaphernes should have taken no measures to stop this invasion: and that Aristagoras did not join it in person.

but the Phoenikians did not arrive in time to stop the combined fleet of Ionians, Athenians and Eretrians (who sent five ships) sailing to the territory of Ephesos and landing at Koressos. Under the command of Histiaios' brother Charophinos the men marched to Sardis. They easily entered it; and by accident, or the wilful mischief of some soldier, the town, consisting of thatched houses, was burnt. But the inhabitants collected in the agora, and offered so stout a resistance, that the Ionians and Athenians retired to Mt Tmolus, bivouacked there for the night, and then proceeded on their return march to Ephesos.

Little after all had been done. The affair was not more than a night's raid: and the retreating army was pursued by the Persian forces, which had been summoned to the rescue, and suffered severely. Moreover the citadel of Sardis, on an almost impregnable height, was intact. The Greeks had gained no permanent advantage; and had intensely irritated the Persian king, who on hearing of the burning of the town, solemnly vowed to be avenged on the Athenians and Eretrians[1]. The result, in short, was to make it a settled purpose of Darius to reduce European Greece to his obedience. The ships sent by Athens in fact were, as Herodotos says, a beginning of the mischief which arose between the Greeks and barbarians [5, 97].

B.C. 500—495. *Ionians continue to struggle, but everywhere unsuccessfully.*

But this failure of the expedition against Sardis did not end the revolt. Miletos still held out. The Ionian fleet went sailing from place to place, gaining over various cities to their cause. As far north as Byzantium, and as far south as Kypros, state after state broke out into rebellion: so that for five years the Persians were not only engaged in

[1] For the story of the vow, and the slave who was to remind him three times at dinner to be avenged on the Athenians, see 5, 105.

besieging Miletos, but had to turn their arms now in this direction and now in that. But the result on the whole was nearly everywhere in favour of the Persians. The Ionians had indeed gained an advantage in a sea-fight off Kypros, but they had been weakened by desertion, as had also the land forces in the island, and before a year was out the whole island was reduced again to obedience [5, 112—3]. On the mainland things had gone equally badly for the rebels. One Ionian city after another was subdued, as well as five of the Greek colonies on the Hellespont. In Karia, indeed, there was for a time some resistance maintained, the native Karians having made common cause with the Ionians. But a fiercely contested battle on the Maeander, in which the Greeks lost 10,000 men, was followed by another at Labranda, in the valley of the Marsyas, in which, though they were reinforced by some Milesians, they were beaten still more decisively [5, 119—120]. And though they partially repaired this disaster by cutting off a large Persian force near Pedasos [5, 121], their resistance had no effect in staying the stream of Persian successes in all parts of Asia: and at length Aristagoras, becoming thoroughly alarmed, determined to quit Miletos, though it had not yet fallen, and to take refuge at Myrkinos, his father-in-law's fortified town in Thrace. He committed his power at Miletos to Pythagoras, and sailing to Thrace seized Myrkinos, but soon fell in battle with the native Thrakians [5, 124—126].

Kypros.

B.C. 498—6. *Aristagoras quits Miletos and falls in Thrace.*

It is at this point that the Sixth Book of Herodotos opens.

Soon after the departure of Aristagoras, Histiaios arrived at Sardis, having persuaded the king to send him down to repress the revolt; and having promised to win Sardinia, the wealth of

Histiaios refused admittance into Miletos.

which was loudly rumoured, to his obedience. Alarmed, however, at discovering that Artaphernes was better informed than his master, he took refuge in Chios, and thence attempted to re-enter Miletos. But the Milesians had lately got rid of Aristagoras much to their satisfaction, and were in no mind to admit their old tyrant; who was consequently repulsed in a night-attempt upon the town, and had to retire wounded. He then appealed once more to the Chians for help, and, failing to obtain it, passed to Mytilene; where obtaining eight triremes he sailed to Byzantium, and supported himself by levying blackmail on the cornships [6, 1—5].

B.C. 497—495.

In his absence the Persian officers resolved upon a closer blockade and a more resolute attack on Miletos, which apparently had been more or less languidly invested throughout all this period.

Attempted combination of Ionian towns.

The Ionian cities, which still held out[1], now made some attempt at combination. And their deputies having resolved on concentrating all resistance on their ships, they mustered, 363 in number, off the Island of Lade. For a time the men submitted to the necessary discipline under a general, jointly elected, Dionysios of Phokaea. But this soon proved too much for them; and they quickly returned to their old habits, each squadron doing as it pleased, and the men spending most of their days and nights on shore [6, 7—8, 11—12].

Meanwhile the Persian fleet of 600 sail, consisting of Phoenikian and Kyprian vessels, had mustered for the attack upon Miletos, which was to be supported by a

[1] Miletos, Priene, Myus, Teos, Chios, Erythrae, Phokaea, Lesbos. Of Ephesos, Kolophon, Lebedos, Klazomenae, the three first held aloof, the last was occupied by the Persians [6, 8].

vigorous assault by the land troops. But first diplomacy was tried, and the Ionian states were addressed each by the mouth of their several expelled tyrants, with a promise of indemnity if they yielded, and a threat of the last severity if they were obdurate. None gave way except the Samians. The rest resolved to fight, though they had not the sense to submit to the necessary training [6, 9, 13].

The natural result followed. The Samian ships, except 11 whose captains refused to do so, deserted as soon as the battle began: and the rest, after a more or less vigorous resistance, were utterly defeated, and scattered in every direction [6, 14—17].

B.C. 495. *Battle of Lade.*

Their power at sea, on which they now almost entirely depended, being thus annihilated, the Ionians had nothing to look for but subjugation and vengeance. The battle of Lade was followed before many months by the fall of Miletos: and when that was complete the Ionian revolt was at an end [6, 18—25]. Histiaios felt himself no longer safe at Byzantium; and, after a series of vain endeavours to secure some place of safety for himself and the remains of his forces, was captured in the territory of Atarneus and promptly put to death [6, 26—30].

Miletos falls, B.C. 494.

The spring of the next year was spent by the Persians in securing the complete submission of the islands and cities, especially the cities on the Hellespont and the Pontus. In the course of which they seized the Thrakian Chersonese, Miltiades having abandoned it in alarm and escaped to Athens [6, 31—33].

B.C. 493.

After a digression on the history of the connexion of Miltiades with the Chersonese [6, 34—41], Herodotos goes on to tell us of the reorganization of Ionia attempted by Artaphernes: his establishment of courts of international arbitration, and his redivision for the purposes of tribute [6, 42].

B.C. 492. Mardonios sent down to the coast.

And now the first step was taken towards exacting that vengeance which Darius meditated against Athens and Eretria; which was to lead to the decisive struggle of European Hellene and Asiatic barbarian. Artaphernes and the other Persian officers were superseded by Mardonios, sent by the king on a double mission; to endeavour to conciliate the Ionians by abolishing their tyrants, and allowing them to enjoy their democratical constitutions; and, secondly, to extend the authority of the Great King round the northern shores of the Aegean, and thence to sail for the punishment of Athens and Eretria. The first part of this commission was fulfilled; but in attempting the latter part Mardonios lost his fleet in a storm, as he was rounding Mt. Athos, and more than twenty thousand men, while his land forces suffered severely in Thrace, though ultimately successful in forcing the natives to submit [6, 43—45].

B.C. 492—1. Darius demands earth and water.

No farther hostile movement was made by the Persians during that and the next year. One act of severity is recorded in regard to the Thasians, but it only extended to the depriving them of ships and fortifications. But Darius was pressing on his preparations for another expedition; and now took an important step, by which he effectually emphasised any latent differences that might exist among the Greeks. His envoys were instructed to demand earth and water, the signs of submission, from all Greek states. All the islanders and many of the continental cities consented. Sparta and Athens however were conspicuous by their refusal, and went so far as to violate the law of nations by killing the heralds[1] [6, 48—9].

[1] Herod. 7, 133. Plut. *Themist.* 6.

One result of this was that the quarrel between Athens and Aegina, which was of long standing, was embittered by the denunciation of the Aeginetans by the Athenians to the Spartan government for having given the earth and water; the consequent seizure of hostages from the Island, and their being entrusted to the Athenians; who refused, on a change of policy at Sparta, to restore them, and were consequently involved in a war with Aegina [cc. 49—50, 61, 73, 85—93].

Four digressions.

In recounting this Herodotos is led into four other digressions: (1) on the origin of the double kingship at Sparta [c. 51—52]: (2) on the functions and honours of the Spartan kings [cc. 56—60]: (3) on the quarrel between Kleomenes and Demaratos, and the means taken by the former to depose the latter and establish Leotychides in his stead [cc. 61—72]: (4) on the career of Kleomenes and especially his invasion of Argos [cc. 74—84].

These digressions, though they break the thread of the story somewhat to the reader's confusion, all have a bearing on the main purpose of the history: to show, that is, the state of Greece in preparation for the coming struggle. Thus the two digressions on the Spartan kingship are designed to illustrate the character of one of the two great states which were to take the lead in the subsequent resistance to Persia, and the difference of spirit in which they respectively undertook their share in that work. The third, on the dissensions of the two kings Kleomenes and Demaratos, helps to explain the uncertainty and halting nature which often characterised the policy of Sparta. The fourth, on the career of Kleomenes and his invasion of Argos, has a direct bearing on the attitude of Argos in the Persian

War of 480—479, which was rendered anti-hellenic greatly by its antagonism to Sparta; and it marks also the period at which the old supremacy once possessed by Argos passed finally into the hands of Sparta.

The quarrel of Athens and Aegina mentioned above did not lead at present to any event of great interest. But its importance lies in its renewal after Marathon; and the consequent effort made by Athens no longer to be so poorly off for ships as to have to borrow from Korinth [c. 89]. This subsequent effort of Athens proved the salvation of Greece at Salamis [7, 144]; and Herodotos is looking forward to that, when he details in this book the earlier stages of the quarrel.

Aug. B.C. 490. *Datis and Artaphernes take Eretria.*

Meanwhile Darius had been preparing another expedition to avenge himself on Eretria and Athens. A vast army mustered on the Aleïan plain in Kilikia, where it was taken on board a fleet of ships brought up from Phoenikia: and under the command of Datis and Artaphernes took the Island course from the coast of Ionia towards Greece. After ravaging Naxos, and touching at Delos, and other islands, the fleet sailed to Karystos on the south of Euboea, and, after subduing its inhabitants, sailed up the coast to the territory of Eretria, where they disembarked their cavalry and infantry; after six days' siege took the town; and removed such of its inhabitants, as had not escaped to the hill country, to the island of Styra, there to await their final removal to Persia [cc. 95—96]. Whilst here they appear to have made some descents upon the coast of Boeotia [c. 118]; but they did not allow anything to interrupt the fulfilment of the next and most important part of their commission, and in a few days put their ships across to the coast of Marathon.

And now we come to the event which closes one chapter in the great contest. The battle of Marathon has enjoyed almost a unique renown in the world's history. The Athenians, justly proud of their almost unaided defence of the common father-land, looked fondly back upon it as the achievement of unmixed glory; and the combatants, the μαραθωνομάχαι, were regarded as the flower of a generation heroic and valiant beyond the ordinary level of mankind. Even to this day the name has something that stirs the heart and fires the imagination as scarcely any other name does; and we kindle, more perhaps than their poetical merits deserve, at the lines of Byron,

Sept., B.C. 490. *The invasion of Marathon.*

> The mountains look on Marathon—
> And Marathon looks on the sea;
> And musing there an hour alone
> I dreamed that Greece might still be free;
> For standing on the Persians' grave
> I could not deem myself a slave.

Yet, neither in the amount of loss inflicted on the enemy, nor in the finality of its issue, nor indeed in the actual difficulty of the achievement, can it compare with many other battles in the history of the world or of Greece. At the time it was a prevalent opinion that it had settled the question of the extension of the Persian kingdom over Europe. But the clearest-sighted of the officers engaged, Themistokles, differed from the view of the majority, and held that it was but the beginning of the struggle [Plutarch, *Themist.* 3]. Herodotos, looking back on the whole war, saw that it was but an episode, yet an important one, as giving the European Greeks what they wanted most,—a confidence in themselves, and in their being able to cope with the forces of the Great King [c. 112].

The Battle of Marathon, 15 *Sept.* B.C. 490.

The account of the battle itself [cc. 108—115], and the circumstances attending it, are somewhat briefly given in Herodotos: and there are certain points in his narrative which appear to require some farther explanation than we can ever perhaps hope to give, even by comparing the notices of other ancient writers that have been preserved to us. It will be necessary, however, to compare these accounts, meagre as they are, with that of our author[1], that we may see what difference of view has long existed on several points. First we may notice that Nepos asserts, probably on the authority of Ephoros, that the Athenian generals were divided on the policy of going to Marathon at all: some of them being for staying at Athens and defending it. This may be a mere confusion founded on the story in Herodotos of the division of opinion as to making the attack at once or waiting [c. 109]. But it is not unlikely in itself, and would be consistent with the attitude of the strategi described by Herodotos. Nepos represents Miltiades as urging the bolder course, as calculated to encourage the citizens and dismay the enemy; Herodotos makes him ground his advice on the fear of a medizing party at Athens. Herodotos does not state the number of the Athenian army: but Nepos says it was 9000, which was increased by 1000 Plataeans. The position occupied by this little army was on the slope of a mountain over which passes the road to Kephisia and Athens, in the precinct of Herakles, the special hero of Marathon, the township close by. Below them was the plain of Marathon, about six miles long, a perfectly flat stretch of treeless land

[1] Nepos, *Miltiades*, 4—6: Plutarch *de Malign. Herod.* c. 26; *vit. Aristidis* 5: Pausan. 1, 15, 3 (the picture in Stoa Poikile); 1, 32, 3; 4, 25, 25 (300,000 slain). Suidas s. v. χωρὶς οἱ ἱππεῖς.

along the bay, of a breadth varying from 1½ miles to about 3 miles. 'Two marshes bound the extremities of the 'plain: the southern is not very large, and is almost dry 'at the conclusion of the great heats; but the northern 'which generally covers considerably more than a square 'mile, offers several parts which are at all seasons im-'passable; but however leave a broad firm, sandy beach 'between them and the sea' (Finlay). On this plain the Persian army had been disembarked and were encamped. Herodotos says that they were guided there by Hippias as being the best place in Attica for cavalry [c. 102]; but he says nothing of the landing of the horses, as he did in the case of the disembarcation at Eretria [c. 101]; and, as we shall see, it seems certain that no cavalry were engaged in the battle; although it seems highly improbable that they were never disembarked. The hesitation of the Athenian generals to descend into the plain was quite natural so long as the cavalry was there; and is exactly reproduced by the tactics of the Greek army in 479 B.C., which clung to the slopes of Kithaeron in fear of the Persian cavalry manoeuvring in the valley of the Asopos [9, 19—24]. Nor is it credible that Miltiades wished to make the attack at once. What he feared was, I think, that the majority of the generals would decide upon not fighting at Marathon at all, but on retiring to Athens; or that at any rate they would prevent the attack being made at the right moment. He was apparently in communication with some of the Ionians in the Persian camp, who were looking out for an opportunity of giving information against the army in which they were compulsorily serving: and he wished to be able to order an advance whenever the right moment arrived. That it was not an immediate attack that he wanted is evident from the fact that, though four of the generals gave up to him their days of command[1], he did not attack until his own day came round. There must therefore have been some days' delay, and we can easily reject the account of Nepos, who seems to make the battle take place on the day after the arrival of the Athenians in the precinct of Herakles. When his day of command came, Miltiades found that the proper time for attack had come too; and this we may account for by combining the story in Suidas of the Ionian signal to Miltiades with the story in Herodotos of the traitorous signal by the flashing shield displayed to the Persians [c. 115]. Probably for some time the Persian commanders were uncertain what the small force which they could see above them was. They would not have felt certain whether they were all Athenians, and whether their presence there meant that Athens was wholly undefended and open to their attack. They therefore waited for communication from the party of medizers, whoever they were. At last, whether from information received, or with the intention of being ready directly such information was signalled to them, they began the re-embarcation. Many of them were already on board, and the cavalry wholly so, when the flashing shield gave them notice that it would be safe to sail to Athens[2]. About the same time the Ionians signalled to Miltiades that the cavalry were all on board (χωρὶς οἱ ἱππεῖς). He could see the movements among the Persians, and knew that they were in all the bustle and disorder of an embarcation. He felt that the moment was come. He was separated by nearly a mile of partly sloping ground from the enemy, and it was vitally important to reach them before they could form their ranks. He therefore gave the word to charge at the double (δρόμῳ); an

[1] Plutarch, *Aristeid.* 5, says all of them: but this is against Herodotos.

[2] Herod. c. 115 seems to place the signal after the battle, or at any rate after the embarcation.

unusual movement, and not to be justified, except by the peculiar circumstance that they were charging men who were not in order, and were bent on reaching them before they could become so. Herodotos uses a significant word in describing the charge: he says the Athenians were 'let go' (ἀπείθησαν), as though they had been straining at the leash, as it were, and with difficulty restrained from attacking before: seeing, perhaps, the enemy preparing to embark, and restless at the idea of their escape. But though the Persian embarcation had begun, their superiority in numbers was still so great, that in order to prevent being out-flanked the Greeks were obliged to extend their line by weakening their centre. Accordingly, though the right and left wing of the Greeks were immediately victorious, the centre was broken by the Persians and Sakae, the best of the barbarian soldiers, and retreated, although Themistokles and Aristides were commanding their tribesmen there, and greatly distinguished themselves[1]. Seeing this, the two victorious wings with admirable prudence, instead of pursuing those whom they had conquered, closed inwards, and made a joint attack upon the Persian centre, which appears to have been making its way back towards the coast. The contest was long and stubborn; but at length the Persians fled towards their ships, and in their headlong haste drove each other into the marsh on the north extremity of the plain as they tried to reach the beach; and were butchered by the victorious Athenians as they struggled helplessly in the bogs, or tried to climb into their ships. Then followed an onslaught upon the ships themselves con-

[1] Plut. *Arist.* 5. Their tribes were the Leontis and Antiochis. The pursuit of the Persians could hardly have lasted long. It was checked apparently by their finding themselves threatened by the closing in of the Greek wings; and fearing to be cut off from the coast.

ceived in the true Homeric spirit. Those who had previously gone on board, and those who now managed to get there, strove to push off the ships; while the Athenians called for fire to burn the vessels, or clung on their sterns to prevent their being pushed off. All however but seven managed to get away; but we hear of no survivors among those Persians who were intercepted and prevented from getting on board. The slaughter was no doubt large, but as usual it was wildly exaggerated by subsequent writers. The moderate reckoning of Herodotos [6400, c. 117] may safely be accepted as approximately true. The Athenians had lost the Polemarch Kallimachos and one of the Strategi, and 190 men besides. The loss was a light price for so glorious a victory; which, above all things, was almost purely Athenian: for it was not till the next day that the belated Spartans arrived, viewed the dead, and returned full of the Athenian praises [c. 120]. It seems certain that the Athenian troops spent the night in their quarters at Marathon. We are told that they hastened back when they saw that the Persian fleet was steering for Sunium. Now the fleet could not have got away until the afternoon of the day of battle. They then had to sail to the small island of Styra, and take on board the Eretrian captives. This must have occupied some hours, and the fleet would not have continued its voyage until daybreak, nor would the Athenians have been able to make out which way it was steering if it did. When we join to this Plutarch's assertion that Miltiades returned to Athens the day after the battle, we shall not have any difficulty in rejecting the notion, in itself highly improbable if not impossible, that the Athenians after a long day's battle marched back 25 miles home,—a good march for fresh troops, that had fought no battle at all.

Next day Aristeides was left with his tribesmen in possession of the field to bury the dead and secure the spoil[1]. The rest marched back in time to deter the Persians from attempting a landing at Phalerum. The ships rested on their oars for a time: but the sight of the very men who had just inflicted such a defeat upon them was too much for the Persian commanders. Without making any further attempt they turned their prows and made for Asia.

The Picture in Stoa Poikile.

The memory of this victory was kept alive by a great festival at Athens. But it was made a more living reality to posterity by a great historical fresco in the Stoa Poikile. Pausanias, who had seen it, has left us a description which may help us to realise the scene [1, 15, 3]:

'The last painting (in the Stoa) represents the men 'who fought at Marathon. The Boeotian contingent from 'Plataea and all the Attic army are advancing to charge 'the Barbarians. In this part of the picture there is no 'sign of superiority on either side. But where the battle 'itself is represented, the Barbarians are seen flying and 'pushing each other into the marsh. Again, the last painting 'represents the Phoenikian ships, and the Greeks slaying 'those of the Barbarians who are trying to get on board 'of them. At that point the hero Marathon is introduced 'into the picture, after whom the plain is called; and 'Theseus is represented as rising out of the earth, as well 'as Athena and Herakles. For the Marathonians them-'selves say that they were the first to regard Herakles as 'a god. Of the combatants the most conspicuous in the 'picture are Kallimachos, the Polemarch, and Miltiades[2] 'of the ten generals, and a hero called Echetlos.'

After thus describing the crowning event of this portion

[1] Plutarch *Arist.* 5. [2] Cheering on his men, Aesch. *in Ctes.* 186.

of his work, Herodotos devotes the remaining part of the book to clearing up certain details. First we have an account of the treatment of the Eretrian prisoners, accompanied by a note on the physical features of the country in which they were settled [c. 119]. Next, after noting the late arrival of the Lakedaemonians at Marathon [c. 120], he proceeds to state the case against the popular belief that the Alkmaeonidae were guilty of raising the treasonable signal of the brazen shield [cc. 121—124]. Plutarch in his essay on the 'Malignity of Herodotos' declares that Herodotos has himself suggested the charge that he might curry favour with the family by refuting it. There are signs in the passage that Herodotos wished, perhaps from personal motives, to speak well of the Alkmaeonidae; but we have no ground for accepting Plutarch's allegation that the belief did not exist independently of Herodotos' narrative.

Discussion on the guilt of the Alkmaeonidae.

This mention of the Alkmaeonidae leads him on to the curious story of the origin of the wealth of the family [c. 125], and the marriage of Megakles with the daughter of Kleisthenes of Sikyon [cc. 126—131]; all part apparently of the family traditions of the Alkmaeonidae, and inserted partly perhaps to please a patron, and partly from Herodotos' native love of a good story. It is slightly connected with what follows by incidentally tracing the pedigree of Perikles son of Xanthippos, the impeacher of Miltiades.

The remaining chapters are devoted to the subsequent career of Miltiades, his trial and death [132—136]; which leads again to an account of an earlier achievement of his in taking Lemnos, and a curious account of the connexion of Athens with that island [137—140].

Miltiades.

The crime for which Miltiades was impeached and fined was technically deceiving the people (ἀπάτη τοῦ δήμου). Nepos [c. 8] declares that the real motive of his condemnation was the jealousy of his commanding position, which possessed the people mindful of the recent tyranny of the Peisistratids. Accordingly a considerable controversy has always existed among Greek historians: some pointing to his fall as a glaring instance of the fickleness and ingratitude of a democracy; others maintaining that it was a noble instance of impartial justice, and a warning to all ministers of state, that previous services cannot be pleaded in extenuation of disloyalty, and a selfish use of official position for the gratification of personal objects. Judgment in the case has gone very much in accordance with the sympathy felt for, or prejudice against, a particular form of government. Not to mention Grote's remarks, in whose eyes the Athenian demos can scarcely do wrong, Bishop Thirlwall, the soberest and most impartial of historians, sums it up in the following weighty sentence: 'If the people conceived 'that nothing he had done for them ought to raise him 'above the laws, if they even thought that his services 'had been sufficiently rewarded by the station which en- 'abled him to perform them, and the glory he had reaped 'from them, they were not ungrateful or unjust; and if 'Miltiades thought otherwise, he had not learnt to live in 'a free state.'

The crime and impeachment of Miltiades.

But admirable as this sentence is, it does not refer to one view of the case which I think ought to be put also. The narrative of Herodotos is briefly this: 'Miltiades asked 'for 70 ships, a force of men, and a grant of money, 'without telling the people what country he was going to 'attack, but merely saying that if they would follow him 'he would enrich them: for he would take them to a land 'whence they would with ease obtain gold without stint. 'That he accordingly got what he asked for, and immedi-'ately sailed to Paros, ostensibly to punish the Parians for 'having taken part in the invasion at Marathon, but really 'to satisfy a private grievance. That having failed to 'take Paros, he returned wounded and without gold to 'Athens.'

Now granting the essential truth of this statement, if we proceed to enquire closely wherein lay the crime of Miltiades against the Athenian people, it will not be easy to fix it very clearly to any definite act. If it is said that it was besieging Paros at all; it may be answered that this crime, if it were one, was condoned and even adopted by the Athenian government. For, seeing that the siege lasted 26 days, and that Paros was within two days' sail from Athens, Miltiades might have been recalled at any time. As he was not, can it be doubted that, had he succeeded, nothing would have been heard of any charge; but that he would have gained still greater reputation for enriching and aggrandising the state? Thus Pausanias attributes his impeachment to his *failing to take* Paros [Πάρου ἁμαρτόντι, 1, 32, 4]. Again, if we attribute the impeachment to the dishonesty of his *motive* in attacking Paros, it may well be asked how Herodotos, who was not born till about four years afterwards, could be informed of his secret intentions? It seems probable from the previous chapters that the historian was intimate with some of the Alkmaeonidae, and would probably have heard their version of the matter, for one of them, Xanthippos, was the prosecutor. To take the assertion of an historian, writing many years after the event, as to the thoughts of any man, can never be safe, unless he can bring conclusive evidence to prove his case:

but when that assertion may with great probability be traced to the political foes of the man in question, it scarcely deserves serious attention. When we consider also that the motive attributed to Miltiades seems exceedingly, not to say absurdly, inadequate; and that though Paros was a prosperous island enough, there was nothing in it which of itself could give Miltiades any chance of fulfilling his liberal promises to the Athenians,—we may be led I think to take a somewhat different view of the expedition altogether, and to incline to supplement Herodotos by the statement of Ephoros[1] [fr. 107] who tells us that the attack upon Paros was one of a series of raids upon the Cyclades. This would at once do away with the idea of a simple use of the public resources to wreak a private vengeance on a single island, and would tend to support the assertion (which Nepos probably took from Ephoros) that the ships were granted with the avowed object of punishing those of the islands that had assisted the Persians[2]. That Miltiades may have indulged in grandiloquent promises of the great riches to be obtained is more than probable: but that the people had not clean hands is shewn by their granting him a roving commission, with little or no limitation, incited to it solely by the hope of plunder. And it was the consciousness of this, I think, as much as the eloquent pleadings of his friends, that induced them to vote against a capital sentence. That Miltiades was ill advised to spend so much time in trying to take Paros is very likely; and that a general returning with nothing but failure must look to encounter the wrath of his masters is also certain: but the gravamen against the Athenian government, as such, is not that they punished Miltiades;

[1] Quoted by Stephanos Byz. s. v. Πάρος.

[2] Just as Themistokles did after Salamis.

but that they gave him a commission which no well-ordered government could ever give any minister with honour or safety. The point may be illustrated by comparing the case of Thrasybulos (mentioned in the note on p. 77, l. 9). The hero of the restored democracy [B.C. 404] occupied a position of high credit at Athens after that event. In B.C. 390 he was despatched with 40 triremes, and a general commission to restore Athenian ascendancy in Asia Minor and the Islands. He interpreted his commission as he chose: went to the Hellespont, and there spent several months in collecting money from the Thrakian cities, and other transactions of the sort. While there an order came out for his recall. But he did not obey at once. He seized Byzantium, got money by letting out the tolls of the cornships, and then coasted along Asia Minor, exacting money from the towns as he went. At Aspendos he was killed, and so escaped the legal consequences of his act. But his subordinate Ergokles was impeached, fined, and condemned to death, though he appears to have escaped the extreme penalty. This open commission to Thrasybulos was of the same nature as that to Miltiades: his abuse of it was also similar. But the Athenian government in the latter case cleared itself from complicity with the wrong doing by recalling Thrasybulos[1]. If it had done the same in the case of Miltiades, there would have been just ground for regarding Miltiades as a disloyal officer, and the state as entitled to exact a righteous punishment. As it is, there is little reason to regard his conduct in any worse light than that of any other general who pays the penalty of failure.

[1] The case forms the subject of a speech of Lysias κατ' Ἐργοκλέους (28). See also Xenophon *Hell.* 4, 8, 25—31. Diodor. Sic. 14, 99.

NOTES ON THE TEXT.

p. 2, l. 23. **βιβλία.** See on V, 58.

p. 4, l. 15. Stein conjectures *ὅσοι Λέσβον*.

p. 4, l. 25. **πάντων.** Some MSS. have *πασέων*.

p. 6, l. 22. **ἐλασσωθήσεσθαι.** Cp. Thucyd. 5, 34. Some MSS. have *ἐλασσώσεσθαι*.

p. 7, l. 18. **οἷα στρατιή.** Dobree conj. *οἷα ἀστρατηΐης*, comparing Arist. *Pax*, 525.

p. 8, l. 1. **τοῦ Δαρείου.** The MSS. have *τὸν Δαρεῖον*. Schweighaeuser corrected it. Stein and Abicht bracket the words, which are certainly unnecessary and may have been a gloss on *τὸ παρεὸν ναυτικὸν*.

p. 8, l. 3. **οὐ βουλομένους.** Two MSS. have *ἀρνευμένους*. Abicht writes *ἀρνεομένους*.

p. 8, l. 14. **ἢ ἀγαθοί.** Cobet would omit these words. *Novae Lect.* p. 420.

p. 10, l. 17. **ὅτε γε.** So MSS. SVR: al. *ὅποτε*. Stein reads *ὄκοτε*, cp. 3, 73.

p. 12, l. 23. **Ἴνυκον.** Cp. p. 13, l. 3, *Ἰνύκου*. Stein reads *Ἴνυκα* and *Ἴνυκος*, though the MSS. do not vary in giving the forms of the text. He does it on the authority of Steph. Byz. who says that Herodotos wrote *Ἴνυχα* as from *Ἴνυξ*, 'but seems to have been wrong' (*ἔοικε δ' ἐσφάλθαι*). But is it not as likely that Stephanos' copy of Herodotos was wrong, as that Herodotos should have made the mistake?

p. 15, l. 4. **ἐκ τοῦ Ἀταρνέος.** Some good MSS. (B[2]R) have *ἐς*, which would connect with *διαβαίνει* in a more natural order. Abicht defends the order of the text as adopted for the sake of emphasis. It is not however evident why the words require emphasis.

p. 15, l. 22. **ἀνήχθη**. An excellent emendation of Bredovius for ἤχθη, cp. p. 16, l. 1; p. 21, l. 22. The MSS. however all have ἤχθη.

p. 16, l. 25. **αὐτοῖσι τοῖς ἱροῖσι**. Bekker here and in p. 17, l. 4, *αὐτοῖσι τοῖσι Πέρσῃσι* omits *τοῖσι*, but I think he does so on a false analogy with the common *αὐτοῖσι ἀνδράσι*.

p. 18, l. 28. **ποιέῃ**. The MSS. vary between *ποιοῖ*, *ποιείη*, *ποιοίη* (which Stein adopts). I have followed Dobree and Cobet in reading *ποιέῃ*. Cp. p. 49, l. 30 *ληΐσηται*, an exactly parallel case. See Cobet *Nov. Lect.* p. 363. Cp. 1, 75, 82, 124; 5, 67.

p. 20, l. 15. **Κίμωνος** is not wanted. Cobet would omit it.

p. 20, l. 18. **ἐπιτιμέων**. This word Cobet declares to be meaningless in this connexion, and he proposes *πενθέων*.

p. 21, l. 6. **ἀπὸ Χερσονήσου**. Two MSS. omit *ἀπὸ* and read *Χερσόνησον* (Stein), two other MSS. omit both. Gaisford thinks both words suspicious.

p. 21, l. 24. **κελεύων**. Cobet would omit this word.

p. 23, l. 25. **κατὰ** is omitted by all but three MSS. It seems more probable that Herodotos should have written it than have ventured on an exact number in such a case. See p. 44, l. 20.

p. 25, l. 1. **Σκαπτησύλης**. So AB[1]. But B[2]CPD have *ἐκ Σκαπτῆς Ὕλης*. Steph. Byz. has *Σκαπτησύλη, τὸ ἐθνικὸν Σκαπτησυλῖται*. The Latin form in Lucr. 6, 810 is on the same side, the *-ens* of *Scaptensula* representing the long *-ησ-*.

p. 27, l. 2. **ἤδη**. Stein with one inferior MS. reads *ἢ δή*. But can this combination of particles with an imperative be justified?

p. 27, l. 26. **βουλομένην**. Cobet (*V. L.* 286, 421) would read *οὐ βουλομένην δὲ*, which is specious, but I think is not in harmony with the context, cp. 9, 14.

p. 28, l. 22. **τῷ δὲ Προκλέα**. Some MSS. have *τῷ δὲ νεωτέρῳ*.

p. 28, l. 28. **κατὰ τὰ λεγόμενα**. MSS. vary between *κατὰ ταῦτα λ.* *κατὰ λεγόμενα*. *καταλεγόμενα*. The source of the corruption is evident.

p. 30, l. 2. **ἐξοδίῃσι**. S_2 *τοῖσιν ἐξοδίοισι*. Cobet would read *τῇσι ἐξόδοισι*, as the only correct form. The evidence for *ἐξοδίη* is very slight, only Polybios 4, 46; 8, 26 [where Dindorf has from Strabo 249 corrected it to *ἐξοδείας* and *ἐξοδείαν*]; and Hesychios, who has *ἐξοδίαν*, quoting apparently a fragment of Polybios. But these instances, coupled with the fact that all our MSS. give it here, make it unsafe to change. Stein suggests *ἐξοδηίῃσι*. A neuter *ἐξοδεῖα* is found on the Rosetta Stone.

p. 30, l. 6. **ἢν θυσίη τις δημοτελής.** PR for *θυσίην...δημοτελέα* of B²PR.

p. 35, l. 8. **δι' ἅ.** *διὰ τὸ* R. *διὰ τὰ* (sc. *ταῦτα*) C. *δι' ἅ* Bekk. **διὰ τὰ** Stein.

p. 40, l. 21. **ὡδώθη.** So three of the best MSS. Others have *εὐοδώθη, εὐωδώθη, ὠρθώθη*, cp. 4, 139.

p. 41, l. 7. **παρατίθενται.** ABCd and others. *κατατίθενται* St. with several MSS.

p. 42, l. 4. **μανιὰς νοῦσος.** Cobet's emendation. Cp. Soph. *Aj.* 59. B²PR₃ have *μανίης*. C *μανίη*.

p. 43, l. 16. **'Ησίπεια** Stein. A *ησίπεια*. B *///σίπεια*. CP₃ *ἡ σίπεια*. R *σήπεια*. SV *Σήπεια*.

p. 50, l. 25. **πεντετηρίς.** The MSS. have *πεντήρης*, but see note to the passage.

p. 51, l. 17. **σφίσι.** The MSS. give *σφι*.

p. 67, l. 14. Cobet would omit *τάχιστα*.

p. 67, l. 22. Cobet would omit *ἐν Μαραθῶνι*.

p. 70. Chapter CXXII omitted in AB¹Cd¹, added in B and d in the margin by another hand. Gaisford's MSS. MKPFb omit it.

p. 77, l. 24. **ἢν μή οἱ δῶσι** B²PR₃. Stein conjectures *ἢν μὲν οὐ δῶσι*.

p. 81, l. 1. **αὐτοῖσι.** Stein reads *ἑωυτοῖσι*.

The MSS. vary as to the forms *πλέω* and *πλώω*, and their compounds and tenses. Stein invariably adopts the former. But as both forms seem undoubtedly to have existed, I have taken the latter wherever there seemed good MS. authority for it.

ΗΡΟΔΟΤΟΥ ΕΡΑΤΩ.

BOOK VI.

Histiaios comes from Susa to Sardis. Artaphernes hints that he knows the origin of the Ionian revolt.

I. Ἀρισταγόρης μέν νυν Ἰωνίην ἀποστήσας οὕτω τελευτᾷ. Ἱστιαῖος δὲ ὁ Μιλήτου τύραννος μεμετιμένος ὑπὸ Δαρείου παρῆν ἐς Σάρδις. ἀπιγμένον δὲ αὐτὸν ἐκ τῶν Σούσων εἴρετο Ἀρταφέρνης ὁ Σαρδίων ὕπαρχος, κατὰ κοῖόν τι δοκέοι Ἴωνας 5 ἀπεστάναι· ὁ δὲ οὔτε εἰδέναι ἔφη, ἐθώυμαζέ τε τὸ γεγονὸς ὡς οὐδὲν δῆθεν τῶν παρεόντων πρηγμάτων ἐπιστάμενος. ὁ δὲ Ἀρταφέρνης ὁρέων αὐτὸν τεχνάζοντα εἶπε, εἰδὼς τὴν ἀτρεκείην τῆς ἀποστάσιος· "Οὕτω τοι, Ἱστιαῖε, ἔχει κατὰ ταῦτα τὰ πρήγματα· 10 "τοῦτο τὸ ὑπόδημα ἔρραψας μὲν σύ, ὑπεδήσατο δὲ "Ἀρισταγόρης."

Histiaios therefore flies to Chios, forfeiting his promise of conquering Sardinia.

II. Ἀρταφέρνης μὲν ταῦτα ἐς τὴν ἀπόστασιν ἔχοντα εἶπε. Ἱστιαῖος δὲ δείσας ὡς συνιέντα Ἀρταφέρνεα ὑπὸ τὴν πρώτην ἐπελθοῦσαν νύκτα ἀπέδρη 15 ἐπὶ θάλασσαν, βασιλέα Δαρεῖον ἐξηπατηκώς· ὃς

Σαρδὼ νῆσον τὴν μεγίστην ὑποδεξάμενος κατεργάσεσθαι ὑπέδυνε τῶν Ἰώνων τὴν ἡγεμονίην τοῦ πρὸς Δαρεῖον πολέμου. διαβὰς δὲ ἐς Χίον ἐδέθη ὑπὸ Χίων, καταγνωσθεὶς πρὸς αὐτῶν νεώτερα πρήσσειν
5 πρήγματα ἐς ἑωυτοὺς ἐκ Δαρείου. μαθόντες μέντοι οἱ Χῖοι τὸν πάντα λόγον, ὡς πολέμιος εἴη βασιλέϊ, ἔλυσαν αὐτόν.

He falsely reports a scheme of the king's to transfer the Ionians to Phoenikia.

III. Ἐνθαῦτα δὴ εἰρωτεόμενος ὑπὸ τῶν Ἰώνων ὁ Ἱστιαῖος, κατ' ὅ τι προθύμως οὕτω ἐπέστειλε τῷ
10 Ἀρισταγόρῃ ἀπίστασθαι ἀπὸ βασιλέος καὶ κακὸν τοσοῦτο εἴη Ἴωνας ἐξεργασμένος, τὴν μὲν γενομένην αὐτοῖσι αἰτίην οὐ μάλα ἐξέφαινε, ὁ δὲ ἔλεγέ σφι, ὡς βασιλεὺς Δαρεῖος ἐβουλεύσατο Φοίνικας μὲν ἐξαναστήσας ἐν τῇ Ἰωνίῃ κατοικίσαι, Ἴωνας δὲ ἐν τῇ
15 Φοινίκῃ, καὶ τούτων εἵνεκεν ἐπιστείλειε. οὐδέν τι πάντως ταῦτα βασιλέος βουλευσαμένου ἐδειμάτου τοὺς Ἴωνας.

A plot discovered in Sardis.

IV. Μετὰ δὲ ὁ Ἱστιαῖος δι' ἀγγέλου ποιεύμενος Ἑρμίππου ἀνδρὸς Ἀταρνείτεω τοῖσι ἐν Σάρδισι
20 ἐοῦσι Περσέων ἔπεμπε βιβλία ὡς προλελεσχηνευμένων αὐτῷ ἀποστάσιος πέρι. ὁ δὲ Ἕρμιππος, πρὸς τοὺς μὲν ἀπεπέμφθη, οὐ διδοῖ, φέρων δὲ ἐνεχείρισε τὰ βιβλία Ἀρταφέρνεϊ. ὁ δὲ μαθὼν ἅπαν τὸ γινόμενον ἐκέλευε τὸν Ἕρμιππον τὰ μὲν παρὰ τοῦ
25 Ἱστιαίου δοῦναι φέροντα τοῖσί περ ἔφερε, τὰ δὲ ἀμοιβαῖα τὰ παρὰ τῶν Περσέων ἀντιπεμπόμενα

Ἱστιαίῳ ἑωυτῷ δοῦναι. τούτων δὲ γενομένων φανερῶν ἀπέκτεινε ἐνθαῦτα πολλοὺς Περσέων ὁ Ἀρταφέρνης.

Histiaios after vainly trying to recover Miletos goes to Byzantium.

V. Περὶ Σάρδις μὲν δὴ ἐγένετο ταραχή. Ἱστιαῖον δὲ ταύτης ἀποσφαλέντα τῆς ἐλπίδος Χῖοι κατῆγον ἐς 5 Μίλητον αὐτοῦ Ἱστιαίου δεηθέντος. οἱ δὲ Μιλήσιοι, ἄσμενοι ἀπαλλαχθέντες καὶ Ἀρισταγόρεω, οὐδαμῶς πρόθυμοι ἦσαν ἄλλον τύραννον δέκεσθαι ἐς τὴν χώρην, οἷα ἐλευθερίης γευσάμενοι. καὶ δή, νυκτὸς γὰρ ἐούσης βίῃ ἐπειρᾶτο κατιὼν ὁ Ἱστιαῖος ἐς τὴν 10 Μίλητον, τιτρώσκεται τὸν μηρὸν ὑπό τευ τῶν Μιλησίων. ὁ μὲν δὴ ὡς ἀπωστὸς τῆς ἑωυτοῦ γίνεται, ἀπικνέεται ὀπίσω ἐς τὴν Χίον· ἐνθεῦτεν δέ, οὐ γὰρ ἔπειθε τοὺς Χίους ὥστε ἑωυτῷ δοῦναι νέας, διέβη ἐς Μυτιλήνην καὶ ἔπεισε Λεσβίους δοῦναί οἱ νέας. οἱ 15 δὲ πληρώσαντες ὀκτὼ τριήρεας ἔπλωον ἅμα Ἱστιαίῳ ἐς Βυζάντιον, ἐνθαῦτα δὲ ἰζόμενοι τὰς ἐκ τοῦ Πόντου ἐκπλωούσας τῶν νεῶν ἐλάμβανον, πλὴν ἢ ὅσοι αὐτῶν Ἱστιαίῳ ἔφασαν ἑτοῖμοι εἶναι πείθεσθαι.

B.C. 495. *The Persians prepare to attack Miletos.*

VI. Ἱστιαῖος μέν νυν καὶ Μυτιληναῖοι ἐποίευν 20 ταῦτα, ἐπὶ δὲ Μίλητον αὐτὴν ναυτικὸς πολλὸς καὶ πεζὸς ἦν στρατὸς προσδόκιμος· συστραφέντες γὰρ οἱ στρατηγοὶ τῶν Περσέων καὶ ἓν ποιήσαντες στρατόπεδον ἤλαυνον ἐπὶ τὴν Μίλητον, τὰ ἄλλα πολίσματα περὶ ἐλάσσονος ποιησάμενοι. τοῦ δὲ ναυτικοῦ 25 Φοίνικες μὲν ἦσαν προθυμότατοι, συνεστρατεύοντο δὲ καὶ Κύπριοι νεωστὶ κατεστραμμένοι καὶ Κίλικές τε καὶ Αἰγύπτιοι.

The Ionians in council decide not to resist by land, but to prepare a large fleet.

VII. Οἱ μὲν δὴ ἐπὶ τὴν Μίλητον καὶ τὴν ἄλλην Ἰωνίην ἐστράτευον, Ἴωνες δὲ πυνθανόμενοι ταῦτα
5 ἔπεμπον προβούλους σφέων αὐτῶν ἐς Πανιώνιον. ἀπικομένοισι δὲ τούτοισι ἐς τοῦτον τὸν χῶρον καὶ βουλευομένοισι ἔδοξε πεζὸν μὲν στρατὸν μὴ συλλέγειν ἀντίξοον Πέρσῃσι, ἀλλὰ τὰ τείχεα ῥύεσθαι αὐτοὺς Μιλησίους, τὸ δὲ ναυτικὸν πληροῦν ὑπολει-
10 πομένους μηδεμίαν τῶν νεῶν, πληρώσαντας δὲ συλλέγεσθαι τὴν ταχίστην ἐς Λάδην, προναυμαχήσοντας Μιλήτου· ἡ δὲ Λάδη ἐστὶ νῆσος σμικρὴ ἐπὶ τῇ πόλι τῇ Μιλησίων κειμένη.

The Ionian forces.

VIII. Μετὰ δὲ ταῦτα πεπληρωμένῃσι τῇσι νηυσὶ
15 παρῆσαν οἱ Ἴωνες, σὺν δέ σφι καὶ Αἰολέων οἳ Λέσβον νέμονται· ἐτάσσοντο δὲ ὧδε· τὸ μὲν πρὸς τὴν ἠῶ εἶχον κέρας αὐτοὶ Μιλήσιοι, νέας παρεχόμενοι ὀγδώκοντα, εἴχοντο δὲ τούτων Πριηνέες δυώδεκα νηυσὶ καὶ Μυούσιοι τρισὶ νηυσί, Μυουσίων δὲ Τήϊοι εἴχοντο
20 ἑπτακαίδεκα νηυσί, Τηΐων δὲ εἴχοντο Χῖοι ἑκατὸν νηυσί· πρὸς δὲ τούτοισι Ἐρυθραῖοί τε ἐτάσσοντο καὶ Φωκαιέες, Ἐρυθραῖοι μὲν ὀκτὼ νέας παρεχόμενοι, Φωκαιέες δὲ τρεῖς· Φωκαιέων δὲ εἴχοντο Λέσβιοι νηυσὶ ἑβδομήκοντα· τελευταῖοι δὲ ἐτάσσοντο ἔχοντες
25 τὸ πρὸς ἑσπέρην κέρας Σάμιοι ἑξήκοντα νηυσί. πάντων δὲ τούτων ὁ σύμπας ἀριθμὸς ἐγένετο τρεῖς καὶ πεντήκοντα καὶ τριηκόσιαι τριήρεες.

The expelled Ionian tyrants urged to detach their several countrymen from the insurgents.

IX. Αὗται μὲν Ἰώνων ἦσαν· τῶν δὲ βαρβάρων τὸ πλῆθος τῶν νεῶν ἦσαν ἑξακόσιαι. ὡς δὲ καὶ αὗται ἀπίκατο πρὸς τὴν Μιλησίην καὶ ὁ πεζός σφι ἅπας παρῆν, ἐνθαῦτα οἱ Περσέων στρατηγοὶ πυθόμενοι τὸ πλῆθος τῶν Ἰάδων νεῶν καταρρώδησαν, μὴ οὐ δυνατοὶ 5 γένωνται ὑπερβαλέσθαι, καὶ οὕτω οὔτε τὴν Μίλητον οἷοί τε ἔωσι ἐξελεῖν μὴ οὐκ ἐόντες ναυκράτορες, πρός τε Δαρείου κινδυνεύσωσι κακόν τι λαβεῖν. ταῦτα ἐπιλεγόμενοι, συλλέξαντες τῶν Ἰώνων τοὺς τυράννους, οἳ ὑπ' Ἀρισταγόρεω μὲν τοῦ Μιλησίου καταλυ- 10 θέντες τῶν ἀρχέων ἔφευγον ἐς Μήδους, ἐτύγχανον δὲ τότε συστρατευόμενοι ἐπὶ τὴν Μίλητον,—τούτων τῶν ἀνδρῶν τοὺς παρεόντας συγκαλέσαντες ἔλεγόν σφι τάδε· "Ἄνδρες Ἴωνες, νῦν τις ὑμέων εὖ ποιήσας "φανήτω τὸν βασιλέος οἶκον· τοὺς γὰρ ἑωυτοῦ ἕκαστος 15 "ὑμέων πολιήτας πειράσθω ἀποσχίζων ἀπὸ τοῦ λοιποῦ "συμμαχικοῦ. προϊσχόμενοι δὲ ἐπαγγείλασθε τάδε, "ὡς πείσονταί τε ἄχαρι οὐδὲν διὰ τὴν ἀπόστασιν, οὐδέ "σφι οὔτε τὰ ἱρὰ οὔτε τὰ ἴδια ἐμπεπρήσεται, οὐδὲ "βιαιότερον ἕξουσι οὐδὲν, ἢ πρότερον εἶχον· εἰ δὲ 20 "ταῦτα μὲν οὐ ποιήσουσι, οἱ δὲ πάντως διὰ μάχης "ἐλεύσονται, τάδε σφι λέγετε ἐπηρεάζοντες, τά πέρ "σφεας κατέξει, ὡς ἑσσωθέντες τῇ μάχῃ ἐξανδρα- "ποδιεῦνται, καὶ ὥς σφεων τοὺς παῖδας ἐκτομίας "ποιήσομεν, τὰς δὲ παρθένους ἀνασπάστους ἐς Βάκ- 25 "τρα, καὶ ὡς τὴν χώρην ἄλλοισι παραδώσομεν."

They fail to do so.

X. Οἱ μὲν δὴ ἔλεγον ταῦτα, τῶν δὲ Ἰώνων οἱ

τύραννοι διέπεμπον νυκτὸς ἕκαστος ἐς τοὺς ἑωυτοῦ
ἐξαγγελλόμενος. οἱ δὲ Ἴωνες, ἐς τοὺς καὶ ἀπίκοντο
αὗται αἱ ἀγγελίαι, ἀγνωμοσύνῃ τε διεχρέοντο καὶ οὐ
προσίεντο τὴν προδοσίην, ἑωυτοῖσί τε ἕκαστοι ἐδόκεον
5 μούνοισι ταῦτα τοὺς Πέρσας ἐξαγγέλλεσθαι. ταῦτα
μέν νυν ἰθέως ἀπικομένων ἐς τὴν Μίλητον τῶν
Περσέων ἐγίνετο.

Dionysios of Phokaia put in command of the Ionian fleet.

XI. Μετὰ δὲ τῶν Ἰώνων συλλεχθέντων ἐς τὴν
Λάδην ἐγίνοντο ἀγοραί. καὶ δή κού σφι καὶ ἄλλοι
10 ἠγορεύοντο, ἐν δὲ δὴ καὶ ὁ Φωκαιεὺς στρατηγὸς
Διονύσιος λέγων τάδε· "Ἐπὶ ξυροῦ γὰρ ἀκμῆς ἔχεται
"ἡμῖν τὰ πρήγματα, ἄνδρες Ἴωνες, ἢ εἶναι ἐλευθέροισι
"ἢ δούλοισι, καὶ τούτοισι ὡς δρηπέτῃσι· νῦν ὦν ὑμεῖς,
"ἢν μὲν βούλησθε ταλαιπωρίας ἐνδέκεσθαι, τὸ παρα-
15 "χρῆμα μὲν πόνος ὑμῖν ἔσται, οἷοί τε δὲ ἔσεσθε ὑπερ-
"βαλόμενοι τοὺς ἐναντίους εἶναι ἐλεύθεροι· εἰ δὲ
"μαλακίῃ τε καὶ ἀταξίῃ διαχρήσεσθε, οὐδεμίαν ὑμέων
"ἔχω ἐλπίδα μὴ οὐ δώσειν ὑμέας δίκην βασιλέϊ τῆς
"ἀποστάσιος. ἀλλ' ἐμοί τε πείθεσθε καὶ ἐμοὶ ὑμέας
20 "αὐτοὺς ἐπιτρέψατε· καὶ ὑμῖν ἐγὼ θεῶν τὰ ἶσα νεμόν-
"των ὑποδέκομαι ἢ οὐ συμμίξειν τοὺς πολεμίους ἢ
"συμμίσγοντας πολλὸν ἐλασσωθήσεσθαι." Ταῦτα
ἀκούσαντες οἱ Ἴωνες ἐπιτράπουσι σφέας αὐτοὺς τῷ
Διονυσίῳ.

Discontent at the severity of the discipline of Dionysios.

25 XII. Ὁ δὲ ἀνάγων ἑκάστοτε ἐπὶ κέρας τὰς νέας,
ὅκως τοῖσι ἐρέτῃσι χρήσαιτο διέκπλοον ποιεύμενος
τῇσι νηυσὶ δι' ἀλληλέων καὶ τοὺς ἐπιβάτας ὁπλίσειε,
τὸ λοιπὸν τῆς ἡμέρης τὰς νέας ἔχεσκε ἐπ' ἀγκυρέων,

παρεῖχέ τε τοῖσι Ἴωσι πόνον δι' ἡμέρης. μέχρι μέν νυν ἡμερέων ἑπτὰ ἐπείθοντό τε καὶ ἐποίευν τὸ κελευόμενον, τῇ δὲ ἐπὶ ταύτῃσι οἱ Ἴωνες, οἷα ἀπαθέες ἐόντες πόνων τοιούτων τετρυμένοι τε ταλαιπωρίῃσί τε καὶ ἡλίῳ, ἔλεξαν πρὸς ἑωυτοὺς τάδε· "Τίνα δαιμόνων 5
"παραβάντες τάδε ἀναπίμπλαμεν; οἵτινες παρα-
"φρονήσαντες καὶ ἐκπλώσαντες ἐκ τοῦ νόου ἀνδρὶ
"Φωκαιέϊ ἀλαζόνι, παρεχομένῳ νέας τρεῖς, ἐπιτρέψ-
"αντες ἡμέας αὐτοὺς ἔχομεν· ὁ δὲ παραλαβὼν ἡμέας
"λυμαίνεται λύμῃσι ἀνηκέστοισι, καὶ δὴ πολλοὶ μὲν 10
"ἡμέων ἐς νούσους πεπτώκασι, πολλοὶ δὲ ἐπίδοξοι
"τὠυτὸ τοῦτο πείσεσθαί εἰσι· πρό τε τούτων τῶν
"κακῶν ἡμῖν γε κρέσσον καὶ ὁτιῶν ἄλλο παθεῖν ἐστὶ,
"καὶ τὴν μέλλουσαν δουληΐην ὑπομεῖναι, ἥ τις ἔσται,
"μᾶλλον ἢ τῇ παρεούσῃ συνέχεσθαι. φέρετε, τοῦ 15
"λοιποῦ μὴ πειθώμεθα αὐτοῦ." Ταῦτα ἔλεξαν, καὶ μετὰ ταῦτα αὐτίκα πείθεσθαι οὐδεὶς ἤθελε, ἀλλ' οἷα στρατιὴ, σκηνάς τε πηξάμενοι ἐν τῇ νήσῳ ἐσκιητροφέοντο καὶ ἐσβαίνειν οὐκ ἐθέλεσκον ἐς τὰς νέας οὐδ' ἀναπειρᾶσθαι. 20

The Samians accept the Persian terms.

XIII. Μαθόντες δὲ ταῦτα τὰ γινόμενα ἐκ τῶν Ἰώνων οἱ στρατηγοὶ τῶν Σαμίων, ἐνθαῦτα δὴ παρ' Αἰάκεος τοῦ Συλοσῶντος ἐκείνους τοὺς πρότερον ἔπεμπε λόγους ὁ Αἰάκης κελευόντων τῶν Περσέων, δεόμενός σφεων ἐκλιπεῖν τὴν Ἰώνων συμμαχίην, οἱ Σάμιοι ὦν 25
ὁρέοντες ἅμα μὲν ἐοῦσαν ἀταξίην πολλὴν ἐκ τῶν Ἰώνων ἐδέκοντο τοὺς λόγους, ἅμα δὲ κατεφαίνετό σφι εἶναι ἀδύνατα τὰ βασιλέος πρήγματα ὑπερβαλέσθαι, εὖ τε ἐπιστάμενοι, ὡς, εἰ καὶ τὸ παρεὸν ναυτικὸν ὑπερβα-

λοίατο τοῦ Δαρείου, ἄλλο σφι παρέσται πενταπλήσιον. προφάσιος ὦν ἐπιλαβόμενοι, ἐπεί τε τάχιστα εἶδον τοὺς Ἴωνας οὐ βουλομένους εἶναι χρηστούς, ἐν κέρδεϊ ἐποιεῦντο περιποιῆσαι τά τε ἱρὰ τὰ σφέτερα καὶ
5 τὰ ἴδια. ὁ δὲ Αἰάκης, παρ' ὅτευ τοὺς λόγους ἐδέκοντο οἱ Σάμιοι, παῖς μὲν ἦν Συλοσῶντος τοῦ Αἰάκεος, τύραννος δὲ ἐὼν Σάμου ὑπὸ τοῦ Μιλησίου Ἀρισταγόρεω ἀπεστέρητο τὴν ἀρχὴν κατά περ οἱ ἄλλοι τῆς Ἰωνίης τύραννοι.

Battle of Lade B.C. 495.

10 XIV. Τότε ὦν ἐπεὶ ἐπέπλωον οἱ Φοίνικες, οἱ Ἴωνες ἀντανῆγον καὶ αὐτοὶ τὰς νέας ἐπὶ κέρας. ὡς δὲ καὶ ἀγχοῦ ἐγένοντο καὶ συνέμισγον ἀλλήλοισι, τὸ ἐνθεῦτεν οὐκ ἔχω ἀτρεκέως συγγράψαι, οἵ τινες τῶν Ἰώνων ἐγένοντο ἄνδρες κακοὶ ἢ ἀγαθοὶ ἐν τῇ ναυμαχίῃ
15 ταύτῃ· ἀλλήλους γὰρ καταιτιῶνται. λέγονται δὲ Σάμιοι ἐνθαῦτα κατὰ τὰ συγκείμενα πρὸς τὸν Αἰάκεα ἀειράμενοι τὰ ἱστία ἀποπλῶσαι ἐκ τῆς τάξιος ἐς τὴν Σάμον, πλὴν ἕνδεκα νεῶν. τούτων δὲ οἱ τριήραρχοι παρέμενον καὶ ἐναυμάχεον ἀνηκουστήσαντες τοῖσι
20 στρατηγοῖσι· καί σφι τὸ κοινὸν τῶν Σαμίων ἔδωκε διὰ τοῦτο τὸ πρῆγμα ἐν στήλῃ ἀναγραφῆναι πατρόθεν ὡς ἀνδράσι ἀγαθοῖσι γενομένοισι, καὶ ἔστι αὕτη ἡ στήλη ἐν τῇ ἀγορῇ. ἰδόμενοι δὲ καὶ Λέσβιοι τοὺς προσεχέας φεύγοντας τὠυτὸ ἐποίευν τοῖσι Σαμίοισι·
25 ὡς δὲ καὶ οἱ πλεῦνες τῶν Ἰώνων ἐποίευν τὰ αὐτὰ ταῦτα. XV. Τῶν δὲ παραμεινάντων ἐν τῇ ναυμαχίῃ περιέφθησαν τρηχύτατα Χῖοι ὡς ἀποδεικνύμενοί τε ἔργα λαμπρὰ καὶ οὐκ ἐθελοκακέοντες· παρείχοντο μὲν γάρ, ὥσπερ καὶ πρότερον εἰρέθη, νέας ἑκατὸν καὶ ἐπ'

ἑκάστης αὐτέων ἄνδρας τεσσεράκοντα τῶν ἀστῶν λογάδας ἐπιβατεύοντας· ὁρέοντες δὲ τοὺς πολλοὺς τῶν συμμάχων προδιδόντας οὐκ ἐδικαίευν γενέσθαι τοῖσι κακοῖσι αὐτῶν ὁμοῖοι, ἀλλὰ μετ' ὀλίγων συμμάχων μεμουνωμένοι διεκπλώοντες ἐναυμάχεον, ἐς ὃ τῶν πο- 5 λεμίων ἑλόντες νέας συχνὰς ἀπέβαλον τῶν σφετέρων νεῶν τὰς πλεῦνας. XVI. Χῖοι μὲν δὴ τῇσι λοιπῇσι τῶν νεῶν ἀποφεύγουσι ἐς τὴν ἑωυτῶν· ὅσοισι δὲ τῶν Χίων ἀδύνατοι ἦσαν αἱ νέες ὑπὸ τρωμάτων, οὗτοι δὲ ὡς ἐδιώκοντο, καταφυγγάνουσι πρὸς τὴν Μυκάλην. 10 νέας μὲν δὴ αὐτοῦ ταύτῃ ἐποκείλαντες κατέλιπον, οἱ δὲ πεζῇ ἐκομίζοντο διὰ τῆς ἠπείρου. ἐπεὶ δὲ ἐσέβαλον ἐς τὴν Ἐφεσίην κομιζόμενοι οἱ Χῖοι, νυκτός τε ἀπίκατο ἐς αὐτὴν καὶ ἐόντων τῇσι γυναιξὶ αὐτόθι θεσμοφορίων· ἐνθαῦτα δὴ οἱ Ἐφέσιοι, οὔτε προακηκοότες ὡς εἶχε περὶ 15 τῶν Χίων, ἰδόντες τε στρατὸν ἐς τὴν χώρην ἐσβεβληκότα, πάγχυ σφέας καταδόξαντες εἶναι κλῶπας καὶ ἰέναι ἐπὶ τὰς γυναῖκας, ἐξεβοήθεον πανδημεὶ καὶ ἔκτεινον τοὺς Χίους. XVII. Οὗτοι μέν νυν τοιαύτῃσι περιέπιπτον τύχῃσι· Διονύσιος δὲ ὁ Φωκαιεὺς ἐπείτε 20 ἔμαθε τῶν Ἰώνων τὰ πρήγματα διεφθαρμένα, νέας ἑλὼν τρεῖς τῶν πολεμίων ἀπέπλωε ἐς μὲν Φώκαιαν οὐκέτι, εὖ εἰδὼς ὡς ἀνδραποδιεῖται σὺν τῇ ἄλλῃ Ἰωνίῃ, ὁ δὲ ἰθέως ὡς εἶχε ἔπλωε ἐς Φοινίκην, γαύλους δὲ ἐνθαῦτα καταδύσας καὶ χρήματα λαβὼν πολλὰ 25 ἔπλωε ἐς Σικελίην, ὁρμεόμενος δὲ ἐνθεῦτεν ληϊστὴς κατεστήκεε Ἑλλήνων μὲν οὐδενός, Καρχηδονίων δὲ καὶ Τυρσηνῶν.

Fall of Miletos.

XVIII. Οἱ δὲ Πέρσαι ἐπείτε τῇ ναυμαχίῃ ἐνίκων τοὺς Ἴωνας, τὴν Μίλητον πολιορκέοντες ἐκ 30

γῆς καὶ θαλάσσης καὶ ὑπορύσσοντες τὰ τείχεα καὶ παντοίας μηχανὰς προσφέροντες αἱρέουσι κατ' ἄκρης ἕκτῳ ἔτεϊ ἀπὸ τῆς ἀποστάσιος τῆς Ἀρισταγόρεω, καὶ ἠνδραποδίσαντο τὴν πόλιν ὥστε συμπεσεῖν τὸ πάθος
5 τῷ χρηστηρίῳ τῷ ἐς Μίλητον γενομένῳ.

An oracle fulfilled by the destruction of the Milesians.

XIX. Χρεομένοισι γὰρ Ἀργείοισι ἐν Δελφοῖσι περὶ σωτηρίης τῆς πόλιος τῆς σφετέρης ἐχρήσθη ἐπίκοινον χρηστήριον, τὸ μὲν ἐς αὐτοὺς τοὺς Ἀργείους φέρον, τὴν δὲ παρενθήκην ἔχρησε ἐς Μιλησίους. τὸ
10 μέν νυν ἐς αὐτοὺς Ἀργείους ἔχον, ἐπεὰν κατὰ τοῦτο γένωμαι τοῦ λόγου, τότε μνησθήσομαι, τὰ δὲ τοῖσι Μιλησίοισι οὐ παρεοῦσι ἔχρησε, ἔχει ὧδε·

Καὶ τότε δή, Μίλητε, κακῶν ἐπιμήχανε ἔργων,
πολλοῖσιν δεῖπνόν τε καὶ ἀγλαὰ δῶρα γενήσῃ,
15 σαὶ δ' ἄλοχοι πολλοῖσι πόδας νίψουσι κομήταις,
νηοῦ δ' ἡμετέρου Διδύμοις ἄλλοισι μελήσει.

τότε δὴ ταῦτα τοὺς Μιλησίους κατελάμβανε, ὅτε γε ἄνδρες μὲν οἱ πλεῦνες ἐκτείνοντο ὑπὸ τῶν Περσέων ἐόντων κομητέων, γυναῖκες δὲ καὶ τέκνα ἐν ἀνδραπόδων
20 λόγῳ ἐγίνοντο, ἱρὸν δὲ τὸ ἐν Διδύμοισι, ὁ νηός τε καὶ τὸ χρηστήριον, συληθέντα ἐνεπίμπρατο. τῶν δ' ἐν τῷ ἱρῷ τούτῳ χρημάτων πολλάκις μνήμην ἑτέρωθι τοῦ λόγου ἐποιησάμην.

Merciful treatment of the survivors.

XX. Ἐνθεῦτεν οἱ ζωγρηθέντες τῶν Μιλησίων
25 ἤγοντο ἐς Σοῦσα. βασιλεὺς δέ σφεας Δαρεῖος κακὸν οὐδὲν ἄλλο ποιήσας κατοίκισε ἐπὶ τῇ Ἐρυθρῇ καλεομένῃ θαλάσσῃ, ἐν Ἄμπῃ πόλι, παρ' ἣν Τίγρης ποταμὸς παραρρέων ἐς θάλασσαν ἐξίει. τῆς δὲ Μιλησίης

χώρης αὐτοὶ μὲν οἱ Πέρσαι εἶχον τὰ περὶ τὴν πόλιν καὶ τὸ πεδίον, τὰ δὲ ὑπεράκρια ἔδοσαν Καρσὶ Πηδασεῦσι ἐκτῆσθαι.

Grief at Athens shewn by the fining of Phrynichos.

XXI. Παθοῦσι δὲ ταῦτα Μιλησίοισι πρὸς Περσέων οὐκ ἀπέδοσαν τὴν ὁμοίην Συβαρῖται, οἳ Λάον 5 τε καὶ Σκίδρον οἴκεον τῆς πόλιος ἀπεστερημένοι. Συβάριος γὰρ ἁλούσης ὑπὸ Κροτωνιητέων Μιλήσιοι πάντες ἡβηδὸν ἀπεκείραντο τὰς κεφαλὰς καὶ πένθος μέγα προσεθήκαντο· πόλιες γὰρ αὗται μάλιστα δὴ τῶν ἡμεῖς ἴδμεν ἀλλήλῃσι ἐξεινώθησαν. οὐδὲν ὁμοίως 10 καὶ Ἀθηναῖοι· Ἀθηναῖοι μὲν γὰρ δῆλον ἐποίησαν ὑπεραχθεσθέντες τῇ Μιλήτου ἁλώσι τῇ τε ἄλλῃ πολλαχῇ, καὶ δὴ καὶ ποιήσαντι Φρυνίχῳ δρᾶμα Μιλήτου ἅλωσιν καὶ διδάξαντι ἐς δάκρυά τε ἔπεσε τὸ θέητρον, καὶ ἐζημίωσάν μιν ὡς ἀναμνήσαντα οἰκήϊα 15 κακὰ χιλίῃσι δραχμῇσι, καὶ ἐπέταξαν μηκέτι μηδένα χρᾶσθαι τούτῳ τῷ δράματι.

The richer Samians, disapproving of the action of their leaders, abandon their country and sail to Sicily, and seize the town of Zankle.

XXII. Μίλητος μέν νυν Μιλησίων ἠρήμωτο, Σαμίων δὲ τοῖσί τι ἔχουσι τὸ μὲν ἐς τοὺς Μήδους ἐκ τῶν στρατηγῶν τῶν σφετέρων ποιηθὲν οὐδαμῶς 20 ἤρεσκε, ἐδόκεε δὲ μετὰ τὴν ναυμαχίην αὐτίκα βουλευομένοισι, πρὶν ἤ σφι ἐς τὴν χώρην ἀπικέσθαι τὸν τύραννον Αἰάκεα, ἐς ἀποικίην ἐκπλώειν μηδὲ μένοντας Μήδοισί τε καὶ Αἰάκεϊ δουλεύειν. Ζαγκλαῖοι γὰρ οἱ ἀπὸ Σικελίης τὸν αὐτὸν χρόνον τοῦτον πέμποντες ἐς 25 τὴν Ἰωνίην ἀγγέλους ἐπεκαλέοντο τοὺς Ἴωνας ἐς

Καλὴν ἀκτήν, βουλόμενοι αὐτόθι πόλιν κτίσαι Ἰώνων·
ἡ δὲ Καλὴ αὕτη ἀκτὴ καλεομένη ἐστὶ μὲν Σικελῶν,
πρὸς δὲ Τυρσηνίην τετραμμένη τῆς Σικελίης· τούτων
ὦν ἐπικαλεομένων οἱ Σάμιοι μοῦνοι Ἰώνων ἐστάλη-
5 σαν, σὺν δέ σφι Μιλησίων οἱ ἐκπεφευγότες. XXIII.
Ἐν ᾧ τοιόνδε δή τι συνήνεικε γενέσθαι· Σάμιοι γὰρ
κομιζόμενοι ἐς Σικελίην ἐγίνοντο ἐν Λοκροῖσι τοῖσι
Ἐπιζεφυρίοισι, καὶ Ζαγκλαῖοι αὐτοί τε καὶ ὁ βασιλεὺς
αὐτῶν, τῷ οὔνομα ἦν Σκύθης, περικατέατο πόλιν τῶν
10 Σικελῶν ἐξελεῖν βουλόμενοι, μαθὼν δὲ ταῦτα ὁ
Ῥηγίου τύραννος Ἀναξίλεως, ὥστε ἐὼν διάφορος
τοῖσι Ζαγκλαίοισι, συμμίξας τοῖσι Σαμίοισι ἀναπείθει, ὡς χρεὸν εἴη Καλὴν μὲν ἀκτήν, ἐπ' ἣν
ἔπλωον, ἐᾶν χαίρειν, τὴν δὲ Ζάγκλην σχεῖν ἐοῦσαν
15 ἐρῆμον ἀνδρῶν. πειθομένων δὲ τῶν Σαμίων καὶ
σχόντων τὴν Ζάγκλην ἐνθαῦτα οἱ Ζαγκλαῖοι ὡς
ἐπύθοντο ἐχομένην τὴν πόλιν ἑωυτῶν, ἐβοήθεον αὐτῇ
καὶ ἐπεκαλέοντο Ἱπποκράτεα τὸν Γέλης τύραννον·
ἦν γὰρ δή σφι οὗτος σύμμαχος. ἐπείτε δὲ αὐτοῖσι
20 καὶ ὁ Ἱπποκράτης σὺν τῇ στρατιῇ ἧκε βοηθέων,
Σκύθην μὲν τὸν μούναρχον τῶν Ζαγκλαίων ὡς ἀπο-
βαλόντα τὴν πόλιν ὁ Ἱπποκράτης πεδήσας, καὶ τὸν
ἀδελφεὸν αὐτοῦ Πυθογένεα, ἐς Ἴνυκον πόλιν ἀπέ-
πεμψε, τοὺς δὲ λοιποὺς Ζαγκλαίους κοινολογησάμενος
25 τοῖσι Σαμίοισι καὶ ὅρκους δοὺς καὶ δεξάμενος προέ-
δωκε. μισθὸς δέ οἱ ἦν εἰρημένος ὅδε ὑπὸ τῶν Σαμίων,
πάντων τῶν ἐπίπλων καὶ ἀνδραπόδων τὰ ἡμίσεα
λαβεῖν τῶν ἐν τῇ πόλι, τὰ δ' ἐπὶ τῶν ἀγρῶν πάντα
Ἱπποκράτεα λαγχάνειν. τοὺς μὲν δὴ πλεῦνας τῶν
30 Ζαγκλαίων αὐτὸς ἐν ἀνδραπόδων λόγῳ εἶχε δήσας,
τοὺς δὲ κορυφαίους αὐτῶν τριηκοσίους ἔδωκε τοῖσι

Σαμίοισι κατασφάξαι. οὐ μέντοι οἵ γε Σάμιοι ἐποίησαν ταῦτα. XXIV. Σκύθης δὲ ὁ τῶν Ζαγκλαίων μούναρχος ἐκ τῆς Ἰνύκου ἐκδιδρήσκει ἐς Ἱμέρην, ἐκ δὲ ταύτης παρῆν ἐς τὴν Ἀσίην καὶ ἀνέβη παρὰ βασιλέα Δαρεῖον. καί μιν ἐνόμισε Δαρεῖος πάντων ἀνδρῶν 5 δικαιότατον εἶναι, ὅσοι ἐκ τῆς Ἑλλάδος παρ' ἑωυτὸν ἀνέβησαν. καὶ γὰρ παραιτησάμενος βασιλέα ἐς Σικελίην ἀπίκετο καὶ αὖτις ἐκ τῆς Σικελίης ὀπίσω παρὰ βασιλέα, ἐς ὃ γήραϊ μέγα ὄλβιος ἐὼν ἐτελεύτησε ἐν Πέρσῃσι. Σάμιοι δὲ ἀπαλλαχθέντες Μήδων ἀπο- 10 νητὶ πόλιν καλλίστην Ζάγκλην περιεβεβλέατο.

Results of the fall of Miletos.

XXV. Μετὰ δὲ τὴν ναυμαχίην τὴν ὑπὲρ Μιλήτου γενομένην Φοίνικες κελευσάντων Περσέων κατῆγον ἐς Σάμον Αἰάκεα τὸν Συλοσῶντος ὡς πολλοῦ τε ἄξιον γενόμενον σφίσι καὶ μεγάλα κατεργασάμενον. καὶ 15 Σαμίοισι μούνοισι τῶν ἀποστάντων ἀπὸ Δαρείου διὰ τὴν ἔκλειψιν τῶν νεῶν τὴν ἐν τῇ ναυμαχίῃ οὔτε ἡ πόλις οὔτε τὰ ἱρὰ ἐνεπρήσθη. Μιλήτου δὲ ἁλούσης αὐτίκα Καρίην ἔσχον οἱ Πέρσαι, τὰς μὲν ἐθελοντὴν τῶν πολίων ὑποκυψάσας, τὰς δὲ ἀνάγκῃ προσηγάγοντο. 20

Histiaios, hearing of the fall of Miletos, leaves Byzantium and seizes Chios.

XXVI. Ταῦτα μὲν δὴ οὕτω ἐγίνετο, Ἱστιαίῳ δὲ τῷ Μιλησίῳ ἐόντι περὶ Βυζάντιον καὶ συλλαμβάνοντι τὰς Ἰώνων ὁλκάδας ἐκπλωούσας ἐκ τοῦ Πόντου ἐξαγγέλλεται τὰ περὶ Μίλητον γενόμενα. τὰ μὲν δὴ περὶ 25 Ἑλλήσποντον ἔχοντα πρήγματα ἐπιτράπει Βισάλτῃ

Ἀπολλοφάνεος παιδὶ Ἀβυδηνῷ, αὐτὸς δὲ ἔχων Λεσβίους ἐς Χίον ἔπλωε, καὶ Χίων φρουρῇ οὐ προσιεμένῃ μιν συνέβαλε ἐν Κοίλοισι καλεομένοισι τῆς Χίης χώρης. τούτων τε δὴ ἐφόνευσε συχνούς, καὶ τῶν
5 λοιπῶν Χίων, οἷα δὴ κεκακωμένων ἐκ τῆς ναυμαχίης, ὁ Ἱστιαῖος ἔχων τοὺς Λεσβίους ἐπεκράτησε, ἐκ Πολίχνης τῆς Χίων ὁρμεόμενος.

Two previous disasters to the Chians.

XXVII. Φιλέει δέ κως προσημαίνειν, εὖτ' ἂν μέλλῃ μεγάλα κακὰ ἢ πόλι ἢ ἔθνεϊ ἔσεσθαι· καὶ γὰρ
10 Χίοισι πρὸ τούτων σημήϊα μεγάλα ἐγένετο. τοῦτο μέν σφι πέμψασι ἐς Δελφοὺς χορὸν νεηνιέων ἑκατὸν δύο μοῦνοι τούτων ἀπενόστησαν, τοὺς δὲ ὀκτώ τε καὶ ἐνενήκοντα αὐτῶν λοιμὸς ὑπολαβὼν ἀπήνεικε· τοῦτο δὲ ἐν τῇ πόλι τὸν αὐτὸν τοῦτον χρόνον, ὀλίγῳ πρὸ
15 τῆς ναυμαχίης, παισὶ γράμματα διδασκομένοισι ἐνέπεσε ἡ στέγη, ὥστε ἀπ' ἑκατὸν καὶ εἴκοσι παίδων εἷς μοῦνος ἀπέφυγε. ταῦτα μέν σφι σημήϊα ὁ θεὸς προέδεξε, μετὰ δὲ ταῦτα ἡ ναυμαχίη ὑπολαβοῦσα ἐς γόνυ τὴν πόλιν ἔβαλε, ἐπὶ δὲ τῇ ναυμαχίῃ ἐπεγένετο
20 Ἱστιαῖος Λεσβίους ἄγων· κεκακωμένων δὲ τῶν Χίων, καταστροφὴν εὐπετέως αὐτῶν ἐποιήσατο.

Histiaios thence goes to Thasos, Lesbos, and the plain of the Kaïkos in Mysia, where he is captured by Harpagos and put to death B.C. 494.

XXVIII. Ἐνθεῦτεν δὲ ὁ Ἱστιαῖος ἐστρατεύετο ἐπὶ Θάσον ἄγων Ἰώνων καὶ Αἰολέων συχνούς. περικατημένῳ δέ οἱ Θάσον ἦλθε ἀγγελίη, ὡς οἱ Φοίνικες
25 ἀναπλώουσι ἐκ τῆς Μιλήτου ἐπὶ τὴν ἄλλην Ἰωνίην.

πυθόμενος δὲ ταῦτα Θάσον μὲν ἀπόρθητον λείπει, αὐτὸς δὲ ἐς τὴν Λέσβον ἠπείγετο ἄγων πᾶσαν τὴν στρατιήν. ἐκ Λέσβου δὲ λιμαινούσης οἱ τῆς στρατιῆς πέρην διαβαίνει, ἐκ τοῦ Ἀταρνέος ὡς ἀμήσων τὸν σῖτον τόν τε ἐνθεῦτεν καὶ τὸν ἐκ Καΐκου πεδίου τὸν 5 τῶν Μυσῶν. ἐν δὲ τούτοισι τοῖσι χωρίοισι ἐτύγχανε ἐὼν Ἅρπαγος ἀνὴρ Πέρσης, στρατηγὸς στρατιῆς οὐκ ὀλίγης, ὅς οἱ ἀποβάντι συμβαλὼν αὐτόν τε Ἱστιαῖον ζωγρίῃ ἔλαβε καὶ τὸν στρατὸν αὐτοῦ τὸν πλέω διέφθειρε. XXIX. Ἐζωγρήθη δὲ ὁ Ἱστιαῖος ὧδε· ὡς 10 ἐμάχοντο οἱ Ἕλληνες τοῖσι Πέρσῃσι ἐν τῇ Μαλήνῃ τῆς Ἀταρνείτιδος χώρης, οἱ μὲν συνέστασαν χρόνον ἐπὶ πολλόν, ἡ δὲ ἵππος ὕστερον ὁρμηθεῖσα ἐπιπίπτει τοῖσι Ἕλλησι· τό τε δὴ ἔργον τῆς ἵππου τοῦτο ἐγένετο, καὶ τετραμμένων τῶν Ἑλλήνων ὁ Ἱστιαῖος 15 ἐλπίζων οὐκ ἀπολέεσθαι ὑπὸ βασιλέος διὰ τὴν παρεοῦσαν ἁμαρτάδα φιλοψυχίην τοιήνδε τινὰ ἀναιρέεται· ὡς φεύγων τε κατελαμβάνετο ὑπ᾽ ἀνδρὸς Πέρσεω καὶ ὡς καταιρεόμενος ὑπ᾽ αὐτοῦ ἔμελλε συγκεντηθήσεσθαι, Περσίδα γλῶσσαν μετεὶς καταμηνύει ἑωυτόν, ὡς εἴη 20 Ἱστιαῖος ὁ Μιλήσιος. XXX. Εἰ μέν νυν, ὡς ἐζωγρήθη, ἀνήχθη ἀγόμενος παρὰ βασιλέα Δαρεῖον, ὁ δὲ οὔτ᾽ ἂν ἔπαθε κακὸν οὐδὲν δοκέειν ἐμοί, ἀπῆκέ τ᾽ ἂν αὐτῷ τὴν αἰτίην· νῦν δέ μιν αὐτῶν τε τούτων εἵνεκεν, καὶ ἵνα μὴ διαφυγὼν αὖτις μέγας παρὰ βασιλέϊ 25 γένηται, Ἀρταφέρνης τε ὁ Σαρδίων ὕπαρχος καὶ ὁ λαβὼν Ἅρπαγος, ὡς ἀπίκετο ἀγόμενος ἐς Σάρδις, τὸ μὲν αὐτοῦ σῶμα αὐτοῦ ταύτῃ ἀνεσταύρωσαν, τὴν δὲ κεφαλὴν ταριχεύσαντες ἀνήνεικαν παρὰ βασιλέα Δαρεῖον ἐς Σοῦσα. Δαρεῖος δὲ πυθόμενος ταῦτα καὶ 30 ἐπαιτιησάμενος τοὺς ταῦτα ποιήσαντας, ὅτι μιν οὐ ζώοντα ἀνήγαγον ἐς ὄψιν τὴν ἑωυτοῦ, τὴν κεφαλὴν τὴν Ἱστιαίου λούσαντάς τε καὶ περιστείλαντας εὖ ἐνετείλατο θάψαι ὡς ἀνδρὸς μεγάλως ἑωυτῷ τε καὶ Πέρσῃσι εὐεργέτεω.

After wintering at Miletos (B.C. 494—493) *the Persian fleet reduce Chios, Lesbos, and Tenedos. The Persian drag-net.*

5 XXXI. Τὰ μὲν περὶ Ἱστιαῖον οὕτω ἔσχε. ὁ δὲ ναυτικὸς στρατὸς ὁ Περσέων χειμερίσας περὶ Μίλητον τῷ δευτέρῳ ἔτεϊ ὡς ἀνέπλωσε, αἱρέει εὐπετέως τὰς νήσους τὰς πρὸς τῇ ἠπείρῳ κειμενας, Χίον καὶ Λέσβον καὶ Τένεδον. ὅκως δὲ λάβοι τινὰ
10 τῶν νήσων, ὡς ἑκάστην αἱρέοντες οἱ βάρβαροι ἐσαγήνευον τοὺς ἀνθρώπους. σαγηνεύουσι δὲ τόνδε τὸν τρόπον· ἀνὴρ ἀνδρὸς ἁψάμενος τῆς χειρὸς ἐκ θαλάσσης τῆς βορηΐης ἐπὶ τὴν νοτίην διήκουσι, καὶ ἔπειτεν διὰ πάσης τῆς νήσου διέρχονται ἐκθηρεύ-
15 οντες τοὺς ἀνθρώπους. αἵρεον δὲ καὶ τὰς ἐν τῇ ἠπείρῳ πόλιας τὰς Ἰάδας κατὰ ταὐτά, πλὴν οὐκ ἐσαγήνευον τοὺς ἀνθρώπους· οὐ γὰρ οἷά τ' ἦν. XXXII. Ἐνθαῦτα Περσέων οἱ στρατηγοὶ οὐκ ἐψεύσαντο τὰς ἀπειλάς, τὰς ἐπηπείλησαν τοῖσι Ἴωσι
20 στρατοπεδευομένοισι ἐναντία σφίσι. ὡς γὰρ δὴ ἐπεκράτησαν τῶν πολίων, παῖδάς τε τοὺς εὐειδεστάτους ἐκλεγόμενοι ἐξέταμνον καὶ ἐποίευν ἀντὶ τοῦ εἶναι ἐνόρχιας εὐνούχους, καὶ παρθένους τὰς καλλιστευούσας ἀνασπάστους παρὰ βασιλέα· ταῦτά τε δὴ ἐποίευν,
25 καὶ τὰς πόλιας ἐνεπίμπρασαν αὐτοῖσι τοῖς ἱροῖσι. οὕτω δὴ τὸ τρίτον Ἴωνες κατεδουλώθησαν, πρῶτον μὲν ὑπὸ Λυδῶν, δὶς δὲ ἐπεξῆς τότε ὑπὸ Περσέων.

B.C. 493. *Then they take the cities of the European coast of the Hellespont up to Byzantium (the Byzantines and Kalchedonians retreating into the Euxine and settling at Mesambria), then Proconnesos and Artake, then the Thracian Chersonese.*

XXXIII. Ἀπὸ δὲ Ἰωνίης ἀπαλλασσόμενος ὁ ναυτικὸς στρατὸς τὰ ἐπ' ἀριστερὰ ἐσπλώοντι τοῦ Ἑλλησπόντου αἵρεε πάντα· τὰ γὰρ ἐπὶ δεξιὰ αὐτοῖσι τοῖσι Πέρσῃσι ὑποχείρια ἦν γεγονότα κατ' ἤπειρον. Εἰσὶ δὲ ἐν τῇ Εὐρώπῃ αἵδε τοῦ Ἑλλησπόντου· Χερ- 5 σόνησός τε, ἐν τῇ πόλιες συχναὶ ἔνεισι, καὶ Πέρινθος καὶ τὰ τείχεα τὰ ἐπὶ Θρηΐκης καὶ Σηλυβρίη τε καὶ Βυζάντιον. Βυζάντιοι μέν νυν καὶ οἱ πέρηθε Καλχηδόνιοι οὐδ' ὑπέμειναν ἐπιπλώοντας τοὺς Φοίνικας, ἀλλ' οἴχοντο ἀπολιπόντες τὴν σφετέρην ἔσω ἐς τὸν 10 Εὔξεινον πόντον, καὶ ἐνθαῦτα πόλιν Μεσαμβρίην οἴκησαν· οἱ δὲ Φοίνικες κατακαύσαντες ταύτας τὰς χώρας τὰς καταλεχθείσας τράπονται ἐπί τε Προκόννησον καὶ Ἀρτάκην, πυρὶ δὲ καὶ ταύτας νείμαντες ἔπλωον αὖτις ἐς τὴν Χερσόνησον ἐξαιρήσοντες τὰς 15 ἐπιλοίπους τῶν πολίων, ὅσας πρότερον προσσχόντες οὐ κατέσυραν. ἐπὶ δὲ Κύζικον οὐδὲ ἔπλωσαν ἀρχήν· αὐτοὶ γὰρ Κυζικηνοὶ ἔτι πρότερον τοῦ Φοινίκων ἐσπλόου ἐγεγόνεσαν ὑπὸ βασιλέϊ Οἰβάρεϊ τῷ Μεγαβάζου ὁμολογήσαντες, τῷ ἐν Δασκυλείῳ ὑπάρχῳ. 20 τῆς δὲ Χερσονήσου, πλὴν Καρδίης πόλιος, τὰς ἄλλας πάσας ἐχειρώσαντο οἱ Φοίνικες.

Miltiades son of Kypselos, invited by the Dolonki, becomes tyrant of the Chersonese [before B.C. 546*], and fortifies it.*

XXXIV. Ἐτυράννευε δὲ αὐτέων μέχρι τότε

Μιλτιάδης ὁ Κίμωνος τοῦ Στησαγόρεω, κτησαμένου
τὴν ἀρχὴν ταύτην πρότερον Μιλτιάδεω τοῦ Κυψέλου
τρόπῳ τοιῷδε· εἶχον Δόλογκοι Θρήϊκες τὴν Χερσόνησον ταύτην. οὗτοι ὦν οἱ Δόλογκοι πιεσθέντες
5 πολέμῳ ὑπὸ Ἀψινθίων ἐς Δελφοὺς ἔπεμψαν τοὺς
βασιλέας περὶ τοῦ πολέμου χρησομένους. ἡ δὲ
Πυθίη σφι ἀνεῖλε οἰκιστὴν ἐπάγεσθαι ἐπὶ τὴν χώρην
τοῦτον, ὃς ἄν σφεας ἀπιόντας ἐκ τοῦ ἱροῦ πρῶτος ἐπὶ
ξείνια καλέσῃ. ἰόντες δὲ οἱ Δόλογκοι τὴν ἱρὴν ὁδὸν
10 διὰ Φωκέων τε καὶ Βοιωτῶν ἤϊσαν καί σφεας ὡς οὐδεὶς
ἐκάλεε, ἐκτράπονται ἐπ' Ἀθηνέων. XXXV. Ἐν δὲ
τῇσι Ἀθήνῃσι τηνικαῦτα εἶχε μὲν τὸ πᾶν κράτος
Πεισίστρατος, ἀτὰρ εδυνάστευε καὶ Μιλτιάδης ὁ Κυψέλου, ἐὼν οἰκίης τεθριπποτρόφου, τὰ μὲν ἀνέκαθεν
15 ἀπ' Αἰακοῦ τε καὶ Αἰγίνης γεγονώς, τὰ δὲ νεώτερα
Ἀθηναῖος, Φιλαίου τοῦ Αἴαντος παιδός, γενομένου
πρώτου τῆς οἰκίης ταύτης Ἀθηναίου. οὗτος ὁ Μιλτιάδης κατήμενος ἐν τοῖσι προθύροισι τοῖσι ἑωυτοῦ,
ὁρέων τοὺς Δολόγκους παριόντας ἐσθῆτα ἔχοντας οὐκ
20 ἐγχωρίην καὶ αἰχμὰς προσεβώσατο, καί σφι προσ-
ελθοῦσι ἐπηγγείλατο καταγωγὴν καὶ ξείνια. οἱ δὲ
δεξάμενοι καὶ ξεινισθέντες ὑπ' αὐτοῦ ἐξέφαινον πᾶν
τὸ μαντήϊον, ἐκφήναντες δὲ ἐδέοντο αὐτοῦ τῷ θεῷ
μιν πείθεσθαι. Μιλτιάδεα δὲ ἀκούσαντα παραυτίκα
25 ἔπεισε ὁ λόγος οἷα ἀχθόμενόν τε τῇ Πεισιστράτου
ἀρχῇ καὶ βουλόμενον ἐκποδὼν εἶναι. αὐτίκα δὲ
ἐστάλη ἐς Δελφοὺς ἐπειρησόμενος τὸ χρηστήριον,
εἰ ποιέῃ τά περ αὐτοῦ οἱ Δόλογκοι προσεδέοντο.
XXXVI. Κελευούσης δὲ καὶ τῆς Πυθίης, οὕτω δὴ
30 Μιλτιάδης ὁ Κυψέλου, Ὀλύμπια ἀναραιρηκὼς πρότερον τούτων τεθρίππῳ, τότε παραλαβὼν Ἀθηναίων

πάντα τὸν βουλόμενον μετέχειν τοῦ στόλου ἔπλωε ἅμα τοῖσι Δολόγκοισι, καὶ ἔσχε τὴν χώρην. καί μιν οἱ ἐπαγαγόμενοι τύραννον κατεστήσαντο. ὁ δὲ πρῶτον μὲν ἀπετείχισε τὸν ἰσθμὸν τῆς Χερσονήσου ἐκ Καρδίης πόλιος ἐς Πακτύην, ἵνα μὴ ἔχοιέν σφεας οἱ 5 Ἀψίνθιοι δηλέεσθαι ἐσβάλλοντες ἐς τὴν χώρην. εἰσὶ δὲ οὗτοι στάδιοι ἕξ τε καὶ τριήκοντα τοῦ ἰσθμοῦ· ἀπὸ δὲ τοῦ ἰσθμοῦ τούτου ἡ Χερσόνησος ἔσω πᾶσά ἐστι σταδίων εἴκοσι καὶ τετρακοσίων τὸ μῆκος. XXXVII. Ἀποτειχίσας ὦν τὸν αὐχένα τῆς Χερ- 10 σονήσου ὁ Μιλτιάδης καὶ τοὺς Ἀψινθίους τρόπῳ τοιούτῳ ὠσάμενος, τῶν λοιπῶν πρώτοισι ἐπολέμησε Λαμψακηνοῖσι. καί μιν οἱ Λαμψακηνοὶ λοχήσαντες αἱρέουσι ζωγρίῃ. ἦν δὲ ὁ Μιλτιάδης Κροίσῳ τῷ Λυδῷ ἐν γνώμῃ γεγονώς· πυθόμενος ὦν ὁ Κροῖσος ταῦτα 15 πέμπων προηγόρευε τοῖσι Λαμψακηνοῖσι μετιέναι Μιλτιάδεα, εἰ δὲ μή, σφέας πίτυος τρόπον ἠπείλεε ἐκτρίψειν. πλανωμένων δὲ τῶν Λαμψακηνῶν ἐν τοῖσι λόγοισι, τί ἐθέλει τὸ ἔπος εἶναι, τό σφι ἠπείλησεν ὁ Κροῖσος, πίτυος τρόπον ἐκτρίψειν, μόγις κοτὲ μαθὼν 20 τῶν τις πρεσβυτέρων εἶπε τὸ ἐόν, ὅτι πίτυς μούνη δενδρέων πάντων ἐκκοπεῖσα βλαστὸν οὐδένα μετίει, ἀλλὰ πανώλεθρος ἐξαπόλλυται. δείσαντες ὦν οἱ Λαμψακηνοὶ Κροῖσον λύσαντες μετῆκαν Μιλτιάδεα.

Miltiades son of Kypselos is succeeded by his nephew Stesagoras son of Kimon, who having been assassinated was succeeded by his brother Miltiades son of Kimon [between B.C. 527—514].

XXXVIII. Οὗτος μὲν δὴ διὰ Κροῖσον ἐκφεύγει, 25 μετὰ δὲ τελευτᾷ ἄπαις, τὴν ἀρχήν τε καὶ τὰ χρήματα

παραδοὺς Στησαγόρῃ τῷ Κίμωνος ἀδελφεοῦ παιδὶ ὁμομητρίου. καί οἱ τελευτήσαντι Χερσονησῖται θύουσι, ὡς νόμος οἰκιστῇ, καὶ ἀγῶνα ἱππικόν τε καὶ γυμνικὸν ἐπιστᾶσι, ἐν τῷ Λαμψακηνῶν οὐδενὶ ἐγγί-
5 νεται ἀγωνίζεσθαι. πολέμου δὲ ἐόντος πρὸς Λαμψακηνοὺς καὶ Στησαγόρην κατέλαβε ἀποθανεῖν ἄπαιδα, πληγέντα τὴν κεφαλὴν πελέκεϊ ἐν τῷ πρυτανηΐῳ πρὸς ἀνδρὸς αὐτομόλου μὲν τῷ λόγῳ, πολεμίου δὲ καὶ ὑποθερμοτέρου τῷ ἔργῳ. XXXIX. Τελευτήσαντος
10 δὲ καὶ Στησαγόρεω τρόπῳ τοιῷδε ἐνθαῦτα Μιλτιάδεα τὸν Κίμωνος, Στησαγόρεω δὲ τοῦ τελευτήσαντος ἀδελφεὸν, καταλαμψόμενον τὰ πρήγματα ἐπὶ Χερσονήσου ἀποστέλλουσι τριήρεϊ οἱ Πεισιστρατίδαι, οἵ μιν καὶ ἐν Ἀθήνῃσι ἐποίευν εὖ, ὡς οὐ συνειδότες
15 δῆθεν τοῦ πατρὸς [Κίμωνος] αὐτοῦ τὸν θάνατον, τὸν ἐγὼ ἐν ἄλλῳ λόγῳ σημανέω ὡς ἐγένετο. Μιλτιάδης δὲ ἀπικόμενος ἐς τὴν Χερσόνησον εἶχε κατ' οἴκους, τὸν ἀδελφεὸν Στησαγόρην δηλαδὴ ἐπιτιμέων. οἱ δὲ Χερσονησῖται πυνθανόμενοι ταῦτα συνελέχθησαν ἀπὸ
20 πασέων τῶν πολίων οἱ δυναστεύοντες πάντοθεν, κοινῷ δὲ στόλῳ ἀπικόμενοι ὡς συλλυπηθησόμενοι ἐδέθησαν ὑπ' αὐτοῦ. Μιλτιάδης τε δὴ ἴσχει τὴν Χερσόνησον πεντακοσίους βόσκων ἐπικούρους, καὶ γαμέει Ὀλόρου τοῦ Θρηΐκων βασιλέος θυγατέρα
25 Ἡγησιπύλην.

In B.C. 495 *Miltiades was expelled from the Chersonese by the Scythians, but returned. In* B.C. 493 *he fled to Athens for fear of the Phoenikian fleet of Darius.*

XL. Οὗτος δὲ ὁ Κίμωνος Μιλτιάδης νεωστὶ μὲν ἐληλύθεε ἐς τὴν Χερσόνησον, κατελάμβανε δέ μιν ἐλθόντα ἄλλα τῶν κατεχόντων πρηγμάτων χαλεπώτερα. τρίτῳ μὲν γὰρ ἔτεϊ τούτων Σκύθας ἐκφεύγει· Σκύθαι γὰρ οἱ νομάδες ἐρεθισθέντες ὑπὸ βασιλέος Δαρείου συνεστράφησαν καὶ ἤλασαν μέχρι τῆς Χερσονήσου ταύτης. τούτους ἐπιόντας οὐκ ὑπομείνας ὁ 5 Μιλτιάδης ἔφευγε ἀπὸ Χερσονήσου, ἐς ὃ οἵ τε Σκύθαι ἀπηλλάχθησαν καί μιν οἱ Δόλογκοι κατήγαγον ὀπίσω. ταῦτα μὲν δὴ τρίτῳ ἔτεϊ πρότερον ἐγεγόνεε τῶν τότε μιν κατεχόντων, XLI. τότε δὲ πυνθανόμενος εἶναι τοὺς Φοίνικας ἐν Τενέδῳ, πληρώσας τριήρεας πέντε 10 χρημάτων τῶν παρεόντων ἀπέπλωε ἐς τὰς Ἀθήνας. καὶ ὥσπερ ὡρμήθη ἐκ Καρδίης πόλιος, ἔπλωε διὰ τοῦ Μέλανος κόλπου παραμείβετό τε τὴν Χερσόνησον, καὶ οἱ Φοίνικές οἱ περιπίπτουσι τῇσι νηυσί. αὐτὸς μὲν δὴ Μιλτιάδης σὺν τῇσι τέσσερσι τῶν νεῶν κατα- 15 φεύγει ἐς Ἴμβρον, τὴν δέ οἱ πέμπτην τῶν νεῶν κατεῖλον διώκοντες οἱ Φοίνικες. τῆς δὲ νεὸς ταύτης ἔτυχε τῶν Μιλτιάδεω παίδων ὁ πρεσβύτατος ἄρχων Μητίοχος, οὐκ ἐκ τῆς Ὀλόρου τοῦ Θρήϊκος ἐὼν θυγατρός, ἀλλ' ἐξ ἄλλης. καὶ τοῦτον ἅμα τῇ νηῒ 20 εἷλον οἱ Φοίνικες, καί μιν πυθόμενοι ὡς εἴη Μιλτιάδεω παῖς, ἀνήγαγον παρὰ βασιλέα, δοκέοντες χάριτα μεγάλην καταθήσεσθαι, ὅτι δὴ Μιλτιάδης γνώμην ἀπεδέξατο ἐν τοῖσι Ἴωσι πείθεσθαι κελεύων τοῖσι Σκύθῃσι, ὅτε οἱ Σκυθαι προσεδέοντο λύσαντας τὴν 25 σχεδίην ἀποπλώειν ἐς τὴν ἑωυτῶν. Δαρεῖος δέ, ὡς οἱ Φοίνικες Μητίοχον τὸν Μιλτιάδεω ἀνήγαγον, ἐποίησε κακὸν μὲν οὐδὲν Μητίοχον, ἀγαθὰ δὲ συχνά· καὶ γὰρ οἶκον καὶ κτῆσιν ἔδωκε καὶ Περσίδα γυναῖκα, ἐκ τῆς οἱ τέκνα ἐγένετο, τὰ ἐς Πέρσας κεκοσμέαται. 30 Μιλτιάδης δὲ ἐξ Ἴμβρου ἀπικνέεται ἐς τὰς Ἀθήνας.

B.C. 493. *Reorganisation of Ionia.*

XLII. Καὶ κατὰ τὸ ἔτος τοῦτο ἐκ τῶν Περσέων οὐδὲν ἔτι πλέον ἐγένετο τούτων ἐς νεῖκος φέρον Ἴωσι, ἀλλὰ τάδε μὲν χρήσιμα κάρτα τοῖσι Ἴωσι ἐγένετο τούτου τοῦ ἔτεος· Ἀρταφέρνης ὁ Σαρδίων ὕπαρχος
5 μεταπεμψάμενος ἀγγέλους ἐκ τῶν πολίων συνθήκας σφίσι αὐτοῖσι τοὺς Ἴωνας ἠνάγκασε ποιέεσθαι, ἵνα δωσίδικοι εἶεν καὶ μὴ ἀλλήλους φέροιέν τε καὶ ἄγοιεν. ταῦτά τε ἠνάγκασε ποιέειν, καὶ τὰς χώρας σφέων μετρήσας κατὰ παρασάγγας, τοὺς καλέουσι οἱ
10 Πέρσαι τὰ τριήκοντα στάδια, κατὰ δὴ τούτους μετρήσας φόρους ἔταξε ἑκάστοισι, οἳ κατὰ χώρην διατελέουσι ἔχοντες ἐκ τούτου τοῦ χρόνου αἰεὶ ἔτι καὶ ἐς ἐμὲ ὡς ἐτάχθησαν ἐξ Ἀρταφέρνεος, ἐτάχθησαν δὲ σχεδὸν κατὰ τὰ αὐτὰ τὰ καὶ πρότερον εἶχον.

B.C. 492. *Mardonius made Satrap of Asia Minor, establishes democracy in the Ionian cities, and proceeds from the Hellespont to coast along the European shore.*

15 XLIII. Καί σφι ταῦτα μὲν εἰρηναῖα ἦν, ἅμα δὲ τῷ ἔαρι τῶν ἄλλων καταλελυμένων στρατηγῶν ἐκ βασιλέος Μαρδόνιος ὁ Γωβρύεω κατέβαινε ἐπὶ θάλασσαν, στρατὸν πολλὸν μὲν κάρτα πεζὸν ἅμα ἀγόμενος, πολλὸν δὲ ναυτικόν, ἡλικίην τε νέος ἐὼν
20 καὶ νεωστὶ γεγαμηκὼς βασιλέος Δαρείου θυγατέρα Ἀρταζώστρην. ἄγων δὲ τὸν στρατὸν τοῦτον ὁ Μαρδόνιος ἐπείτε ἐγένετο ἐν τῇ Κιλικίῃ, αὐτὸς μὲν ἐπιβὰς ἐπὶ νεὸς ἐκομίζετο ἅμα τῇσι ἄλλῃσι νηυσί, στρατιὴν δὲ τὴν πεζὴν ἄλλοι ἡγεμόνες ἦγον ἐπὶ τὸν
25 Ἑλλήσποντον. ὡς δὲ παραπλώων τὴν Ἀσίην ἀπίκετο

ὁ Μαρδόνιος ἐς τὴν Ἰωνίην, ἐνθαῦτα μέγιστον θῶυμα ἐρέω τοῖσι μὴ ἀποδεκομένοισι Ἑλλήνων Περσέων τοῖσι ἑπτὰ Ὀτάνεα γνώμην ἀποδέξασθαι, ὡς χρεὸν εἴη δημοκρατέεσθαι Πέρσας· τοὺς γὰρ τυράννους τῶν Ἰώνων καταπαύσας πάντας ὁ Μαρδόνιος δημο- 5 κρατίας κατίστα ἐς τὰς πόλιας. ταῦτα δὲ ποιήσας ἠπείγετο ἐς τὸν Ἑλλήσποντον. ὡς δὲ συνελέχθη μὲν χρῆμα πολλὸν νεῶν, συνελέχθη δὲ καὶ πεζὸς στρατὸς πολλὸς, διαβάντες τῇσι νηυσὶ τὸν Ἑλλήσποντον ἐπορεύοντο διὰ τῆς Εὐρώπης, ἐπορεύοντο δὲ ἐπί τε 10 Ἐρέτριαν καὶ Ἀθήνας.

The fleet of Mardonius is wrecked on Mt Athos;

XLIV. Αὗται μὲν ὦν σφι πρόσχημα ἦσαν τοῦ στόλου, ἀτὰρ ἐν νόῳ ἔχοντες ὅσας ἂν πλείστας δύναιντο καταστρέφεσθαι τῶν Ἑλληνίδων πολίων, τοῦτο μὲν δὴ τῇσι νηυσὶ Θασίους οὐδὲ χεῖρας ἀντα- 15 ειραμένους κατεστρέψαντο, τοῦτο δὲ τῷ πεζῷ Μακεδόνας πρὸς τοῖσι ὑπάρχουσι δούλους προσεκτήσαντο· τὰ γὰρ ἐντὸς Μακεδόνων ἔθνεα πάντα σφι ἤδη ἦν ὑποχείρια γεγονότα. ἐκ μὲν δὴ Θάσου διαβαλόντες πέρην ὑπὸ τὴν ἤπειρον ἐκομίζοντο μέχρι Ἀκάνθου, 20 ἐκ δὲ Ἀκάνθου ὁρμεόμενοι τὸν Ἄθων περιέβαλλον. ἐπιπεσὼν δέ σφι περιπλώουσι βορῆς ἄνεμος μέγας τε καὶ ἄπορος κάρτα τρηχέως περιέσπε πλήθεϊ πολλὰς τῶν νεῶν ἐκβάλλων πρὸς τὸν Ἄθων. λέγεται γὰρ κατὰ τριηκοσίας μὲν τῶν νεῶν τὰς διαφθαρείσας 25 εἶναι, ὑπὲρ δὲ δύο μυριάδας ἀνθρώπων· ὥστε γὰρ θηριωδεστάτης ἐούσης τῆς θαλάσσης ταύτης τῆς περὶ τὸν Ἄθων οἱ μὲν ὑπὸ τῶν θηρίων διεφθείροντο ἁρπαζόμενοι, οἱ δὲ πρὸς τὰς πέτρας ἀρασσόμενοι· οἱ δὲ αὐτῶν νέειν οὐκ ἠπιστέατο καὶ κατὰ τοῦτο διεφθείροντο, οἱ δὲ ῥίγεϊ.

and the army much damaged by the Brygi, whom however he finally subdues.

XLV. Ὁ μὲν δὴ ναυτικὸς στρατὸς οὕτω ἔπρησσε. Μαρδονίῳ δὲ καὶ τῷ πεζῷ στρατοπεδευομένῳ
5 ἐν Μακεδονίῃ νυκτὸς Βρύγοι Θρήϊκες ἐπεχείρησαν. καί σφεων πολλοὺς φονεύουσι οἱ Βρύγοι, Μαρδόνιον δὲ αὐτὸν τρωματίζουσι. οὐ μέντοι οὐδὲ αὐτοὶ δουλοσύνην διέφυγον πρὸς Περσέων· οὐ γὰρ δὴ πρότερον ἀπανέστη ἐκ τῶν χωρέων τούτων Μαρδόνιος, πρὶν
10 ἤ σφεας ὑποχειρίους ἐποιήσατο. τούτους μέντοι καταστρεψάμενος ἀπῆγε τὴν στρατιὴν ὀπίσω, ἅτε τῷ πεζῷ τε προσπταίσας πρὸς τοὺς Βρύγους καὶ τῷ ναυτικῷ μεγάλως περὶ Ἄθων. οὗτος μέν νυν ὁ στόλος αἰσχρῶς ἀγωνισάμενος ἀπηλλάχθη ἐς τὴν
15 Ἀσίην.

B.C. 491. *The Thasians deprived of their fleet, and ordered to dismantle their fortifications.*

XLVI. Δευτέρῳ δὲ ἔτεϊ τούτων ὁ Δαρεῖος πρῶτα μὲν Θασίους διαβληθέντας ὑπὸ τῶν ἀστυγειτόνων, ὡς ἀπόστασιν μηχανῴατο, πέμψας ἄγγελον ἐκέλευέ σφεας τὸ τεῖχος περιαιρέειν καὶ τὰς νέας ἐς Ἄβδηρα
20 κομίζειν. οἱ γὰρ δὴ Θάσιοι οἷα ὑπὸ Ἱστιαίου τε τοῦ Μιλησίου πολιορκηθέντες καὶ προσόδων ἐουσέων μεγάλων ἐχρέοντο τοῖσι χρήμασι ναῦς τε ναυπηγεύμενοι μακρὰς καὶ τεῖχος ἰσχυρότερον περιβαλλόμενοι. ἡ δὲ πρόσοδός σφι ἐγίνετο ἔκ τε τῆς
25 ἠπείρου καὶ ἀπὸ τῶν μετάλλων. ἐκ μέν γε τῶν ἐκ

Σκαπτησύλης τῶν χρυσέων μετάλλων τὸ ἐπίπαν ὀγδώκοντα τάλαντα προσήϊε, ἐκ δὲ τῶν ἐν αὐτῇ Θάσῳ ἐλάσσω μὲν τούτων, συχνὰ δὲ οὕτω, ὥστε τὸ ἐπίπαν Θασίοισι ἐοῦσι καρπῶν ἀτελέσι προσήϊε ἀπό τε τῆς ἠπείρου καὶ τῶν μετάλλων ἔτεος ἑκάστου 5 διηκόσια τάλαντα, ὅτε δὲ τὸ πλεῖστον προσῆλθε, τριηκόσια. XLVII. Εἶδον δὲ καὶ αὐτὸς τὰ μέταλλα ταῦτα, καὶ μακρῷ ἦν αὐτῶν θωυμασιώτατα τὰ οἱ Φοίνικες ἀνεῦρον οἱ μετὰ Θάσου κτίσαντες τὴν νῆσον ταύτην, ἥτις νῦν ἐπὶ τοῦ Θάσου τούτου τοῦ Φοίνικος 10 τὸ οὔνομα ἔσχε. τὰ δὲ μέταλλα τὰ Φοινικικὰ ταῦτα ἐστὶ τῆς Θάσου μεταξὺ Αἰνύρων τε χώρου καλεομένου καὶ Κοινύρων, ἀντίον δὲ Σαμοθρηΐκης, οὖρος μέγα ἀνεστραμμένον ἐν τῇ ζητήσι.

Darius sends envoys to demand earth and water of the Greek cities.

XLVIII. Τοῦτο μὲν νύν ἐστι τοιοῦτο. οἱ δὲ 15 Θάσιοι τῷ βασιλέϊ κελεύσαντι καὶ τὸ τεῖχος τὸ σφέτερον κατεῖλον καὶ τὰς νέας τὰς πάσας ἐκόμισαν ἐς Ἄβδηρα.

Μετὰ δὲ τοῦτο ἀπεπειρᾶτο ὁ Δαρεῖος τῶν Ἑλλήνων, ὅ τι ἐν νόῳ ἔχοιεν, κότερα πολεμέειν ἑωυτῷ ἢ 20 παραδιδόναι σφέας αὐτούς. διέπεμπε ὦν κήρυκας ἄλλους ἄλλῃ τάξας ἀνὰ τὴν Ἑλλάδα, κελεύων αἰτέειν βασιλέϊ γῆν τε καὶ ὕδωρ. τούτους μὲν δὴ ἐς τὴν Ἑλλάδα ἔπεμπε, ἄλλους δὲ κήρυκας διέπεμπε ἐς τὰς ἑωυτοῦ δασμοφόρους πόλιας τὰς παραθαλασσίους, 25 κελεύων νέας τε μακρὰς καὶ ἱππαγωγὰ πλοῖα ποιέεσθαι.

Many of the continental towns obey, and all the islands. The Athenians use this as a pretext for accusing the Aeginetans to the Spartans.

XLIX. Οὗτοί τε δὴ παρεσκευάζοντο ταῦτα, καὶ τοῖσι ἥκουσι ἐς τὴν Ἑλλάδα κήρυξι πολλοὶ μὲν ἠπειρωτέων ἔδοσαν τὰ προΐσχετο αἰτέων ὁ Πέρσης, πάντες δὲ νησιῶται ἐς τοὺς ἀπικοίατο αἰτήσοντες.
5 οἵ τε δὴ ἄλλοι νησιῶται διδοῦσι γῆν τε καὶ ὕδωρ Δαρείῳ, καὶ δὴ καὶ Αἰγινῆται. ποιήσασι δέ σφι ταῦτα ἰθέως Ἀθηναῖοι ἐπεκέατο, δοκέοντες ἐπὶ σφίσι ἔχοντας τοὺς Αἰγινήτας δεδωκέναι, ὡς ἅμα τῷ Πέρσῃ ἐπὶ σφέας στρατεύωνται. καὶ ἄσμενοι προφάσιος
10 ἐπελάβοντο, φοιτέοντές τε ἐς τὴν Σπάρτην κατηγόρεον τῶν Αἰγινητέων τὰ πεποιήκοιεν προδόντες τὴν Ἑλλάδα.

The Spartan king Kleomenes goes to Aegina to arrest the medizers, but his authority is undermined by the other king Demaratus.

L. Πρὸς ταύτην δὲ τὴν κατηγορίην Κλεομένης ὁ Ἀναξανδρίδεω βασιλεὺς ἐὼν Σπαρτιητέων διέβη ἐς
15 Αἴγιναν, βουλόμενος συλλαβεῖν Αἰγινητέων τοὺς αἰτιωτάτους. ὡς δὲ ἐπειρᾶτο συλλαμβάνων, ἄλλοι τε δὴ αὐτῷ ἐγίνοντο ἀντίξοοι τῶν Αἰγινητέων, ἐν δὲ δὴ καὶ Κρῖος ὁ Πολυκρίτου μάλιστα, ὃς οὐκ ἔφη αὐτὸν οὐδένα ἄξειν χαίροντα Αἰγινητέων· ἄνευ γάρ
20 μιν Σπαρτιητέων τοῦ κοινοῦ ποιέειν ταῦτα ὑπ' Ἀθηναίων ἀναγνωσθέντα χρήμασι· ἅμα γὰρ ἄν μιν τῷ ἑτέρῳ βασιλέϊ ἐλθόντα συλλαμβάνειν. ἔλεγε δὲ ταῦτα ἐξ ἐπιστολῆς τῆς Δημαρήτου. Κλεομένης δὲ ἀπελαυνόμενος ἐκ τῆς Αἰγίνης εἴρετο τὸν Κρῖον ὅ τι

οἱ εἴη τὸ οὔνομα· ὁ δέ οἱ τὸ ἐὸν ἔφρασε. ὁ δὲ Κλεομένης πρὸς αὐτὸν ἔφη· "Ἤδη νῦν καταχαλκοῦ ὦ κριὲ τὰ κέρεα, ὡς συνοισόμενος μεγάλῳ κακῷ."

Origin of the double kingship in Sparta.
[*Digression to* c. 60.]

LI. Ἐν δὲ τῇ Σπάρτῃ τοῦτον τὸν χρόνον ὑπομένων Δημάρητος ὁ Ἀρίστωνος διέβαλλε τὸν 5 Κλεομένεα, ἐὼν βασιλεὺς καὶ οὗτος Σπαρτιητέων, οἰκίης δὲ τῆς ὑποδεεστέρης, κατ' ἄλλο μὲν οὐδὲν ὑποδεεστέρης (ἀπὸ γὰρ τοῦ αὐτοῦ γεγόνασι), κατὰ πρεσβυγενείην δέ κως τετίμηται μᾶλλον ἡ Εὐρυσθένεος. LII. Λακεδαιμόνιοι γὰρ ὁμολογέοντες οὐδενὶ 10 ποιητῇ λέγουσι αὐτὸν Ἀριστόδημον τὸν Ἀριστομάχου τοῦ Κλεοδαίου τοῦ Ὕλλου βασιλεύοντα ἀγαγεῖν σφέας ἐς ταύτην τὴν χώρην, τὴν νῦν ἐκτέαται, ἀλλ' οὐ τοὺς Ἀριστοδήμου παῖδας. μετὰ δὲ χρόνον οὐ πολλὸν Ἀριστοδήμῳ τεκεῖν τὴν γυναῖκα, τῇ οὔνομα 15 εἶναι Ἀργείην· θυγατέρα δὲ αὐτὴν λέγουσι εἶναι Αὐτεσίωνος τοῦ Τισαμενοῦ τοῦ Θερσάνδρου τοῦ Πολυνείκεος· ταύτην δὲ τεκεῖν δίδυμα, ἐπιδόντα δὲ τὸν Ἀριστόδημον τὰ τέκνα νούσῳ τελευτᾶν. Λακεδαιμονίους δὲ τοὺς τότε ἐόντας βουλεῦσαι κατὰ νόμον 20 βασιλέα τῶν παίδων τὸν πρεσβύτερον ποιήσασθαι· οὐκ ὦν δή σφεας ἔχειν, ὁκότερον ἕλωνται, ὥστε καὶ ὁμοίων καὶ ἴσων ἐόντων· οὐ δυναμένους δὲ γνῶναι, ἢ καὶ πρὸ τούτου, ἐπειρωτᾶν τὴν τεκοῦσαν. τὴν δὲ οὐδὲ αὐτὴν φάναι διαγινώσκειν· εἰδυῖαν μὲν καὶ τὸ 25 κάρτα λέγειν ταῦτα, βουλομένην δὲ εἴ κως ἀμφότεροι γενοίατο βασιλέες. τοὺς ὦν δὴ Λακεδαιμονίους ἀπορέειν, ἀπορέοντας δὲ πέμπειν ἐς Δελφοὺς ἐπειρη-

σομένους, ὅ τι χρήσωνται τῷ πρήγματι. τὴν δὲ Πυθίην κελεύειν σφέας ἀμφότερα τὰ παιδία ἡγήσασθαι βασιλέας, τιμᾶν δὲ μᾶλλον τὸν γεραίτερον. τὴν μὲν δὴ Πυθίην ταῦτά σφι ἀνελεῖν, τοῖσι δὲ Λακε-
5 δαιμονίοισι ἀπορέουσι οὐδὲν ἔσσον, ὅκως ἐξεύρωσι αὐτῶν τὸν πρεσβύτερον, ὑποθέσθαι ἄνδρα Μεσσήνιον, τῷ οὔνομα εἶναι Πανίτην. ὑποθέσθαι δὲ τοῦτον τὸν Πανίτην τάδε τοῖσι Λακεδαιμονίοισι, φυλάξαι τὴν γειναμένην, ὁκότερον τῶν παίδων
10 πρότερον λούει καὶ σιτίζει· καὶ ἢν μὲν κατὰ ταὐτὰ φαίνηται αἰεὶ ποιεῦσα, τοὺς δὲ πᾶν ἕξειν ὅσον τι καὶ δίζηνται καὶ ἐθέλουσι ἐξευρεῖν, ἢν δὲ πλανᾶται καὶ ἐκείνη ἐναλλὰξ ποιεῦσα, δῆλά σφι ἔσεσθαι, ὡς οὐδὲ ἐκείνη πλέον οὐδὲν οἶδε, ἐπ' ἄλλην τέ σφεας τρά-
15 πεσθαι ὁδόν. ἐνθαῦτα δὴ τοὺς Σπαρτιήτας κατὰ τὰς τοῦ Μεσσηνίου ὑποθήκας φυλάξαντας τὴν μητέρα τῶν Ἀριστοδήμου παίδων λαβεῖν κατὰ τὰ αὐτὰ τιμῶσαν τὸν πρότερον καὶ σίτοισι καὶ λουτροῖσι, οὐκ εἰδυῖαν τῶν εἵνεκεν ἐφυλάσσετο. λαβόντας δὲ τὸ
20 παιδίον τὸ τιμώμενον πρὸς τῆς γειναμένης ὡς ἐὸν πρότερον τρέφειν ἐν τῷ δημοσίῳ· καί οἱ οὔνομα τεθῆναι Εὐρυσθένεα, τῷ δὲ Προκλέα. τούτους ἀνδρωθέντας αὐτούς τε ἀδελφεοὺς ἐόντας λέγουσι διαφόρους εἶναι τὸν πάντα χρόνον τῆς ζόης ἀλλή-
25 λοισι, καὶ τοὺς ἀπὸ τούτων γενομένους ὡσαύτως διατελέειν.

Variations of the legend.

LIII. Ταῦτα μὲν Λακεδαιμόνιοι λέγουσι μοῦνοι Ἑλλήνων· τάδε δὲ κατὰ τὰ λεγόμενα ὑπ' Ἑλλήνων ἐγὼ γράφω· τούτους γὰρ δὴ τοὺς Δωριέων βασιλέας μέχρι μὲν Περσέος τοῦ Δανάης, τοῦ θεοῦ ἀπεόντος, καταλεγομένους ὀρθῶς ὑπ' Ἑλλήνων καὶ ἀποδεικνυμένους ὥς εἰσι Ἕλληνες· ἤδη γὰρ τηνικαῦτα ἐς Ἕλληνας οὗτοι ἐτέλεον. ἔλεξα δὲ μέχρι Περσέος τοῦδε εἵνεκεν, ἀλλ' οὐκ ἀνέκαθεν ἔτι ἔλαβον, ὅτι οὐκ 5 ἔπεστι ἐπωνυμίη Περσέϊ οὐδεμία πατρὸς θνητοῦ, ὥσπερ Ἡρακλέϊ Ἀμφιτρύων· ἤδη ὦν ὀρθῷ λόγῳ χρεομένῳ μέχρι τοῦ Περσέος ὀρθῶς εἴρηταί μοι, ἀπὸ δὲ Δανάης τῆς Ἀκρισίου καταλέγοντι τοὺς ἄνω αἰεὶ πατέρας αὐτῶν φαινοίατο ἂν ἐόντες οἱ τῶν Δωριέων 10 ἡγεμόνες Αἰγύπτιοι ἰθαγενέες. LIV. Ταῦτα μέν νυν κατὰ τὰ Ἕλληνες λέγουσι γεγενεηλόγηται· ὡς δὲ ὁ Περσέων λόγος λέγεται, αὐτὸς ὁ Περσεὺς ἐὼν Ἀσσύριος ἐγένετο Ἕλλην, ἀλλ' οὐκ οἱ Περσέος πρόγονοι· τοὺς δὲ Ἀκρισίου γε πατέρας ὁμολο- 15 γέοντας κατ' οἰκηϊότητα Περσέϊ οὐδέν, τούτους δὲ εἶναι, κατά περ Ἕλληνες λέγουσι, Αἰγυπτίους. LV. Καὶ ταῦτα μέν νυν περὶ τούτων εἰρήσθω. ὅ τι δὲ ἐόντες Αἰγύπτιοι, καὶ ὅ τι ἀποδεξάμενοι ἔλαβον τὰς Δωριέων βασιληΐας, ἄλλοισι γὰρ περὶ αὐτῶν εἴρηται, 20 ἐάσομεν αὐτά· τὰ δὲ ἄλλοι οὐ κατελάβοντο, τούτων μνήμην ποιήσομαι.

Functions and honours of the Spartan kings: (1) in war,

LVI. Γέρεά τε δὴ τάδε τοῖσι βασιλεῦσι Σπαρτιῆται δεδώκασι· ἱρωσύνας δύο, Διός τε Λακεδαίμονος καὶ Διὸς οὐρανίου, καὶ πόλεμόν γε ἐκφέρειν 25 ἐπ' ἣν ἂν βούλωνται χώρην, τούτου δὲ μηδένα εἶναι Σπαρτιητέων διακωλυτήν, εἰ δὲ μή, αὐτὸν ἐν τῷ ἄγεϊ ἐνέχεσθαι· στρατευομένων δὲ πρώτους ἰέναι τοὺς βασιλέας, ὑστάτους δὲ ἀπιέναι· ἑκατὸν δὲ ἄνδρας λογάδας ἐπὶ στρατιῆς φυλάσσειν αὐτούς· προβάτοισι δὲ χρᾶσθαι ἐν τῇσι ἐξοδίῃσι, ὁκόσοισι ἂν ἐθέλωσι, τῶν δὲ θυομένων ἁπάντων τὰ δέρματά τε καὶ τὰ νῶτα λαμβάνειν σφέας.

(2) *in peace,*

5 LVII. Ταῦτα μὲν τὰ ἐμπολέμια, τὰ δὲ ἄλλα τὰ εἰρηναῖα κατὰ τάδε σφι δέδοται· ἢν θυσίη τις δημοτελὴς ποιέηται, πρώτους ἐπὶ τὸ δεῖπνον ἵζειν τοὺς βασιλέας καὶ ἀπὸ τούτων πρῶτον ἄρχεσθαι, διπλήσια νέμοντας ἑκατέρῳ τὰ πάντα ἢ τοῖσι 10 ἄλλοισι δαιτυμόνεσι· καὶ σπονδαρχίας εἶναι τούτων, καὶ τῶν τυθέντων τὰ δέρματα. νεομηνίας δὲ ἀνὰ πάσας καὶ ἑβδόμας ἱσταμένου τοῦ μηνὸς δίδοσθαι ἐκ τοῦ δημοσίου ἱρήϊον τέλεον ἑκατέρῳ ἐς Ἀπόλλωνος καὶ μέδιμνον ἀλφίτων καὶ οἴνου τετάρτην 15 Λακωνικήν, καὶ ἐν τοῖσι ἀγῶσι πᾶσι προεδρίας ἐξαιρέτους· καὶ προξείνους ἀποδεικνύναι τούτοισι προσκεῖσθαι τοὺς ἂν ἐθέλωσι τῶν ἀστῶν, καὶ Πυθίους αἱρέεσθαι δύο ἑκάτερον· οἱ δὲ Πύθιοί εἰσι θεοπρόποι ἐς Δελφούς, σιτεόμενοι μετὰ τῶν βασιλέων 20 τὰ δημόσια· μὴ ἐλθοῦσι δὲ τοῖσι βασιλεῦσι ἐπὶ τὸ δεῖπνον ἀποπέμπεσθαί σφι ἐς τὰ οἰκία ἀλφίτων τε δύο χοίνικας ἑκατέρῳ καὶ οἴνου κοτύλην, παρεοῦσι δὲ διπλήσια πάντα δίδοσθαι· τὠυτὸ δὲ τοῦτο καὶ πρὸς ἰδιωτέων κληθέντας ἐπὶ δεῖπνον τιμᾶσθαι· τὰς δὲ 25 μαντηΐας τὰς γινομένας τούτους φυλάσσειν, συνειδέναι δὲ καὶ τοὺς Πυθίους· δικάζειν δὲ μούνους τοὺς βασιλέας τοσάδε μοῦνα· πατρούχου τε παρθένου πέρι, ἐς τὸν ἱκνέεται ἔχειν, ἢν μή περ ὁ πατὴρ αὐτὴν ἐγγυήσῃ, καὶ ὁδῶν δημοσιέων πέρι· καὶ ἤν τις θετὸν

παῖδα ποιέεσθαι ἐθέλῃ, βασιλέων ἐναντίον ποιέεσθαι· καὶ παρίζειν βουλεύουσι τοῖσι γέρουσι, ἐοῦσι δυῶν δέουσι τριήκοντα. ἢν δὲ μὴ ἔλθωσι, τοὺς μάλιστά σφι τῶν γερόντων προσήκοντας ἔχειν τὰ τῶν βασιλέων γέρεα, δύο ψήφους τιθεμένους, τρίτην δὲ 5 τὴν ἑωυτῶν.

(3) *honours after death.*

LVIII. Ταῦτα μὲν ζώουσι τοῖσι βασιλεῦσι δέδοται ἐκ τοῦ κοινοῦ τῶν Σπαρτιητέων, ἀποθανοῦσι δὲ τάδε· ἱππέες περιαγγέλλουσι τὸ γεγονὸς κατὰ πᾶσαν τὴν Λακωνικήν, κατὰ δὲ τὴν πόλιν γυναῖκες 10 περιιοῦσαι λέβητα κροτέουσι. ἐπεὰν ὦν τοῦτο γένηται τοιοῦτο, ἀνάγκη ἐξ οἰκίης ἑκάστης ἐλευθέρους δύο καταμιαίνεσθαι, ἄνδρα τε καὶ γυναῖκα· μὴ ποιήσασι δὲ τοῦτο ζημίαι μεγάλαι ἐπικέαται. νόμος δὲ τοῖσι Λακεδαιμονίοισι κατὰ τῶν βασιλέων τοὺς 15 θανάτους ἐστὶ ὡυτὸς καὶ τοῖσι βαρβάροισι τοῖσι ἐν τῇ Ἀσίῃ· τῶν γὰρ ὦν βαρβάρων οἱ πλεῦνες τὠυτῷ νόμῳ χρέονται κατὰ τοὺς θανάτους τῶν βασιλέων. ἐπεὰν γὰρ ἀποθάνῃ βασιλεὺς Λακεδαιμονίων, ἐκ πάσης δέει Λακεδαίμονος, χωρὶς Σπαρτιητέων, ἀριθμῷ 20 τῶν περιοίκων ἀναγκαστοὺς ἐς τὸ κῆδος ἰέναι· τούτων ὦν καὶ τῶν εἱλωτέων καὶ αὐτῶν Σπαρτιητέων ἐπεὰν συλλεχθέωσι ἐς τὠυτὸ πολλαὶ χιλιάδες, σύμμιγα τῇσι γυναιξὶ κόπτονταί τε προθύμως καὶ οἰμωγῇ διαχρέονται ἀπλέτῳ, φάμενοι τὸν ὕστατον αἰεὶ ἀπο- 25 γενόμενον τῶν βασιλέων, τοῦτον δὴ γενέσθαι ἄριστον. ὃς δ' ἂν ἐν πολέμῳ τῶν βασιλέων ἀποθάνῃ, τούτῳ δὲ εἴδωλον σκευάσαντες ἐν κλίνῃ εὖ ἐστρωμένῃ ἐκφέρουσι. ἐπεὰν δὲ θάψωσι, ἀγορὴ δέκα ἡμερέων οὐκ ἵσταταί σφι, οὐδ' ἀρχαιρεσίη συνίζει, ἀλλὰ πενθέουσι ταύτας τὰς ἡμέρας.

Analogies with the Persians,

LIX. Συμφέρονται δὲ ἄλλο τόδε τοῖσι Πέρσῃσι· ἐπεὰν ἀποθανόντος τοῦ βασιλέος ἄλλος ἐνίστηται 5 βασιλεύς, οὗτος ὁ ἐσιὼν ἐλευθεροῖ ὅστις τι Σπαρτιητέων τῷ βασιλέϊ ἢ τῷ δημοσίῳ ὤφειλε. ἐν δ' αὖ Πέρσῃσι ὁ κατιστάμενος βασιλεὺς τὸν προοφειλόμενον φόρον μετίει τῇσι πόλισι πάσῃσι.

and with the Egyptians.

LX. Συμφέρονται δὲ καὶ τάδε Αἰγυπτίοισι 10 Λακεδαιμόνιοι· οἱ κήρυκες αὐτῶν καὶ αὐληταὶ καὶ μάγειροι ἐκδέκονται τὰς πατρωΐας τέχνας, καὶ αὐλητής τε αὐλητέω γίνεται καὶ μάγειρος μαγείρου καὶ κῆρυξ κήρυκος· οὐ κατὰ λαμπροφωνίην ἐπιτιθέμενοι ἄλλοι σφέας παρακληΐουσι, ἀλλὰ κατὰ τὰ πάτρια 15 ἐπιτελέουσι.

[*Resuming from* c. 50. B.C. 491.] *On his return from Aegina Kleomenes determines to depose Demaratos.*

LXI. Ταῦτα μὲν δὴ οὕτω γίνεται. τότε δὲ τὸν Κλεομένεα ἐόντα ἐν τῇ Αἰγίνῃ καὶ κοινὰ τῇ Ἑλλάδι ἀγαθὰ προεργαζόμενον ὁ Δημάρητος διέβαλε, οὐκ Αἰγινητέων οὕτω κηδόμενος, ὡς φθόνῳ καὶ ἄγῃ 20 χρεόμενος. Κλεομένης δὲ νοστήσας ἀπ' Αἰγίνης ἐβούλευε τὸν Δημάρητον παῦσαι τῆς βασιληΐης, διὰ πρῆγμα τοιόνδε ἐπίβασιν ἐς αὐτὸν ποιεύμενος·

The story of King Ariston and the beautiful wife of his friend Agetos.

Ἀρίστωνι βασιλεύοντι ἐν Σπάρτῃ καὶ γήμαντι

γυναῖκας δύο παῖδες οὐκ ἐγίνοντο. καὶ οὐ γὰρ συνεγινώσκετο αὐτὸς τούτων εἶναι αἴτιος, γαμέει τρίτην γυναῖκα. ὧδε δὲ γαμέει. ἦν οἱ φίλος τῶν Σπαρτιητέων ἀνήρ, τῷ προσεκέετο τῶν ἀστῶν μάλιστα ὁ Ἀρίστων. τούτῳ τῷ ἀνδρὶ ἐτύγχανε 5 ἐοῦσα γυνὴ καλλίστη μακρῷ τῶν ἐν Σπάρτῃ γυναικῶν, καὶ ταῦτα μέντοι καλλίστη ἐξ αἰσχίστης γενομένη. ἐοῦσαν γάρ μιν τὸ εἶδος φλαύρην ἡ τροφὸς αὐτῆς, οἷα ἀνθρώπων τε ὀλβίων θυγατέρα καὶ δυσειδέα ἐοῦσαν, πρὸς δὲ καὶ ὁρέουσα τοὺς 10 γονέας συμφορὴν τὸ εἶδος αὐτῆς ποιευμένους, ταῦτα ἕκαστα μαθοῦσα ἐπιφράζεται τοιάδε· ἐφόρεε αὐτὴν ἀνὰ πᾶσαν ἡμέρην ἐς τὸ τῆς Ἑλένης ἱρόν· τὸ δ' ἐστὶ ἐν τῇ Θεράπνῃ καλευμένῃ, ὕπερθε τοῦ Φοιβηίου ἱροῦ· ὅκως δὲ ἐνείκειε ἡ τροφός, πρός τε 15 τὤγαλμα ἵστα καὶ ἐλίσσετο τὴν θεὸν ἀπαλλάξαι τῆς δυσμορφίης τὸ παιδίον. καὶ δή κοτε ἀπιούσῃ ἐκ τοῦ ἱροῦ τῇ τροφῷ γυναῖκα λέγεται ἐπιφανῆναι, ἐπιφανεῖσαν δὲ ἐπείρεσθαί μιν, ὅ τι φέρει ἐν τῇ ἀγκάλῃ, καὶ τὴν φράσαι, ὡς παιδίον φορέει· τὴν δὲ κελεῦσαί 20 οἱ δέξαι· τὴν δὲ οὐ φάναι· ἀπειρῆσθαι γάρ οἱ ἐκ τῶν γειναμένων μηδενὶ ἐπιδεικνύναι· τὴν δὲ πάντως ἑωυτῇ κελεύειν ἐπιδέξαι· ὁρέουσαν δὲ τὴν γυναῖκα περὶ πολλοῦ ποιευμένην ἰδέσθαι, οὕτω δὴ τὴν τροφὸν δέξαι τὸ παιδίον· τὴν δὲ καταψῶσαν τοῦ 25 παιδίου τὴν κεφαλὴν εἶπαι, ὡς καλλιστεύσει πασέων τῶν ἐν Σπάρτῃ γυναικῶν. ἀπὸ μὲν δὴ ταύτης τῆς ἡμέρης μεταπεσεῖν τὸ εἶδος. γαμέει δὲ δή μιν ἐς γάμου ὥρην ἀπικομένην Ἄγητος ὁ Ἀλκείδεω, οὗτος δὴ ὁ τοῦ Ἀρίστωνος φίλος. LXII. Τὸν δὲ Ἀρί- 30 στωνα ἔκνιζε ἄρα τῆς γυναικὸς ταύτης ὁ ἔρως·

μηχανᾶται δὴ τοιάδε· αὐτός τε τῷ ἑταίρῳ, τοῦ ἦν ἡ γυνὴ αὕτη, ὑποδέκεται δωτίνην δώσειν τῶν ἑωυτοῦ πάντων ἕν, τὸ ἂν αὐτὸς ἐκεῖνος ἕληται, καὶ τὸν ἑταῖρον ἑωυτῷ ἐκέλευε ὡσαύτως τὴν ὁμοίην διδόναι.
5 ὁ δὲ οὐδὲν φοβηθεὶς ἀμφὶ τῇ γυναικί, ὁρέων ἐοῦσαν καὶ Ἀρίστωνι γυναῖκα, καταινέει ταῦτα· ἐπὶ τούτοισι δὲ ὅρκους ἐπήλασαν. μετὰ δὲ αὐτός τε ὁ Ἀρίστων ἔδωκε τοῦτο, ὅ τι δὴ ἦν, τὸ εἵλετο τῶν κειμηλίων τῶν Ἀρίστωνος ὁ Ἄγητος, καὶ αὐτὸς τὴν ὁμοίην ζητέων
10 φέρεσθαι παρ' ἐκείνου, ἐνθαῦτα δὴ τοῦ ἑταίρου τὴν γυναῖκα ἐπειρᾶτο ἀπάγεσθαι. ὁ δὲ πλὴν τούτου μούνου τὰ ἄλλα ἔφη καταινέσαι. ἀναγκαζόμενος μέντοι τῷ τε ὅρκῳ καὶ τῆς ἀπάτης τῇ παραγωγῇ ἀπίει ἀπάγεσθαι.

Ariston marries the woman as his third wife. She bears Demaratos, whose paternity is doubted.

15 LXIII. Οὕτω μὲν δὴ τὴν τρίτην ἐσηγάγετο γυναῖκα ὁ Ἀρίστων, τὴν δευτέρην ἀποπεμψάμενος, ἐν δέ οἱ χρόνῳ ἐλάσσονι καὶ οὐ πληρώσασα τοὺς δέκα μῆνας ἡ γυνὴ αὕτη τίκτει τοῦτον δὴ τὸν Δημάρητον. καί τίς οἱ τῶν οἰκετέων ἐν θώκῳ κατημένῳ
20 μετὰ τῶν ἐφόρων ἐξαγγέλλει, ὥς οἱ παῖς γέγονε. ὁ δὲ ἐπιστάμενός τε τὸν χρόνον, τῷ ἠγάγετο τὴν γυναῖκα, καὶ ἐπὶ δακτύλων συμβαλλόμενος τοὺς μῆνας εἶπε ἀπομόσας "οὐκ ἂν ἐμὸς εἴη·" τοῦτο ἤκουσαν μὲν οἱ ἔφοροι, πρῆγμα μέντοι οὐδὲν ἐποιή-
25 σαντο τὸ παραυτίκα, ὁ δὲ παῖς αὔξετο, καὶ τῷ Ἀρίστωνι τὸ εἰρημένον μετέμελε· παῖδα γὰρ τὸν Δημάρητον ἐς τὰ μάλιστά οἱ ἐνόμισε εἶναι. Δημάρητον δὲ αὐτῷ οὔνομα ἔθετο διὰ τόδε· πρότερον

τούτων πανδημεὶ Σπαρτιῆται Ἀρίστωνι, ὡς ἀνδρὶ εὐδοκιμέοντι διὰ πάντων δὴ τῶν βασιλέων τῶν ἐν τῇ Σπάρτῃ γενομένων, ἀρὴν ἐποιήσαντο παῖδα γενέσθαι· διὰ τοῦτο μέν οἱ τὸ οὔνομα Δημάρητος ἐτέθη. LXIV. Χρόνου δὲ προϊόντος Ἀρίστων μὲν ἀπέθανε, 5 Δημάρητος δὲ ἔσχε τὴν βασιληΐην. ἔδεε δέ, ὡς οἶκε, ἀνάπυστα γενόμενα ταῦτα καταπαῦσαι Δημάρητον τῆς βασιληΐης, δι' ἃ Κλεομένεϊ διεβλήθη μεγάλως πρότερόν τε ὁ Δημάρητος ἀπαγαγὼν τὴν στρατιὴν ἐξ Ἐλευσῖνος καὶ δὴ καὶ τότε ἐπ' Αἰγινητέων τοὺς 10 μηδίσαντας διαβάντος Κλεομένεος.

Kleomenes agrees with Leotychides to make him king, in place of his cousin Demaratos, whom Leotychides had also a private reason for hating.

LXV. Ὁρμηθεὶς ὦν ἀποτίνυσθαι ὁ Κλεομένης συντίθεται Λευτυχίδῃ τῷ Μενάρεος τοῦ Ἄγιος, ἐόντι οἰκίης τῆς αὐτῆς Δημαρήτῳ, ἐπ' ᾧ τε, ἢν αὐτὸν καταστήσῃ βασιλέα ἀντὶ Δημαρήτου, ἕψεταί οἱ ἐπ' 15 Αἰγινήτας. ὁ δὲ Λευτυχίδης ἦν ἐχθρὸς τῷ Δημαρήτῳ μάλιστα γεγονὼς διὰ πρῆγμα τοιόνδε· ἁρμοσαμένου Λευτυχίδεω Πέρκαλον τὴν Χίλωνος τοῦ Δημαρμένου θυγατέρα ὁ Δημάρητος ἐπιβουλεύσας ἀποστερέει Λευτυχίδην τοῦ γάμου, φθάσας αὐτὸς τὴν Πέρκαλον 20 ἁρπάσας καὶ σχὼν γυναῖκα· κατὰ τοῦτο μὲν τῷ Λευτυχίδῃ ἡ ἔχθρη ἡ ἐς τὸν Δημάρητον ἐγεγόνεε, τότε δὲ ἐκ τῆς Κλεομένεος προθυμίης ὁ Λευτυχίδης κατόμνυται Δημαρήτου, φὰς αὐτὸν οὐκ ἱκνεομένως βασιλεύειν Σπαρτιητέων, οὐκ ἐόντα παῖδα Ἀρίστωνος. μετὰ δὲ 25 τὴν κατωμοσίην ἐδίωκε ἀνασώζων ἐκεῖνο τὸ ἔπος, το εἶπε Ἀρίστων τότε, ὅτε οἱ ἐξήγγειλε ὁ οἰκέτης παῖδα

γεγονέναι, ὁ δὲ συμβαλόμενος τοὺς μῆνας ἀπώμοσε, φὰς οὐκ ἑωυτοῦ εἶναι. τούτου δὴ ἐπιβατεύων τοῦ ῥήματος ὁ Λευτυχίδης ἀπέφαινε τὸν Δημάρητον οὔτε ἐξ Ἀρίστωνος γεγονότα οὔτε ἱκνεομένως βασι-
5 λεύοντα Σπάρτης, τοὺς ἐφόρους μάρτυρας παρεχόμενος κείνους, οἳ τότε ἔτυχον πάρεδροί τε ἐόντες καὶ ἀκούσαντες ταῦτα Ἀρίστωνος.

The Spartans agree to refer the matter of the paternity of Demaratos to the oracle at Delphi. Kleomenes secures a decision against Demaratos by an intrigue, which cost the Pythia her office.

LXVI. Τέλος δὲ ἐόντων περὶ αὐτῶν νεικέων ἔδοξε Σπαρτιήτῃσι ἐπείρεσθαι τὸ χρηστήριον τὸ ἐν
10 Δελφοῖσι, εἰ Ἀρίστωνος εἴη παῖς ὁ Δημάρητος. ἀνοίστου δὲ γενομένου ἐκ προνοίης τῆς Κλεομένεος ἐς τὴν Πυθίην ἐνθαῦτα προσποιέεται Κλεομένης Κόβωνα τὸν Ἀριστοφάντου, ἄνδρα ἐν Δελφοῖσι δυναστεύοντα μέγιστον, ὁ δὲ Κόβων Περίαλλαν τὴν πρόμαντιν
15 ἀναπείθει, τὰ Κλεομένης ἐβούλετο λέγεσθαι, λέγειν. οὕτω δὴ ἡ Πυθίη ἐπειρωτώντων τῶν θεοπρόπων ἔκρινε μὴ Ἀρίστωνος εἶναι Δημάρητον παῖδα. ὑστέρῳ μέντοι χρόνῳ ἀνάπυστα ἐγένετο ταῦτα, καὶ Κόβων τε ἔφυγε ἐκ Δελφῶν καὶ Περίαλλα ἡ πρόμαντις
20 ἐπαύσθη τῆς τιμῆς.

Demaratos remained in Sparta for a time; but, on receiving an insult from Leotychides, determines to put an end to his uncertainty.

LXVII. Κατὰ μὲν δὴ τὴν Δημαρήτου κατάπαυσιν τῆς βασιληίης οὕτω ἐγένετο, ἔφευγε δὲ Δημάρητος ἐκ Σπάρτης ἐς Μήδους ἐκ τοιοῦδε ὀνείδεος·

μετὰ τῆς βασιληΐης τὴν κατάπαυσιν ὁ Δημάρητος ἦρχε αἱρεθεὶς ἀρχήν. ἦσαν μὲν δὴ γυμνοπαιδίαι, θηωμένου δὲ τοῦ Δημαρήτου ὁ Λευτυχίδης, γεγονὼς ἤδη αὐτὸς βασιλεὺς ἀντ' ἐκείνου, πέμψας τὸν θεράποντα ἐπὶ γέλωτί τε καὶ λάσθῃ εἰρώτα τὸν Δημάρητον, ὁκοῖόν τι εἴη τὸ ἄρχειν μετὰ τὸ βασιλεύειν· 5 ὁ δὲ ἀλγήσας τῷ ἐπειρωτήματι εἶπε φὰς "αὐτὸς μὲν "ἀμφοτέρων ἤδη πεπειρῆσθαι, ἐκεῖνον δὲ οὔ, τὴν "μέντοι ἐπειρώτησιν ταύτην ἄρξειν Λακεδαιμονίοισι "ἢ μυρίης κακότητος ἢ μυρίης εὐδαιμονίης." ταῦτα 10 δὲ εἴπας καὶ κατακαλυψάμενος ἤϊε ἐκ τοῦ θεήτρου ἐς τὰ ἑωυτοῦ οἰκία, αὐτίκα δὲ παρασκευασάμενος ἔθυε τῷ Διὶ βοῦν, θύσας δὲ τὴν μητέρα ἐκάλεσε.

He therefore solemnly appeals to his mother to tell him the truth.

LXVIII. Ἀπικομένῃ δὲ τῇ μητρὶ ἐσθεὶς ἐς τὰς χεῖράς οἱ τῶν σπλάγχνων κατικέτευε λέγων τοιάδε· 15 "Ὦ μῆτερ, θεῶν σε τῶν τε ἄλλων καταπτόμενος "ἱκετεύω καὶ τοῦ ἑρκείου Διὸς τοῦδε φράσαι μοι τὴν "ἀληθείην, τίς μεύ ἐστι πατὴρ ὀρθῷ λόγῳ. Λευτυ"χίδης μὲν γὰρ ἔφη ἐν τοῖσι νείκεσι λέγων κυέουσάν "σε ἐκ τοῦ προτέρου ἀνδρὸς οὕτω ἐλθεῖν παρὰ Ἀρίσ- 20 "τωνα, οἱ δὲ καὶ τὸν ματαιότερον λόγον λέγοντες φασί "σε ἐλθεῖν παρὰ τῶν οἰκετέων τὸν ὀνοφορβόν, καὶ ἐμὲ "εἶναι ἐκείνου παῖδα. ἐγώ ὦν σε μετέρχομαι τῶν "θεῶν εἰπεῖν τὠληθές· οὔτε γάρ, εἴ περ πεποίηκάς "τι τῶν λεγομένων, μούνη δὴ πεποίηκας, μετὰ πολ- 25 "λέων δέ, ὅ τε λόγος πολλὸς ἐν Σπάρτῃ, ὡς Ἀρίστωνι "σπέρμα παιδοποιὸν οὐκ ἐνῆν· τεκεῖν γὰρ ἄν οἱ καὶ "τὰς προτέρας γυναῖκας."

His mother's explanation. He is the son of the Hero Astrabakos, or of Ariston.

LXIX. Ὁ μὲν δὴ τοιαῦτα ἔλεγε, ἡ δὲ ἀμείβετο τοισίδε· "Ὦ παῖ, ἐπείτε με λιτῇσι μετέρχεαι εἰπεῖν "τὴν ἀληθείην, πᾶν ἐς σὲ κατειρήσεται τὠληθές. ὥς "με ἠγάγετο Ἀρίστων ἐς ἑωυτοῦ, νυκτὶ τρίτῃ ἀπὸ τῆς
5 "πρώτης ἦλθέ μοι φάσμα εἰδόμενον Ἀρίστωνι, συνευ- "νηθὲν δὲ τοὺς στεφάνους τοὺς εἶχε ἐμοὶ περιετίθει. "καὶ τὸ μὲν οἰχώκεε, ἧκε δὲ μετὰ ταῦτα Ἀρίστων. "ὡς δέ με εἶδε ἔχουσαν στεφάνους, εἰρώτα, τίς εἴη ὁ "μοι δούς. ἐγὼ δὲ ἐφάμην ἐκεῖνον· ὁ δὲ οὐκ ὑπεδέκετο·
10 "ἐγὼ δὲ κατωμνύμην φαμένη αὐτὸν οὐ καλῶς ποιέειν "ἀπαρνεύμενον· ὀλίγῳ γάρ τι πρότερον ἐλθόντα καὶ "συνευνηθέντα δοῦναί μοι τοὺς στεφάνους. ὁρέων δέ "με κατομνυμένην ὁ Ἀρίστων ἔμαθε, ὡς θεῖον εἴη τὸ "πρῆγμα. καὶ τοῦτο μὲν οἱ στέφανοι ἐφάνησαν ἐόντες
15 "ἐκ τοῦ ἡρωΐου τοῦ παρὰ τῇσι θύρῃσι τῇσι αὐλείῃσι "ἱδρυμένου, τὸ καλέουσι Ἀστραβάκου, τοῦτο δὲ οἱ "μάντιες τὸν αὐτὸν τοῦτον ἥρωα ἀναίρεον εἶναι. οὕτω, "ὦ παῖ, ἔχεις πᾶν, ὅσον τι καὶ βούλεαι πυθέσθαι. "ἢ γὰρ ἐκ τοῦ ἥρωος τούτου γέγονας, καί τοι πατήρ
20 "ἐστι Ἀστράβακος ὁ ἥρως, ἢ Ἀρίστων· ἐν γάρ σε τῇ "νυκτὶ ταύτῃ ἀναιρέομαι. τῇ δέ σευ μάλιστα κατάπ- "τονται οἱ ἐχθροὶ, λέγοντες, ὡς αὐτὸς ὁ Ἀρίστων, ὅτε "αὐτῷ σὺ ἠγγέλθης γεγεννημένος, πολλῶν ἀκουόντων "οὐ φήσειέ σε ἑωυτοῦ εἶναι, τὸν χρόνον γὰρ, τοὺς δέκα
25 "μῆνας, οὐδέκω ἐξήκειν, ἀϊδρείῃ τῶν τοιούτων ἐκεῖνος "τοῦτο ἀπέρριψε τὸ ἔπος. τίκτουσι γὰρ γυναῖκες καὶ "ἐννεάμηνα καὶ ἑπτάμηνα, καὶ οὐ πᾶσαι δέκα μῆνας "ἐκτελέσασαι· ἐγὼ δὲ σὲ, ὦ παῖ, ἑπτάμηνον ἔτεκον. "ἔγνω δὲ καὶ αὐτὸς ὁ Ἀρίστων οὐ μετὰ πολλὸν

"χρόνον, ὡς ἀγνοίη τὸ ἔπος ἐκβάλοι τοῦτο. λόγους δὲ
"ἄλλους περὶ γενέσιος τῆς σεωυτοῦ μὴ δέκεο· τὰ γὰρ
"ἀληθέστατα πάντα ἀκήκοας. ἐκ δὲ ὀνοφορβῶν αὐτῷ
"τε Λευτυχίδῃ καὶ τοῖσι ταῦτα λέγουσι τίκτοιεν αἱ
"γυναῖκες παῖδας." 5

He flies to Elis, thence to Zakynthos, and thence to the Court of Darius, who receives him with great liberality.

LXX. Ἡ μὲν δὴ ταῦτα ἔλεγε, ὁ δὲ πυθόμενός
τε τὰ ἐβούλετο καὶ ἐπόδια λαβὼν ἐπορεύετο ἐς Ἦλιν,
τῷ λόγῳ φὰς, ὡς ἐς Δελφοὺς χρησόμενος τῷ χρηστηρίῳ πορεύεται. Λακεδαιμόνιοι δὲ ὑποτοπηθέντες
Δημάρητον δρησμῷ ἐπιχειρέειν ἐδίωκον. καί κως 10
ἔφθη ἐς Ζάκυνθον διαβὰς ὁ Δημάρητος ἐκ τῆς
Ἤλιδος. ἐπιδιαβάντες δὲ οἱ Λακεδαιμόνιοι αὐτοῦ τε
ἅπτοντο καὶ τοὺς θεράποντας αὐτὸν ἀπαιρέονται.
μετὰ δὲ, οὐ γὰρ ἐξεδίδοσαν αὐτὸν οἱ Ζακύνθιοι,
ἐνθεῦτεν διαβαίνει ἐς τὴν Ἀσίην παρὰ βασιλέα 15
Δαρεῖον. ὁ δὲ ὑπεδέξατό τε αὐτὸν μεγαλωστὶ καὶ
γῆν τε καὶ πόλις ἔδωκε. οὕτω ἀπίκετο ἐς τὴν Ἀσίην
Δημάρητος καὶ τοιαύτῃ χρησάμενος τύχῃ, ἄλλα τε
Λακεδαιμονίοισι συχνὰ ἔργοισί τε καὶ γνώμῃσι ἀπολαμπρυνθεὶς, ἐν δὲ δὴ καὶ Ὀλυμπιάδα σφι ἀνελό- 20
μενος τεθρίππῳ προσέβαλε, μοῦνος τοῦτο πάντων δὴ
τῶν γενομένων βασιλέων ἐν Σπάρτῃ ποιήσας.

Leotychides succeeded Demaratos at Sparta. Zeuxidemos died in the lifetime of his father Leotychides, leaving a son Archidamos. Leotychides married again and had a daughter, Lampito, who married her nephew Archidamos.

LXXI. Λευτυχίδης δὲ ὁ Μενάρεος Δημαρήτου

καταπαυθέντος διεδέξατο τὴν βασιληΐην. καί οἱ γίνεται παῖς Ζευξίδημος, τὸν δὴ Κυνίσκον μετεξέτεροι Σπαρτιητέων ἐκάλεον. οὗτος ὁ Ζευξίδημος οὐκ ἐβασίλευσε Σπάρτης· πρὸ Λευτυχίδεω γὰρ τελευτᾷ,
5 λιπὼν παῖδα Ἀρχίδημον. Λευτυχίδης δὲ στερηθεὶς Ζευξιδήμου γαμέει δευτέρην γυναῖκα Εὐρυδάμην, ἐοῦσαν Μενίου μὲν ἀδελφεὴν, Διακτορίδεω δὲ θυγατέρα, ἐκ τῆς οἱ ἔρσεν μὲν γίνεται οὐδὲν, θυγάτηρ δὲ Λαμπιτὼ, τὴν Ἀρχίδημος ὁ Ζευξιδήμου γαμέει δόντος
10 αὐτῷ Λευτυχίδεω.

At a later period (about B.C. 478) *Leotychides was convicted of taking a bribe in Thessaly and banished, and died in Tegea.*

LXXII. Οὐ μὲν οὐδὲ Λευτυχίδης κατεγήρα ἐν Σπάρτῃ, ἀλλὰ τίσιν τοιήνδε τινὰ Δημαρήτῳ ἐξέτισε· ἐστρατήγησε Λακεδαιμονίοισι ἐς Θεσσαλίην, παρεὸν δέ οἱ ὑποχείρια πάντα ποιήσασθαι ἐδωροδόκησε
15 ἀργύριον πολλόν. ἐπ' αὐτοφώρῳ δὲ ἁλοὺς αὐτοῦ ἐν τῷ στρατοπέδῳ, ἐπικατήμενος χειρίδι πλέῃ ἀργυρίου, ἔφυγε ἐκ Σπάρτης ὑπὸ δικαστήριον ὑπαχθεὶς, καὶ τὰ οἰκία οἱ κατεσκάφη· ἔφυγε δὲ ἐς Τεγέην καὶ ἐτελεύτησε ἐν ταύτῃ.

B.C. 491. *Kleomenes and Leotychides make a joint expedition into Aegina. The Aeginetans thereupon give ten hostages who are depositea in Attica.*

20 LXXIII. Ταῦτα μὲν δὴ ἐγένετο χρόνῳ ὕστερον· τότε δὲ ὡς τῷ Κλεομένεϊ ὡδώθη τὸ ἐς τὸν Δημάρητον πρῆγμα, αὐτίκα παραλαβὼν Λευτυχίδην ἤιε ἐπὶ τοὺς Αἰγινήτας, δεινόν τινά σφι ἔγκοτον διὰ τὸν προπηλακισμὸν ἔχων. οὕτω δὴ οὔτε οἱ Αἰγινῆται

ἀμφοτέρων τῶν βασιλέων ἡκόντων ἐπ' αὐτοὺς ἐδικαίευν ἔτι ἀντιβαίνειν, ἐκεῖνοί τε ἐπιλεξάμενοι ἄνδρας δέκα Αἰγινητέων τοὺς πλείστου ἀξίους καὶ πλούτῳ καὶ γένεϊ ἦγον, καὶ ἄλλους καὶ δὴ καὶ Κριόν τε τὸν Πολυκρίτου καὶ Κάσαμβον τὸν Ἀριστοκράτεος, οἵ 5 περ εἶχον μέγιστον κράτος· ἀγαγόντες δέ σφεας ἐς γῆν τὴν Ἀττικὴν παραθήκην παρατίθενται ἐς τοὺς ἐχθίστους Αἰγινήτῃσι Ἀθηναίους.

The falseness of Kleomenes' dealing in regard to Demaratos becoming known, Kleomenes fled, and after a time raised a party in Arcadia against Sparta.

LXXIV. Μετὰ δὲ ταῦτα Κλεομένεα ἐπάϊστον γενόμενον κακοτεχνήσαντα ἐς Δημάρητον δεῖμα ἔλαβε 10 Σπαρτιητέων, καὶ ὑπεξέσχε ἐς Θεσσαλίην. ἐνθεῦτεν δὲ ἀπικόμενος ἐς τὴν Ἀρκαδίην νεώτερα ἔπρησσε πρήγματα, συνιστὰς τοὺς Ἀρκάδας ἐπὶ τῇ Σπάρτῃ, ἄλλους τε ὅρκους προσάγων σφι ἦ μὲν ἕψεσθαί σφεας αὐτῷ τῇ ἂν ἐξηγῆται, καὶ δὴ καὶ ἐς Νώνακριν 15 πόλιν πρόθυμος ἦν τῶν Ἀρκάδων τοὺς προεστεῶτας ἀγινέων ἐξορκοῦν τὸ Στυγὸς ὕδωρ. ἐν δὲ ταύτῃ τῇ πόλι λέγεται εἶναι ὑπ' Ἀρκάδων τὸ Στυγὸς ὕδωρ, καὶ δὴ καὶ ἔστι τοιόνδε τι· ὕδωρ ὀλίγον φαινόμενον ἐκ πέτρης στάζει ἐς ἄγκος, τὸ δὲ ἄγκος αἱμασιῆς τις 20 περιθέει κύκλος. ἡ δὲ Νώνακρις, ἐν τῇ ἡ πηγὴ αὕτη τυγχάνει ἐοῦσα, πόλις ἐστὶ τῆς Ἀρκαδίης πρὸς Φενεῷ.

The Spartans in terror restore him to his office; but he presently became insane, and whilst in confinement mangled himself in a horrible manner: which the various Greek states accounted for as a divine visitation for acts of sacrilege.

LXXV. Μαθόντες δὲ Λακεδαιμόνιοι Κλεομένεα ταῦτα πρήσσοντα κατῆγον αὐτὸν δείσαντες ἐπὶ τοῖσι αὐτοῖσι ἐς Σπάρτην, τοῖσι καὶ πρότερον ἦρχε. κατελθόντα δὲ αὐτὸν αὐτίκα ὑπέλαβε μανιὰς νοῦσος, ἐόντα
5 καὶ πρότερον ὑπομαργότερον· ὅκως γάρ τεῳ ἐντύχοι Σπαρτιητέων, ἐνέχραυε ἐς τὸ πρόσωπον τὸ σκῆπτρον. ποιεῦντα δὲ αὐτὸν ταῦτα καὶ παραφρονήσαντα ἔδησαν οἱ προσήκοντες ἐν ξύλῳ· ὁ δὲ δεθεὶς τὸν φύλακον μουνωθέντα ἰδὼν τῶν ἄλλων αἴτεε μάχαιραν, οὐ
10 βουλομένου δὲ τὰ πρῶτα τοῦ φυλάκου διδόναι ἠπείλεε τά μιν λυθεὶς ποιήσει, ἐς ὃ δείσας τὰς ἀπειλὰς ὁ φύλακος (ἦν γὰρ τῶν τις εἱλωτέων) διδοῖ οἱ μάχαιραν. Κλεομένης δὲ παραλαβὼν τὸν σίδηρον ἤρχετο ἐκ τῶν κνημέων ἑωυτὸν λωβώμενος· ἐπιτάμνων
15 γὰρ κατὰ μῆκος τὰς σάρκας προέβαινε ἐκ τῶν κνημέων ἐς τοὺς μηρούς, ἐκ δὲ τῶν μηρῶν ἔς τε τὰ ἰσχία καὶ τὰς λαπάρας, ἐς ὃ ἐς τὴν γαστέρα ἀπίκετο καὶ ταύτην καταχορδεύων ἀπέθανε τρόπῳ τοιούτῳ, ὡς μὲν οἱ πολλοὶ λέγουσι Ἑλλήνων, ὅτι τὴν Πυθίην
20 ἀνέγνωσε τὰ περὶ Δημάρητον γενόμενα λέγειν, ὡς δὲ Ἀθηναῖοι λέγουσι, διότι ἐς Ἐλευσῖνα ἐσβαλὼν ἔκειρε τὸ τέμενος τῶν θεῶν, ὡς δὲ Ἀργεῖοι, ὅτι ἐξ ἱροῦ αὐτῶν τοῦ Ἄργου Ἀργείων τοὺς καταφυγόντας ἐκ τῆς μάχης καταγινέων κατέκοπτε καὶ αὐτὸ τὸ ἄλσος
25 ἐν ἀλογίῃ ἔχων ἐνέπρησε.

His impieties in the invasion of Argos [*about* B.C. 510].

LXXVI. Κλεομένεϊ γὰρ μαντευομένῳ ἐν Δελφοῖσι ἐχρήσθη Ἄργος αἱρήσειν. ἐπείτε δὲ Σπαρτιήτας ἄγων ἀπίκετο ἐπὶ ποταμὸν Ἐρασῖνον, ὃς λέγεται ῥέειν ἐκ τῆς Στυμφηλίδος λίμνης (τὴν γὰρ δὴ λίμνην ταύτην ἐς χάσμα ἀφανὲς ἐκδιδοῦσαν ἀναφαίνεσθαι ἐν Ἄργεϊ, τὸ ἐνθεῦτεν δὲ τὸ ὕδωρ ἤδη τοῦτο ὑπ' Ἀργείων Ἐρασῖνον καλέεσθαι), ἀπικόμενος 5 δ' ὦν ὁ Κλεομένης ἐπὶ τὸν ποταμὸν τοῦτον ἐσφαγιάζετο αὐτῷ. καὶ οὐ γὰρ ἐκαλλιέρεε οὐδαμῶς διαβαίνειν μιν, ἄγασθαι μὲν ἔφη τοῦ Ἐρασίνου οὐ προδιδόντος τοὺς πολιήτας, Ἀργείους μέντοι οὐδ' ὣς χαιρήσειν. μετὰ δὲ ταῦτα ἐξαναχωρήσας τὴν στρατιὴν κατήγαγε 10 ἐς Θυρέην, σφαγιασάμενος δὲ τῇ θαλάσσῃ ταῦρον πλοίοισί σφεας ἤγαγε ἔς τε τὴν Τιρυνθίην χώρην καὶ Ναυπλίην.

He kills a number of Argives at Tiryns by a ruse;

LXXVII. Ἀργεῖοι δ' ἐβοήθεον πυνθανόμενοι ταῦτα ἐπὶ θάλασσαν. ὡς δὲ ἀγχοῦ μὲν ἐγίνοντο τῆς 15 Τίρυνθος, χώρῳ δὲ ἐν τούτῳ τῷ κέεται Ἡσίπεια οὔνομα, μεταίχμιον οὐ μέγα ἀπολιπόντες ἵζοντο ἀντίοι τοῖσι Λακεδαιμονίοισι. ἐνθαῦτα δὴ οἱ Ἀργεῖοι τὴν μὲν ἐκ τοῦ φανεροῦ μάχην οὐκ ἐφοβέοντο, ἀλλὰ μὴ δόλῳ αἱρεθέωσι. καὶ γὰρ δή σφι ἐς τοῦτο τὸ 20 πρῆγμα εἶχε τὸ χρηστήριον, τὸ ἐπίκοινα ἔχρησε ἡ Πυθίη τούτοισί τε καὶ Μιλησίοισι, λέγον ὧδε·

Ἀλλ' ὅταν ἡ θήλεια τὸν ἄρσενα νικήσασα
ἐξελάσῃ, καὶ κῦδος ἐν Ἀργείοισιν ἄρηται,
πολλὰς Ἀργείων ἀμφιδρυφέας τότε θήσει. 25
ὥς ποτέ τις ἐρέει καὶ ἐπεσσομένων ἀνθρώπων·
δεινὸς ὄφις τριέλικτος ἀπώλετο δουρὶ δαμασθείς.

Ταῦτα δὴ πάντα συνελθόντα τοῖσι Ἀργείοισι φόβον

παρεῖχε. καὶ δή σφι πρὸς ταῦτα ἔδοξε τῷ κήρυκι τῶν πολεμίων χρᾶσθαι, δόξαν δέ σφι ἐποίευν τοιόνδε· ὅκως ὁ Σπαρτιήτης κῆρυξ προσημαίνοι τι Λακεδαιμονίοισι, ἐποίευν καὶ οἱ Ἀργεῖοι τὠυτὸ τοῦτο.
5 LXXVIII. Μαθὼν δὲ ὁ Κλεομένης ποιεῦντας τοὺς Ἀργείους ὁκοῖόν τι ὁ σφέτερος κῆρυξ σημήνειε, παραγγέλλει σφι, ὅταν σημήνῃ ὁ κῆρυξ ποιέεσθαι ἄριστον, τότε ἀναλαβόντας τὰ ὅπλα χωρέειν ἐς τοὺς Ἀργείους. ταῦτα καὶ ἐγένετο ἐπιτελέα ἐκ τῶν Λακε-
10 δαιμονίων· ἄριστον γὰρ ποιευμένοισι τοῖσι Ἀργείοισι ἐκ τοῦ κηρύγματος ἐπεκέατο, καὶ πολλοὺς μὲν ἐφόνευσαν αὐτῶν, πολλῷ δ' ἔτι πλεῦνας ἐς τὸ ἄλσος τοῦ Ἄργου καταφυγόντας περιιζόμενοι ἐφύλασσον.

and massacres a large number who had taken refuge in the sacred enclosure of Argos, and burnt the Grove.

LXXIX. Ἐνθεῦτεν δὲ ὁ Κλεομένης ἐποίεε
15 τοιόνδε· ἔχων αὐτομόλους ἄνδρας καὶ πυνθανόμενος τούτων ἐξεκάλεε πέμπων κήρυκα, οὐνομαστὶ λέγων τῶν Ἀργείων τοὺς ἐν τῷ ἱρῷ ἀπεργμένους, ἐξεκάλεε δὲ φὰς αὐτῶν ἔχειν τὰ ἄποινα· ἄποινα δέ ἐστι Πελοποννησίοισι δύο μνέαι τεταγμέναι κατ' ἄνδρα αἰχμάλωτον
20 ἐκτίνειν. κατὰ πεντήκοντα δὴ ὦν τῶν Ἀργείων ὡς ἑκάστους ἐκκαλεύμενος ὁ Κλεομένης ἔκτεινε. ταῦτα δέ κως γινόμενα ἐλελήθεε τοὺς λοιποὺς τοὺς ἐν τῷ τεμένεϊ· ἅτε γὰρ πυκνοῦ ἐόντος τοῦ ἄλσεος οὐκ ὥρων οἱ ἐντὸς τοὺς ἐκτὸς ὅ τι ἔπρησσον, πρίν γε δὴ αὐτῶν
25 τις ἀναβὰς ἐπὶ δένδρος κατεῖδε τὸ ποιεύμενον. οὐκ ὦν δὴ ἔτι καλεόμενοι ἐξήϊσαν. LXXX. Ἐνθαῦτα δὴ ὁ Κλεομένης ἐκέλευε πάντα τινὰ τῶν εἱλωτέων περινέειν ὕλῃ τὸ ἄλσος, τῶν δὲ πειθομένων ἐνέπρησε

τὸ ἄλσος. καιομένου δὲ ἤδη ἐπείρετο τῶν τινὰ αὐτομόλων, τίνος εἴη θεῶν τὸ ἄλσος, ὁ δὲ ἔφη Ἄργου εἶναι· ὁ δὲ ὡς ἤκουσε, ἀναστενάξας μέγα εἶπε· "Ὦ "Ἄπολλον χρηστήριε, ἦ μεγάλως με ἠπάτηκας φά-"μενος Ἄργος αἱρήσειν· συμβάλλομαι δ' ἐξήκειν μοι 5 "τὸ χρηστήριον."

He then sent his army back to Sparta, and went to the temple of Here, between Mycenae and Argos; had the priest dragged out; and offered sacrifice himself.

LXXXI. Μετὰ δὲ ταῦτα ὁ Κλεομένης τὴν μὲν πλέω στρατιὴν ἀπῆκε ἀπιέναι ἐς Σπάρτην, χιλίους δὲ αὐτὸς λαβὼν τοὺς ἀριστέας ἤιε ἐς τὸ Ἡραῖον θύσων. βουλόμενον δὲ αὐτὸν θύειν ἐπὶ τοῦ βωμοῦ ὁ 10 ἱρεὺς ἀπηγόρευε, φὰς οὐκ ὅσιον εἶναι ξείνῳ αὐτόθι θύειν. ὁ δὲ Κλεομένης τὸν ἱρέα ἐκέλευε τοὺς εἵλωτας ἀπὸ τοῦ βωμοῦ ἀπάγοντας μαστιγῶσαι, καὶ αὐτὸς ἔθυσε· ποιήσας δὲ ταῦτα ἀπήιε ἐς τὴν Σπάρτην.

On his return to Sparta he is accused of having spared Argos for a bribe.

LXXXII. Νοστήσαντα δέ μιν ὑπῆγον οἱ ἐχθροὶ 15 ὑπὸ τοὺς ἐφόρους, φάμενοί μιν δωροδοκήσαντα οὐκ ἑλεῖν τὸ Ἄργος, παρεὸν εὐπετέως μιν ἑλεῖν. ὁ δέ σφι ἔλεξε, οὔτε εἰ ψευδόμενος οὔτε εἰ ἀληθέα λέγων, ἔχω σαφηνέως εἶπαι, ἔλεξε δ' ὦν φάμενος, ἐπείτε δὴ τὸ τοῦ Ἄργου ἱρὸν εἷλε, δοκέειν οἱ ἐξεληλυθέναι τὸν 20 τοῦ θεοῦ χρησμὸν, πρὸς ὧν ταῦτα οὐ δικαιοῦν πειρᾶν τῆς πόλιος, πρίν γε δὴ ἱροῖσι χρήσηται καὶ μάθῃ, εἴτε οἱ ὁ θεὸς παραδιδοῖ εἴτε οἱ ἐμποδὼν ἕστηκε·

καλλιερευμένῳ δὲ ἐν τῷ Ἡραίῳ ἐκ τοῦ ἀγάλματος τῶν στηθέων φλόγα πυρὸς ἐκλάμψαι, μαθεῖν δὲ αὐτὸς οὕτω τὴν ἀτρεκείην, ὅτι οὐκ αἱρέει τὸ Ἄργος· εἰ μὲν γὰρ ἐκ τῆς κεφαλῆς τοῦ ἀγάλματος ἐξέλαμψε, 5 αἱρέειν ἂν κατ' ἄκρης τὴν πόλιν, ἐκ τῶν στηθέων δὲ λάμψαντος πᾶν οἱ πεποιῆσθαι, ὅσον ὁ θεὸς ἐβούλετο γενέσθαι. ταῦτα δὲ λέγων πιστά τε καὶ οἰκότα ἐδόκεε Σπαρτιήτῃσι λέγειν, καὶ διέφυγε πολλὸν τοὺς διώκοντας.

The effect of the invasion upon Argos.

10 LXXXIII. Ἄργος δὲ ἀνδρῶν ἐχηρώθη οὕτω, ὥστε οἱ δοῦλοι αὐτῶν ἔσχον πάντα τὰ πρήγματα ἄρχοντές τε καὶ διέποντες, ἐς ὃ ἐπήβησαν οἱ τῶν ἀπολομένων παῖδες. ἔπειτά σφεας οὗτοι ἀνακτώμενοι ὀπίσω ἐς ἑωυτοὺς τὸ Ἄργος ἐξέβαλον· ἐξω-15 θεύμενοι δὲ οἱ δοῦλοι μάχῃ ἔσχον Τίρυνθα. τέως μὲν δή σφι ἦν ἄρθμια ἐς ἀλλήλους, ἔπειτα δὲ ἐς τοὺς δούλους ἦλθε ἀνὴρ μάντις Κλέανδρος, γένος ἐὼν Φιγαλεὺς ἀπ' Ἀρκαδίης· οὗτος τοὺς δούλους ἀνέγνωσε ἐπιθέσθαι τοῖσι δεσπότῃσι. ἐκ τούτου δὲ πόλεμός 20 σφι ἦν ἐπὶ χρόνον συχνὸν, ἐς ὃ δὴ μόγις οἱ Ἀργεῖοι ἐπεκράτησαν.

Another account of the origin of Kleomenes' madness.

LXXXIV. Ἀργεῖοι μέν νυν διὰ ταῦτα Κλεομένεά φασι μανέντα ἀπολέσθαι κακῶς, αὐτοὶ δὲ Σπαρτιῆταί φασι ἐκ δαιμονίου μὲν οὐδενὸς μανῆναι 25 Κλεομένεα, Σκύθῃσι δὲ ὁμιλήσαντά μιν ἀκρητοπότην γενέσθαι καὶ ἐκ τούτου μανῆναι. Σκύθας γὰρ τοὺς νομάδας, ἐπείτε σφι Δαρεῖον ἐσβαλεῖν ἐς τὴν χώρην, μετὰ ταῦτα μεμονέναι μιν τίσασθαι, πέμψαντας δὲ ἐς Σπάρτην συμμαχίην τε ποιέεσθαι, καὶ συντίθεσθαι, ὡς χρεὸν εἴη αὐτοὺς μὲν τοὺς Σκύθας παρὰ Φᾶσιν ποταμὸν πειρᾶν ἐς τὴν Μηδικὴν ἐσβαλεῖν, σφέας δὲ τοὺς Σπαρτιήτας κελεύειν ἐξ Ἐφέσου ὁρμεομένους 5 ἀναβαίνειν καὶ ἔπειτα ἐς τὠυτὸ ἀπαντᾶν. Κλεομένεα δὲ λέγουσι ἡκόντων τῶν Σκυθέων ἐπὶ ταῦτα ὁμιλέειν σφι μεζόνως, ὁμιλέοντα δὲ μᾶλλον τοῦ ἱκνεομένου μαθεῖν τὴν ἀκρητοποσίην παρ' αὐτῶν· ἐκ τούτου δὲ μανῆναί μιν νομίζουσι Σπαρτιῆται. ἔκ τε τοῦ, ὡς 10 αὐτοὶ λέγουσι, ἐπεὰν ζωρότερον βούλωνται πιεῖν, "ἐπισκύθισον" λέγουσι. οὕτω δὴ Σπαρτιῆται τὰ περὶ Κλεομένεα λέγουσι· ἐμοὶ δὲ δοκέει τίσιν ταύτην ὁ Κλεομένης Δημαρήτῳ ἐκτῖσαι.

Resuming from c. 75. *The Aeginetans appeal to the Spartans against the forcible taking of the ten hostages by Leotychides. The Spartans decide to give up Leotychides; as a compromise he is sent to Athens to demand back the hostages.*

LXXXV. Τελευτήσαντος δὲ Κλεομένεος, ὡς 15 ἐπύθοντο Αἰγινῆται, ἔπεμπον ἐς Σπάρτην ἀγγέλους καταβωσομένους Λευτυχίδεω περὶ τῶν ἐν Ἀθήνῃσι ὁμήρων ἐχομένων. Λακεδαιμόνιοι δὲ δικαστήριον συναγαγόντες ἔγνωσαν περιυβρίσθαι Αἰγινήτας ὑπὸ Λευτυχίδεω, καί μιν κατέκριναν ἔκδοτον ἄγεσθαι ἐς 20 Αἴγιναν ἀντὶ τῶν ἐν Ἀθήνῃσι ἐχομένων ἀνδρῶν. μελλόντων δὲ ἄγειν τῶν Αἰγινητέων τὸν Λευτυχίδην εἶπέ σφι Θεασίδης ὁ Λεωπρέπεος, ἐὼν ἐν τῇ Σπάρτῃ δόκιμος ἀνήρ· "Τί βούλεσθε ποιέειν, ἄνδρες Αἰγι-"νῆται; τὸν βασιλέα τῶν Σπαρτιητέων ἔκδοτον 25

"γενόμενον ὑπὸ τῶν πολιητέων ἄγειν; εἰ νῦν ὀργῇ
"χρεόμενοι ἔγνωσαν οὕτω Σπαρτιῆται, ὅκως ἐξ ὑστέ-
"ρης μή τι ὑμῖν, ἢν ταῦτα πρήσσητε, πανώλεθρον
"κακὸν ἐς τὴν χώρην ἐσβάλωσι." ταῦτα ἀκούσαντες
5 οἱ Αἰγινῆται ἔσχοντο τῆς ἀγωγῆς, ὁμολογίῃ δὲ ἐχρήσαντο τοιῇδε, ἐπισπόμενον Λευτυχίδην ἐς Ἀθήνας
ἀποδοῦναι Αἰγινήτῃσι τοὺς ἄνδρας.

The Athenians refuse to give them up.

LXXXVI. Ὡς δὲ ἀπικόμενος Λευτυχίδης ἐς
τὰς Ἀθήνας ἀπαίτεε τὴν παραθήκην, οἱ Ἀθηναῖοι
10 προφάσιας εἷλκον οὐ βουλόμενοι ἀποδοῦναι, φάντες
δύο σφέας ἐόντας βασιλέας παραθέσθαι καὶ οὐ
δικαιοῦν τῷ ἑτέρῳ ἄνευ τοῦ ἑτέρου ἀποδιδόναι. οὐ
φαμένων δὲ ἀποδώσειν τῶν Ἀθηναίων ἔλεξέ σφι
Λευτυχίδης τάδε·

The Speech of Leotychides. Story of Glaukos.

15 "Ὦ Ἀθηναῖοι, ποιέετε μὲν ὁκότερα βούλεσθε
"αὐτοί· καὶ γὰρ ἀποδιδόντες ποιέετε ὅσια, καὶ μὴ
"ἀποδιδόντες τὰ ἐναντία τούτων· ὁκοῖον μέντοι τι ἐν
"τῇ Σπάρτῃ συνηνείχθη γενέσθαι περὶ παρακαταθή-
"κης, βούλομαι ὑμῖν εἶπαι. λέγομεν ἡμεῖς οἱ Σπαρτι-
20 "ῆται γενέσθαι ἐν τῇ Λακεδαίμονι κατὰ τρίτην γενεὴν
"τὴν ἀπ' ἐμέο Γλαῦκον Ἐπικύδεος παῖδα. τοῦτον
"τὸν ἄνδρα φαμὲν τά τε ἄλλα πάντα περιήκειν τὰ
"πρῶτα, καὶ δὴ καὶ ἀκούειν ἄριστα δικαιοσύνης πέρι
"πάντων, ὅσοι τὴν Λακεδαίμονα τοῦτον τὸν χρόνον
25 "οἴκεον. συνενειχθῆναι δέ οἱ ἐν χρόνῳ ἱκνεομένῳ τάδε
"λέγομεν, ἄνδρα Μιλήσιον ἀπικόμενον ἐς Σπάρτην
"βούλεσθαί οἱ ἐλθεῖν ἐς λόγους, προϊσχόμενον τοιάδε·

"'Εἰμὶ μὲν Μιλήσιος, ἥκω δὲ τῆς σῆς, Γλαῦκε, δικαιο-
"'σύνης βουλόμενος ἀπολαῦσαι. ὡς γὰρ δὴ ἀνὰ πᾶσαν
"'μὲν τὴν ἄλλην Ἑλλάδα, ἐν δὲ καὶ περὶ Ἰωνίην τῆς
"'σῆς δικαιοσύνης ἦν λόγος πολλὸς, ἐμεωυτῷ λόγους
"'ἐδίδουν καὶ ὅτι ἐπικίνδυνός ἐστι αἰεί κοτε ἡ Ἰωνίη, 5
"'ἡ δὲ Πελοπόννησος ἀσφαλέως ἱδρυμένη, καὶ διότι
"'χρήματα οὐδαμὰ τοὺς αὐτοὺς ἔστι ὁρᾶν ἔχοντας.
"'ταῦτά τε ὦν ἐπιλεγομένῳ καὶ βουλευομένῳ ἔδοξέ
"'μοι τὰ ἡμίσεα πάσης τῆς οὐσίης ἐξαργυρώσαντα
"'θέσθαι παρὰ σὲ, εὖ ἐξεπισταμένῳ, ὥς μοι κείμενα 10
"'ἔσται παρὰ σοὶ σόα. συ δή μοι καὶ τὰ χρήματα δέξαι
"'καὶ τάδε τὰ σύμβολα σῶζε λαβών· ὃς δ' ἂν ἔχων
"'ταῦτα ἀπαιτέῃ, τούτῳ ἀποδοῦναι.' Ὁ μὲν δὴ ἀπὸ
"Μιλήτου ἥκων ξεῖνος τοσαῦτα ἔλεξε, Γλαῦκος δὲ
"ἐδέξατο τὴν παρακαταθήκην ἐπὶ τῷ εἰρημένῳ λόγῳ. 15
"χρόνου δὲ πολλοῦ διελθόντος ἦλθον ἐς τὴν Σπάρτην
"τούτου τοῦ παραθεμένου τὰ χρήματα οἱ παῖδες, ἐλ-
"θόντες δὲ ἐς λόγους τῷ Γλαύκῳ καὶ ἀποδεικνύντες τὰ
"σύμβολα ἀπαίτεον τὰ χρήματα. ὁ δὲ διωθέετο ἀντυ-
"ποκρινόμενος τοιάδε· 'Οὔτε μέμνημαι τὸ πρῆγμα, οὔτε 20
"'με περιφέρει οὐδὲν εἰδέναι τούτων τῶν ὑμεῖς λέγετε,
"'βούλομαί τε ἀναμνησθεὶς ποιέειν πᾶν τὸ δίκαιον·
"'καὶ γὰρ εἰ ἔλαβον, ὀρθῶς ἀποδοῦναι, καὶ εἴ γε ἀρχὴν
"'μὴ ἔλαβον, νόμοισι τοῖσι Ἑλλήνων χρήσομαι ἐς
"'ὑμέας. ταῦτα ὦν ὑμῖν ἀναβάλλομαι κυρώσειν ἐς 25
"'τέταρτον μῆνα ἀπὸ τοῦδε.' Οἱ μὲν δὴ Μιλήσιοι συμ-
"φορὴν ποιεύμενοι ἀπαλλάσσοντο ὡς ἀπεστερημένοι
"τῶν χρημάτων, Γλαῦκος δὲ ἐπορεύετο ἐς Δελφοὺς
"χρησόμενος τῷ χρηστηρίῳ. ἐπειρωτέοντα δὲ αὐτὸν
"τὸ χρηστήριον, εἰ ὅρκῳ τὰ χρήματα ληΐσηται, ἡ 30
"Πυθίη μετέρχεται τοισίδε τοῖσι ἔπεσι·

"'Γλαῦκ' Ἐπικυδείδη, τὸ μὲν αὐτίκα κέρδιον οὕτως,
"'ὅρκῳ νικῆσαι καὶ χρήματα ληΐσσασθαι.
"'ὄμνυ', ἐπεὶ θάνατός γε καὶ εὔορκον μένει ἄνδρα.
"'ἀλλ' Ὅρκου πάϊς ἐστὶν ἀνώνυμος, οὐδ' ἔπι χεῖρες,
5 "'οὐδὲ πόδες· κραιπνὸς δὲ μετέρχεται, εἰσόκε πᾶσαν
"'συμμάρψας ὀλέσει γενεὴν καὶ οἶκον ἅπαντα.
"'ἀνδρὸς δ' εὐόρκου γενεὴ μετόπισθεν ἀμείνων.'
"ταῦτα ἀκούσας ὁ Γλαῦκος συγγνώμην τὸν θεὸν
"παραιτέετο αὐτῷ σχεῖν τῶν ῥηθέντων. ἡ δὲ Πυθίη
10 "ἔφη τὸ πειρηθῆναι τοῦ θεοῦ καὶ τὸ ποιῆσαι ἴσον
"δύνασθαι. Γλαῦκος μὲν δὴ μεταπεμψάμενος τοὺς
"Μιλησίους ξείνους ἀποδιδοῖ σφι τὰ χρήματα. τοῦ
"δὲ εἵνεκεν ὁ λόγος ὅδε, ὦ Ἀθηναῖοι, ὡρμήθη λέγεσθαι
"ἐς ὑμέας, εἰρήσεται· Γλαύκου νῦν οὔτε τι ἀπόγονόν
15 "ἐστι οὐδὲν οὔτ' ἱστίη οὐδεμία νομιζομένη εἶναι
"Γλαύκου, ἐκτέτριπταί τε πρόρριζος ἐκ Σπάρτης.
"οὕτω ἀγαθὸν μηδὲ διανοέεσθαι περὶ παρακαταθήκης
"ἄλλο γε ἢ ἀποδιδόναι." Λευτυχίδης μὲν εἴπας ταῦτα,
ὥς οἱ οὐδὲ οὕτω ἐσήκουον οἱ Ἀθηναῖοι, ἀπαλλάσσετο.

The Aeginetans retaliate on Athens by seizing the Sacred vessel off Sunium.

20 LXXXVII. Οἱ δὲ Αἰγινῆται, πρὶν τῶν πρότερον ἀδικημάτων δοῦναι δίκας, τῶν ἐς Ἀθηναίους ὕβρισαν Θηβαίοισι χαριζόμενοι, ἐποίησαν τοιόνδε· μεμφόμενοι τοῖσι Ἀθηναίοισι καὶ ἀξιοῦντες ἀδικέεσθαι, ὡς τιμωρησόμενοι τοὺς Ἀθηναίους παρεσκευάζοντο. καὶ ἦν 25 γὰρ δὴ τοῖσι Ἀθηναίοισι πεντετηρὶς ἐπὶ Σουνίῳ, λοχήσαντες ὦν τὴν θεωρίδα νέα εἷλον πλήρεα ἀνδρῶν τῶν πρώτων Ἀθηναίων, λαβόντες δὲ τοὺς ἄνδρας ἔδησαν.

The Athenians intrigue with Nikodromos who was heading a popular movement in Aegina to betray the island to them,

LXXXVIII. Ἀθηναῖοι δὲ παθόντες ταῦτα πρὸς Αἰγινητέων οὐκέτι ἀνεβάλλοντο μὴ οὐ τὸ πᾶν μηχανήσασθαι ἐπ' Αἰγινήτῃσι. καὶ ἦν γὰρ Νικόδρομος Κνοίθου καλεόμενος ἐν τῇ Αἰγίνῃ ἀνὴρ δόκιμος, οὗτος μεμφόμενος μὲν τοῖσι Αἰγινήτῃσι 5 προτέρην ἑωυτοῦ ἐξέλασιν ἐκ τῆς νήσου, μαθὼν δὲ τότε τοὺς Ἀθηναίους ἀναρτημένους ἔρδειν Αἰγινήτας κακῶς, συντίθεται Ἀθηναίοισι προδοσίην Αἰγίνης, φράσας ἐν τῇ τε ἡμέρῃ ἐπιχειρήσει, καὶ ἐκείνους ἐς τὴν ἥκειν δεήσει βοηθέοντας. 10

but fail because of the difficulty of getting ships in time.

LXXXIX. Μετὰ ταῦτα καταλαμβάνει μὲν κατὰ συνεθήκατο Ἀθηναίοισι ὁ Νικόδρομος τὴν παλαιὴν καλεομένην πόλιν, Ἀθηναῖοι δὲ οὐ παραγίνονται ἐς δέον· οὐ γὰρ ἔτυχον ἐοῦσαι νέες σφι ἀξιόμαχοι τῇσι Αἰγινητέων συμβαλεῖν. ἐν ᾧ ὦν Κορινθίων ἐδέοντο 15 χρῆσαί σφι νέας, ἐν τούτῳ διεφθάρη τὰ πρήγματα. οἱ δὲ Κορίνθιοι, ἦσαν γάρ σφίσι τοῦτον τὸν χρόνον φίλοι ἐς τὰ μάλιστα, Ἀθηναίοισι διδοῦσι δεομένοισι εἴκοσι νέας, διδοῦσι δὲ πενταδράχμους ἀποδόμενοι· δωρεὴν γὰρ ἐν τῷ νόμῳ οὐκ ἐξῆν δοῦναι. ταύτας τε 20 δὴ λαβόντες οἱ Ἀθηναῖοι καὶ τὰς σφετέρας, πληρώσαντες ἑβδομήκοντα νέας τὰς ἁπάσας, ἔπλωον ἐπὶ τὴν Αἴγιναν καὶ ὑστέρησαν ἡμέρῃ μιῇ τῆς συγκειμένης.

Nikodromos escapes, and is settled at Sunium.

XC. Νικόδρομος δέ, ὡς οἱ Ἀθηναῖοι ἐς τὸν καιρὸν

οὐ παρεγίνοντο, ἐς πλοῖον ἐσβὰς ἐκδιδρήσκει ἐκ τῆς Αἰγίνης, σὺν δέ οἱ καὶ ἄλλοι ἐκ τῶν Αἰγινητέων ἕσποντο, τοῖσι Ἀθηναῖοι Σούνιον οἰκῆσαι ἔδοσαν. ἐνθεῦτεν δὲ οὗτοι ὁρμεόμενοι ἔφερόν τε καὶ ἦγον τοὺς 5 ἐν τῇ νήσῳ Αἰγινήτας.

Sacrilege of the Aeginetan oligarchical party.

XCI. Ταῦτα μὲν δὴ ὕστερον ἐγίνετο, Αἰγινητέων δὲ οἱ παχέες ἐπαναστάντος τοῦ δήμου σφι ἅμα Νικοδρόμῳ ἐπεκράτησαν, καὶ ἔπειτά σφεας χειρωσάμενοι ἐξῆγον ἀπολέοντες. ἀπὸ τούτου δὲ καὶ ἄγος 10 σφι ἐγένετο, τὸ ἐκθύσασθαι οὐκ οἷοί τε ἐγίνοντο ἐπιμηχανεόμενοι, ἀλλ' ἔφθησαν ἐκπεσόντες πρότερον ἐκ τῆς νήσου ἤ σφι ἵλεων γενέσθαι τὴν θεόν. ἑπτακοσίους γὰρ δὴ τοῦ δήμου ζωγρήσαντες ἐξῆγον ὡς ἀπολέοντες, εἷς δέ τις τούτων ἐκφυγὼν τὰ δεσμὰ 15 καταφεύγει πρὸς πρόθυρα Δήμητρος θεσμοφόρου, ἐπιλαβόμενος δὲ τῶν ἐπισπαστήρων εἴχετο. οἱ δὲ ἐπεί τέ μιν ἀποσπάσαι οὐκ οἷοί τε ἀπέλκοντες ἐγίνοντο, ἀποκόψαντες αὐτοῦ τὰς χεῖρας ἦγον οὕτω, χεῖρες δὲ ἐκεῖναι ἐμπεφυκυῖαι ἦσαν τοῖσι ἐπισπάσ-20 τροισι.

Sea fight between the Athenians and Aeginetans. The Aeginetans being beaten vainly apply for help to Argos.

XCII. Ταῦτα μέν νυν σφέας αὐτοὺς οἱ Αἰγινῆται ἐργάσαντο, Ἀθηναίοισι δὲ ἥκουσι ἐναυμάχησαν νηυσὶ ἑβδομήκοντα, ἑσσωθέντες δὲ τῇ ναυμαχίῃ ἐπεκαλέοντο τοὺς αὐτοὺς καὶ πρότερον, Ἀργείους. καὶ δή σφι 25 οὗτοι μὲν οὐκέτι βοηθέουσι, μεμφόμενοι, ὅτι Αἰγιναῖαι νέες ἀνάγκῃ λαμφθεῖσαι ὑπὸ Κλεομένεος ἔσχον τε ἐς τὴν Ἀργολίδα χώρην καὶ συναπέβησαν Λακεδαιμονίοισι· συναπέβησαν δὲ καὶ ἀπὸ Σικυωνιέων νεῶν ἄνδρες τῇ αὐτῇ ταύτῃ ἐσβολῇ. καί σφι ὑπ' Ἀργείων ἐπεβλήθη ζημίη, χίλια τάλαντα ἐκτῖσαι, πεντακόσια ἑκατέρους. Σικυώνιοι μέν νυν συγγνόντες ἀδικῆσαι 5 ὡμολόγησαν ἑκατὸν τάλαντα ἐκτίσαντες ἀζήμιοι εἶναι, Αἰγινῆται δὲ οὔτε συνεγινώσκοντο, ἦσάν τε αὐθαδέστεροι. διὰ δὴ ὦν σφι ταῦτα δεομένοισι ἀπὸ μὲν τοῦ δημοσίου οὐδεὶς Ἀργείων ἔτι ἐβοήθεε, ἐθελονταὶ δὲ ἐς χιλίους· ἦγε δὲ αὐτοὺς στρατηγὸς Εὐρυβάτης, 10 πεντάεθλον ἐπασκήσας. τούτων οἱ πλεῦνες οὐκ ἀπενόστησαν ὀπίσω, ἀλλ' ἐτελεύτησαν ὑπ' Ἀθηναίων ἐν Αἰγίνῃ· αὐτὸς δὲ ὁ στρατηγὸς Εὐρυβάτης μουνομαχίην ἐπασκέων τρεῖς μὲν ἄνδρας τρόπῳ τοιούτῳ κτείνει, ὑπὸ δὲ τοῦ τετάρτου Σωφάνεος τοῦ Δεκελέος 15 ἀποθνήσκει.

They however defeat the Athenian fleet.

XCIII. Αἰγινῆται δὲ ἐοῦσι ἀτάκτοισι Ἀθηναίοισι συμβαλόντες τῇσι νηυσὶ ἐνίκησαν, καί σφεων νέας τέσσερας αὐτοῖσι ἀνδράσι εἷλον.

B.C. 490. *Darius pushes on his design of invading Greece, urged on by the Peisistratidae. He deposes Mardonius from the command, and appoints Datis and Artaphernes. The object is the destruction of Athens and Eretria.*

XCIV. Ἀθηναίοισι μὲν δὴ πόλεμος συνῆπτο 20 πρὸς Αἰγινήτας· ὁ δὲ Πέρσης τὸ ἑωυτοῦ ἐποίεε ὥστε ἀναμιμνήσκοντός τε αἰεὶ τοῦ θεράποντος μεμνῆσθαί μιν τῶν Ἀθηναίων καὶ Πεισιστρατιδέων προσκατη-

μένων καὶ διαβαλλόντων Ἀθηναίους, ἅμα δὲ βουλόμενος ὁ Δαρεῖος ταύτης ἐχόμενος τῆς προφάσιος καταστρέφεσθαι τῆς Ἑλλάδος τοὺς μὴ δόντας αὐτῷ γῆν τε καὶ ὕδωρ. Μαρδόνιον μὲν δὴ φλαύρως πρή-
5 ξαντα τῷ στόλῳ παραλύει τῆς στρατηγίης, ἄλλους δὲ στρατηγοὺς ἀποδέξας ἀπέστελλε ἐπί τε Ἐρέτριαν καὶ Ἀθήνας, Δᾶτίν τε ἐόντα Μῆδον γένος καὶ Ἀρταφέρνεα τὸν Ἀρταφέρνεος παῖδα, ἀδελφιδέον ἑωυτοῦ· ἐντειλάμενος δὲ ἀπέπεμπε ἐξανδραποδί-
10 σαντας Ἀθήνας καὶ Ἐρέτριαν ἀγαγεῖν ἑωυτῷ ἐς ὄψιν τὰ ἀνδράποδα.

The Persian army musters on the Aleïan plain in Kilikia. The fleet takes all on board and sails to Naxos, where they burn the temples and town.

XCV. Ὡς δὲ οἱ στρατηγοὶ οὗτοι οἱ ἀποδεχθέντες πορευόμενοι παρὰ βασιλέος ἀπίκοντο τῆς Κιλικίης ἐς τὸ Ἀλήϊον πεδίον, ἅμα ἀγόμενοι πεζὸν
15 στρατὸν πολλόν τε καὶ εὖ ἐσκευασμένον, ἐνθαῦτα στρατοπεδευομένοισι ἐπῆλθε μὲν ὁ ναυτικὸς πᾶς στρατὸς ὁ ἐπιταχθεὶς ἑκάστοισι, παρεγένοντο δὲ καὶ αἱ ἱππαγωγοὶ νέες, τὰς τῷ προτέρῳ ἔτεϊ προεῖπε τοῖσι ἑωυτοῦ δασμοφόροισι Δαρεῖος ἑτοιμάζειν. ἐσβα-
20 λόμενοι δὲ τοὺς ἵππους ἐς ταύτας καὶ τὸν πεζὸν στρατὸν ἐσβιβάσαντες ἐς τὰς νέας, ἔπλωον ἑξακοσίῃσι τριήρεσι ἐς τὴν Ἰωνίην. ἐνθεῦτεν δὲ οὐ παρὰ τὴν ἤπειρον εἶχον τὰς νέας ἰθὺ τοῦ τε Ἑλλησπόντου καὶ τῆς Θρηΐκης, ἀλλ' ἐκ Σάμου ὁρμεόμενοι παρά τε
25 Ἰκάριον καὶ διὰ νήσων τὸν πλόον ἐποιεῦντο, ὡς μὲν ἐμοὶ δοκέειν, δείσαντες μάλιστα τὸν περίπλοον τοῦ Ἄθω, ὅτι τῷ προτέρῳ ἔτεϊ ποιεύμενοι ταύτῃ τὴν

κομιδὴν μεγάλως προσέπταισαν· πρὸς δὲ καὶ ἡ Νάξος σφέας ἠνάγκαζε, πρότερον οὐκ ἁλοῦσα. XCVI. Ἐπεὶ δὲ ἐκ τοῦ Ἰκαρίου πελάγεος προσφερόμενοι προσέμιξαν τῇ Νάξῳ (ἐπὶ ταύτην γὰρ δὴ πρώτην ἐπεῖχον στρατεύεσθαι οἱ Πέρσαι), μεμνημένοι 5 τῶν πρότερον οἱ Νάξιοι πρὸς τὰ οὔρεα οἴχοντο φεύγοντες· οὐδὲ ὑπέμειναν. οἱ δὲ Πέρσαι ἀνδραποδισάμενοι τοὺς κατέλαβον αὐτῶν, ἐνέπρησαν καὶ τὰ ἱρὰ καὶ τὴν πόλιν, ταῦτα δὲ ποιήσαντες ἐπὶ τὰς ἄλλας νήσους ἀνήγοντο. 10

The Delians fly to Tenos. The Persians anchor at Rhenaea, and endeavour to induce the Delians to return. They treat the sacred places with reverence.

XCVII. Ἐν ᾧ δὲ οὗτοι ταῦτα ἐποίευν, οἱ Δήλιοι ἐκλιπόντες καὶ αὐτοὶ τὴν Δῆλον οἴχοντο φεύγοντες ἐς Τῆνον. τῆς δὲ στρατιῆς καταπλωούσης ὁ Δᾶτις προπλώσας οὐκ ἔα τὰς νέας πρὸς τὴν Δῆλον προσορμίζεσθαι, ἀλλὰ πέρην ἐν τῇ Ῥηναίῃ, αὐτὸς δὲ πυθόμενος ἵνα ἦσαν οἱ Δήλιοι, πέμπων κήρυκα ἠγόρευέ 15 σφι τάδε· "Ἄνδρες ἱροί, τί φεύγοντες οἴχεσθε, οὐκ "ἐπιτήδεα καταγνόντες κατ' ἐμεῦ; ἐγὼ γὰρ καὶ αὐτὸς "ἐπὶ τοσοῦτό γε φρονέω καί μοι ἐκ βασιλέος ὧδε "ἐπέσταλται, ἐν τῇ χώρῃ οἱ δύο θεοὶ ἐγένοντο, ταύτην 20 "μηδὲν σίνεσθαι, μήτε αὐτὴν τὴν χώρην μήτε τοὺς "οἰκήτορας αὐτῆς. νῦν ὦν καὶ ἄπιτε ἐπὶ τὰ ὑμέτερα "αὐτῶν καὶ τὴν νῆσον νέμεσθε." Ταῦτα μὲν ἐπεκηρυκεύσατο τοῖσι Δηλίοισι, μετὰ δὲ λιβανωτοῦ τριηκόσια τάλαντα κατανήσας ἐπὶ τοῦ βωμοῦ ἐθυμίησε. 25

An earthquake at Delos,—an omen of the troubles to come on Greece.

XCVIII. Δᾶτις μὲν δὴ ταῦτα ποιήσας ἔπλεε ἅμα τῷ στρατῷ ἐπὶ τὴν Ἐρέτριαν πρῶτα, ἅμα ἀγόμενος καὶ Ἴωνας καὶ Αἰολέας, μετὰ δὲ τοῦτον ἐνθεῦτεν ἐξαναχθέντα Δῆλος ἐκινήθη, ὡς ἔλεγον
5 Δήλιοι, καὶ πρῶτα καὶ ὕστατα μέχρι ἐμεῦ σεισθεῖσα. καὶ τοῦτο μέν κου τέρας ἀνθρώποισι τῶν μελλόντων ἔσεσθαι κακῶν ἔφαινε ὁ θεός. ἐπὶ γὰρ Δαρείου τοῦ Ὑστάσπεος καὶ Ξέρξεω τοῦ Δαρείου καὶ Ἀρταξέρξεω τοῦ Ξέρξεω, τριῶν τούτων ἐπεξῆς γενεέων ἐγένετο
10 πλέω κακὰ τῇ Ἑλλάδι ἢ ἐπὶ εἴκοσι ἄλλας γενεὰς τὰς πρὸ Δαρείου γενομένας, τὰ μὲν ἀπὸ τῶν Περσέων αὐτῇ γενόμενα, τὰ δὲ ἀπ' αὐτῶν τῶν κορυφαίων περὶ τῆς ἀρχῆς πολεμεόντων. οὕτω οὐδὲν ἦν ἀεικὲς κινηθῆναι Δῆλον τὸ πρὶν ἐοῦσαν ἀκίνητον. [καὶ ἐν
15 χρησμῷ ἦν γεγραμμένον περὶ αὐτῆς ὧδε·

κινήσω καὶ Δῆλον, ἀκίνητόν περ ἐοῦσαν.]

δύναται δὲ κατὰ Ἑλλάδα γλῶσσαν ταῦτα τὰ οὐνόματα, Δαρεῖος ἐρξίης, Ξέρξης ἀρήϊος, Ἀρταξέρξης μέγας ἀρήϊος. τούτους μὲν δὴ τοὺς βασιλέας ὧδε
20 ἂν ὀρθῶς κατὰ γλῶσσαν τὴν σφετέρην Ἕλληνες καλέοιεν.

The Persian fleet touches at Karystos on the South of Euboea. The Karystians yield.

XCIX. Οἱ δὲ βάρβαροι ὡς ἀπῆραν ἐκ τῆς Δήλου, προσίσχον πρὸς τὰς νήσους, ἐνθεῦτεν δὲ στρατιήν τε παρελάμβανον καὶ ὁμήρους τῶν νησιω-
25 τέων παῖδας ἐλάμβανον. ὡς δὲ περιπλέοντες τὰς

νήσους προσέσχον καὶ ἐς Κάρυστον, οὐ γὰρ δή σφι οἱ Καρύστιοι οὔτε ὁμήρους ἐδίδοσαν οὔτε ἔφασαν ἐπὶ πόλις ἀστυγείτονας στρατεύεσθαι, λέγοντες Ἐρέτριάν τε καὶ Ἀθήνας, ἐνθαῦτα τούτους ἐπολιόρκεόν τε καὶ τὴν γῆν σφέων ἔκειρον, ἐς ὃ καὶ οἱ Καρύστιοι 5 παρέστησαν ἐς τῶν Περσέων τὴν γνώμην.

The Eretrians send to Athens for help. There is a division of opinion in Eretria, which is betrayed on the 7th day by Euphorbos and Philagros.

C. Ἐρετριέες δὲ πυνθανόμενοι τὴν στρατιὴν τὴν Περσικὴν ἐπὶ σφέας ἐπιπλέουσαν Ἀθηναίων ἐδεήθησαν σφίσι βοηθοὺς γενέσθαι. Ἀθηναῖοι δὲ οὐκ ἀπείπαντο τὴν ἐπικουρίην, ἀλλὰ τοὺς τετρακισχι- 10 λίους κληρουχέοντας τῶν ἱπποβοτέων Χαλκιδέων τὴν χώρην, τούτους σφι διδοῦσι τιμωρούς. τῶν δὲ Ἐρετριέων ἦν ἄρα οὐδὲν ὑγιὲς βούλευμα, οἳ μετεπέμποντο μὲν Ἀθηναίους, ἐφρόνεον δὲ διφασίας ἰδέας. οἱ μὲν γὰρ αὐτῶν ἐβουλεύοντο ἐκλιπεῖν τὴν πόλιν ἐς 15 τὰ ἄκρα τῆς Εὐβοίης, ἄλλοι δὲ αὐτῶν ἴδια κέρδεα προσδεκόμενοι παρὰ τοῦ Πέρσεω οἴσεσθαι προδοσίην ἐσκευάζοντο. μαθὼν δὲ τούτων ἑκάτερα ὡς εἶχε Αἰσχίνης ὁ Νόθωνος, ἐὼν τῶν Ἐρετριέων τὰ πρῶτα, φράζει τοῖσι ἥκουσι τῶν Ἀθηναίων πάντα τὰ 20 παρεόντα σφι πρήγματα, προσεδέετό τε ἀπαλλάσσεσθαί σφεας ἐς τὴν σφετέρην, ἵνα μὴ προσαπόλωνται. οἱ δὲ Ἀθηναῖοι ταῦτα Αἰσχίνῃ συμβουλεύσαντι πείθονται. CI. Καὶ οὗτοι μὲν διαβάντες ἐς Ὠρωπὸν ἔσωζον σφέας αὐτούς, οἱ δὲ Πέρσαι 25 πλέοντες κατέσχον τὰς νέας τῆς Ἐρετρικῆς χώρης κατὰ Ταμύνας καὶ Χοιρέας καὶ Αἰγίλια, κατασχόντες δὲ ἐς ταῦτα τὰ χωρία αὐτίκα ἵππους τε ἐξεβάλλοντο καὶ παρεσκευάζοντο ὡς προσοισόμενοι τοῖσι ἐχθροῖσι. οἱ δὲ Ἐρετριέες ἐπεξελθεῖν μὲν καὶ μαχέσασθαι οὐκ ἐποιεῦντο βουλήν, εἴ κως δὲ διαφυλάξαιεν τὰ τείχεα, 5 τούτου σφι ἔμελε πέρι, ἐπεί τε ἐνίκα μὴ ἐκλιπεῖν τὴν πόλιν. προσβολῆς δὲ γινομένης καρτερῆς πρὸς τὸ τεῖχος ἔπιπτον ἐπὶ ἓξ ἡμέρας πολλοὶ μὲν ἀμφοτέρων, τῇ δὲ ἑβδόμῃ Εὔφορβός τε ὁ Ἀλκιμάχου καὶ Φίλαγρος ὁ Κυνέω, ἄνδρες τῶν ἀστῶν δόκιμοι, προδι-10 δοῦσι τοῖσι Πέρσῃσι. οἱ δὲ ἐσελθόντες ἐς τὴν πόλιν τοῦτο μὲν τὰ ἱρὰ συλήσαντες ἐνέπρησαν, ἀποτινύμενοι τῶν ἐν Σάρδισι κατακαυθέντων ἱρῶν, τοῦτο δὲ τοὺς ἀνθρώπους ἠνδραποδίσαντο κατὰ τὰς Δαρείου ἐντολάς.

The Persian fleet then sails to Marathon on the Attic coast.

15 CII. Χειρωσάμενοι δὲ τὴν Ἐρέτριαν καὶ ἐπισχόντες ὀλίγας ἡμέρας ἔπλεον ἐς τὴν Ἀττικήν, κατέργοντές τε πολλὸν καὶ δοκέοντες ταὐτὰ τοὺς Ἀθηναίους ποιήσειν, τὰ καὶ τοὺς Ἐρετριέας ἐποίησαν, καὶ ἦν γὰρ ὁ Μαραθὼν ἐπιτηδεώτατον χωρίον τῆς 20 Ἀττικῆς ἐνιππεῦσαι καὶ ἀγχοτάτω τῆς Ἐρετρίης, ἐς τοῦτό σφι κατηγέετο Ἱππίης ὁ Πεισιστράτου.

The Athenians send out an army to defend their country under their ten Strategi, the tenth being Miltiades. Previous history of Miltiades.

CIII. Ἀθηναῖοι δὲ ὡς ἐπύθοντο ταῦτα, ἐβοήθεον καὶ αὐτοὶ ἐς τὸν Μαραθῶνα. ἦγον δέ σφεας στρατηγοὶ δέκα, τῶν ὁ δέκατος ἦν Μιλτιάδης, τοῦ τὸν 25 πατέρα Κίμωνα τὸν Στησαγόρεω κατέλαβε φυγεῖν ἐξ Ἀθηνέων Πεισίστρατον τὸν Ἱπποκράτεος. καὶ αὐτῷ φεύγοντι Ὀλυμπιάδα ἀνελέσθαι τεθρίππῳ συνέβη, καὶ ταύτην μὲν τὴν νίκην ἀνελόμενόν μιν τὠυτὸ ἐξενείκασθαι τῷ ὁμομητρίῳ ἀδελφεῷ Μιλτιάδῃ. μετὰ δὲ τῇ ὑστέρῃ Ὀλυμπιάδι τῇσι αὐτῇσι 5 ἵπποισι νικῶν παραδιδοῖ Πεισιστράτῳ ἀνακηρυχθῆναι, καὶ τὴν νίκην παρεὶς τούτῳ κατῆλθε ἐπὶ τὰ ἑωυτοῦ ὑπόσπονδος. καί μιν ἀνελόμενον τῇσι αὐτῇσι ἵπποισι ἄλλην Ὀλυμπιάδα κατέλαβε ἀποθανεῖν ὑπὸ τῶν Πεισιστράτου παίδων οὐκέτι περιεόντος αὐτοῦ 10 Πεισιστράτου· κτείνουσι δὲ οὗτοί μιν κατὰ τὸ πρυτανήϊον νυκτὸς ὑπείσαντες ἄνδρας. τέθαπται δὲ Κίμων πρὸ τοῦ ἄστεος, πέρην τῆς διὰ Κοίλης καλεομένης ὁδοῦ, καταντίον δ' αὐτοῦ αἱ ἵπποι τετάφαται αὗται αἱ τρεῖς Ὀλυμπιάδας ἀνελόμεναι. ἐποίη- 15 σαν δὲ καὶ ἄλλαι ἵπποι ἤδη τὠυτὸ τοῦτο Εὐαγόρεω Λάκωνος, πλέω δὲ τούτων οὐδαμαί. Ὁ μὲν δὴ πρεσβύτερος τῶν παίδων τῷ Κίμωνι Στησαγόρης ἦν τηνικαῦτα παρὰ τῷ πάτρῳ Μιλτιάδῃ τρεφόμενος ἐν τῇ Χερσονήσῳ, ὁ δὲ νεώτερος παρ' αὐτῷ Κίμωνι ἐν 20 Ἀθήνῃσι, τοὔνομα ἔχων ἀπὸ τοῦ οἰκιστέω τῆς Χερσονήσου Μιλτιάδεω Μιλτιάδης. CIV. Οὗτος δὴ ὢν τότε ὁ Μιλτιάδης ἥκων ἐκ τῆς Χερσονήσου καὶ ἐκπεφευγὼς διπλόον θάνατον ἐστρατήγεε Ἀθηναίων. ἅμα μὲν γὰρ οἱ Φοίνικες αὐτὸν οἱ ἐπιδιώξαντες μέχρι 25 Ἴμβρου περὶ πολλοῦ ἐποιεῦντο λαβεῖν τε καὶ ἀναγαγεῖν παρὰ βασιλέα, ἅμα δὲ ἐκφυγόντα τε τούτους καὶ ἀπικόμενον ἐς τὴν ἑωυτοῦ, δοκέοντά τε εἶναι ἐν σωτηρίῃ ἤδη, τὸ ἐνθεῦτέν μιν οἱ ἐχθροὶ ὑποδεξάμενοι καὶ ὑπὸ δικαστήριον αὐτὸν ἀγαγόντες ἐδίωξαν τυραν- 30 νίδος τῆς ἐν Χερσονήσῳ. ἀποφυγὼν δὲ καὶ τούτους στρατηγὸς οὕτω Ἀθηναίων ἀπεδέχθη, αἱρεθεὶς ὑπὸ τοῦ δήμου.

Pheidippides sent to Sparta, who is met by the God Pan on his road.

CV. Καὶ πρῶτα μὲν ἐόντες ἔτι ἐν τῷ ἄστεϊ οἱ στρατηγοὶ ἀποπέμπουσι ἐς Σπάρτην κήρυκα Φειδιπ-
5 πίδην, Ἀθηναῖον μὲν ἄνδρα, ἄλλως δὲ ἡμεροδρόμον τε καὶ τοῦτο μελετῶντα, τῷ δή, ὡς αὐτός τε ἔλεγε Φειδιππίδης καὶ Ἀθηναίοισι ἀπήγγελλε, περὶ τὸ Παρθένιον οὖρος τὸ ὑπὲρ Τεγέης ὁ Πὰν περιπίπτει. βώσαντα δὲ τὸ οὔνομα τοῦ Φειδιππίδεω τὸν Πᾶνα
10 Ἀθηναίοισι κελεῦσαι ἀπαγγεῖλαι, δι' ὅ τι ἑωυτοῦ οὐδεμίαν ἐπιμέλειαν ποιεῦνται, ἐόντος εὐνόου Ἀθηναίοισι καὶ πολλαχῇ γενομένου ἤδη σφι χρησίμου, τὰ δ' ἔτι καὶ ἐσομένου. καὶ ταῦτα μὲν Ἀθηναῖοι, καταστάντων σφίσι εὖ ἤδη τῶν πρηγμάτων, πιστεύσαντες εἶναι
15 ἀληθέα ἱδρύσαντο ὑπὸ τῇ ἀκροπόλι Πανὸς ἱρόν, καὶ αὐτὸν ἀπὸ ταύτης τῆς ἀγγελίης θυσίῃσι ἐπετέῃσι καὶ λαμπάδι ἱλάσκονται.

The Spartans will come when it is full moon.

CVI. Τότε δὲ πεμφθεὶς ὑπὸ τῶν στρατηγῶν ὁ Φειδιππίδης οὗτος, ὅτε πέρ οἱ ἔφη καὶ τὸν Πᾶνα
20 φανῆναι, δευτεραῖος ἐκ τοῦ Ἀθηναίων ἄστεος ἦν ἐν Σπάρτῃ, ἀπικόμενος δὲ ἐπὶ τοὺς ἄρχοντας ἔλεγε· "Ὦ "Λακεδαιμόνιοι, Ἀθηναῖοι ὑμέων δέονται σφίσι βοη-"θῆσαι καὶ μὴ περιιδεῖν πόλιν ἀρχαιοτάτην ἐν τοῖσι "Ἕλλησι δουλοσύνῃ περιπεσοῦσαν πρὸς ἀνδρῶν βαρ-
25 "βάρων· καὶ γὰρ νῦν Ἐρέτριά τε ἠνδραπόδισται καὶ "πόλι λογίμῳ ἡ Ἑλλὰς γέγονε ἀσθενεστέρη." Ὁ μὲν

δή σφι τὰ ἐντεταλμένα ἀπήγγελλε, τοῖσι δὲ ἕαδε μὲν βοηθέειν Ἀθηναίοισι, ἀδύνατα δέ σφι ἦν τὸ παραυτίκα ποιέειν ταῦτα οὐ βουλομένοισι λύειν τὸν νόμον· ἦν γὰρ ἱσταμένου τοῦ μηνὸς εἰνάτη, εἰνάτῃ δὲ οὐκ ἐξελεύσεσθαι ἔφασαν μὴ οὐ πλήρεος ἐόντος τοῦ 5 κύκλου.

Hippias' dream; and his lost tooth.

CVII. Οὗτοι μέν νυν τὴν πανσέληνον ἔμενον, τοῖσι δὲ βαρβάροισι κατηγέετο Ἱππίης ὁ Πεισιστράτου ἐς τὸν Μαραθῶνα, τῆς παροιχομένης νυκτὸς ὄψιν ἰδὼν ἐν τῷ ὕπνῳ τοιήνδε· ἐδόκεε ὁ Ἱππίης τῇ 10 μητρὶ τῇ ἑωυτοῦ συνευνηθῆναι. συνεβάλετο ὦν ἐκ τοῦ ὀνείρου κατελθὼν ἐς τὰς Ἀθήνας καὶ ἀνασωσάμενος τὴν ἀρχὴν τελευτήσειν ἐν τῇ ἑωυτοῦ γηραιός. ἐκ μὲν δὴ τῆς ὄψιος συνεβάλετο ταῦτα, τότε δὲ κατηγεόμενος τοῦτο μὲν τὰ ἀνδράποδα τὰ ἐξ Ἐρε- 15 τρίης ἀπέβησε ἐς τὴν νῆσον τὴν Στυρέων, καλεομένην δὲ Αἰγίλειαν, τοῦτο δὲ καταγομένας ἐς τὸν Μαραθῶνα τὰς νέας ὅρμιζε οὗτος, ἐκβάντας τε ἐς γῆν τοὺς βαρβάρους διέτασσε. καί οἱ ταῦτα διέποντι ἐπῆλθε πταρεῖν τε καὶ βῆξαι μεζόνως ἢ ὡς ἐώθεε, οἷα δέ οἱ 20 πρεσβυτέρῳ ἐόντι τῶν ὀδόντων οἱ πλεῦνες ἐσείοντο. τούτων ὦν ἕνα τῶν ὀδόντων ἐκβάλλει ὑπὸ βίης βήξας· ἐκπεσόντος δὲ ἐς τὴν ψάμμον αὐτοῦ ἐποιέετο πολλὴν σπουδὴν ἐξευρεῖν. ὡς δὲ οὐκ ἐφαίνετό οἱ ὁ ὀδών, ἀναστενάξας εἶπε πρὸς τοὺς παραστάτας· "Ἡ 25 "γῆ ἥδε οὐκ ἡμετέρη ἐστὶ οὐδέ μιν δυνησόμεθα ὑπο-"χειρίην ποιήσασθαι· ὁκόσον δέ τί μοι μέρος μετῆν, "ὁ ὀδὼν μετέχει."

The Athenians, drawn up in the sacred enclosure of Herakles, are joined by 1000 Plataeans.

CVIII. Ἱππίης μὲν δὴ ταύτῃ τὴν ὄψιν συνεβάλετο ἐξεληλυθέναι, Ἀθηναίοισι δὲ τεταγμένοισι ἐν τεμένεϊ Ἡρακλέος ἐπῆλθον βοηθέοντες Πλαταιέες πανδημεί· καὶ γὰρ καὶ ἐδεδώκεσαν σφέας αὐτοὺς
5 τοῖσι Ἀθηναίοισι οἱ Πλαταιέες, καὶ πόνους ὑπὲρ αὐτῶν οἱ Ἀθηναῖοι συχνοὺς ἤδη ἀναιρέοντο· ἔδοσαν δὲ ὧδε.

The origin of the connexion between Plataea and Athens.

Πιεζόμενοι ὑπὸ Θηβαίων οἱ Πλαταιέες ἐδίδοσαν πρῶτα παρατυχοῦσι Κλεομένεΐ τε τῷ Ἀναξανδρίδεω
10 καὶ Λακεδαιμονίοισι σφέας αὐτούς, οἱ δὲ οὐ δεκόμενοι ἔλεγόν σφι τάδε· "Ἡμεῖς μὲν ἑκαστέρω τε οἰκέομεν, "καὶ ὑμῖν τοιήδε τις γίνοιτ' ἂν ἐπικουρίη ψυχρή· "φθαίητε γὰρ ἂν πολλάκις ἐξανδραποδισθέντες ἤ τινα "πυθέσθαι ἡμέων. συμβουλεύομεν δὲ ὑμῖν δοῦναι
15 "ὑμέας αὐτοὺς Ἀθηναίοισι, πλησιοχώροισί τε ἀνδράσι "καὶ τιμωρέειν ἐοῦσι οὐ κακοῖσι." Ταῦτα συνεβούλευον οἱ Λακεδαιμόνιοι, οὐ κατὰ εὐνοίην οὕτω τῶν Πλαταιέων, ὡς βουλόμενοι τοὺς Ἀθηναίους ἔχειν πόνους συνεστεῶτας Βοιωτοῖσι. Λακεδαιμόνιοι μέν νυν
20 Πλαταιεῦσι ταῦτα συνεβούλευον, οἱ δὲ οὐκ ἠπίστησαν, ἀλλ' Ἀθηναίων ἱρὰ ποιεύντων τοῖσι δυώδεκα θεοῖσι ἱκέται ἱζόμενοι ἐπὶ τὸν βωμὸν ἐδίδοσαν σφέας αὐτούς. Θηβαῖοι δὲ πυθόμενοι ταῦτα ἐστρατεύοντο ἐπὶ τοὺς Πλαταιέας· Ἀθηναῖοι δέ σφι ἐβοήθεον.
25 μελλόντων δὲ συνάπτειν μάχην Κορίνθιοι οὐ περιεῖδον, παρατυχόντες δὲ καὶ καταλλάξαντες ἐπιτοε-

ψάντων ἀμφοτέρων, οὔρισαν τὴν χώρην ἐπὶ τοισίδε, ἐᾶν Θηβαίους Βοιωτῶν τοὺς μὴ βουλομένους ἐς Βοιωτοὺς τελέειν. Κορίνθιοι μὲν δὴ ταῦτα γνόντες ἀπαλλάσσοντο, Ἀθηναίοισι δὲ ἀπιοῦσι ἐπεθήκαντο Βοιωτοί, ἐπιθέμενοι δὲ ἑσσώθησαν τῇ μάχῃ. ὑπερ- 5
βάντες δὲ οἱ Ἀθηναῖοι τοὺς οἱ Κορίνθιοι ἔθηκαν Πλαταιεῦσι εἶναι οὔρους, τούτους ὑπερβάντες τὸν Ἀσωπὸν αὐτὸν ἐποιήσαντο οὖρον Θηβαίοισι πρὸς Πλαταιέας εἶναι καὶ Ὑσιάς. ἔδοσαν μὲν δὴ οἱ Πλαταιέες σφέας αὐτοὺς Ἀθηναίοισι τρόπῳ τῷ 10
εἰρημένῳ, ἧκον δὲ τότε ἐς Μαραθῶνα βοηθέοντες.

The Athenian generals are divided in opinion. Some argue against attack, others with Miltiades wish for an immediate advance.

CIX. Τοῖσι δὲ Ἀθηναίων στρατηγοῖσι ἐγίνοντο δίχα αἱ γνῶμαι, τῶν μὲν οὐκ ἐώντων συμβάλλειν, ὀλίγους γὰρ εἶναι στρατιῇ τῇ Μήδων συμβαλεῖν, τῶν δὲ καὶ Μιλτιάδεω κελευόντων. ὡς δὲ δίχα τε 15
ἐγίνοντο καὶ ἐνίκα ἡ χείρων τῶν γνωμέων, ἐνθαῦτα ἦν γὰρ ἑνδέκατος ψηφιδοφόρος ὁ τῷ κυάμῳ λαχὼν Ἀθηναίων πολεμαρχέειν, (τὸ παλαιὸν γὰρ Ἀθηναῖοι ὁμόψηφον τὸν πολέμαρχον ἐποιεῦντο τοῖσι στρατηγοῖσι,) ἦν τε τότε πολέμαρχος Καλλίμαχος Ἀφιδναῖος, 20
πρὸς τοῦτον ἐλθὼν Μιλτιάδης ἔλεγε τάδε·

Speech of Miltiades.

"Ἐν σοὶ νῦν, Καλλίμαχε, ἐστὶ ἢ καταδουλῶσαι "Ἀθήνας, ἢ ἐλευθέρας ποιήσαντα μνημόσυνα λιπέσθαι "ἐς τὸν ἅπαντα ἀνθρώπων βίον, οἷα οὐδὲ Ἁρμόδιός "τε καὶ Ἀριστογείτων λείπουσι. νῦν γὰρ δή, ἐξ οὗ 25

“ἐγένοντο Ἀθηναῖοι, ἐς κίνδυνον ἥκουσι μέγιστον. καὶ
“ἢν μέν γε ὑποκύψωσι τοῖσι Μήδοισι, δέδοκται τὰ
“πείσονται παραδεδομένοι Ἱππίῃ, ἢν δὲ περιγένηται
“αὕτη ἡ πόλις, οἵη τέ ἐστι πρώτη τῶν Ἑλληνίδων
5 “πολίων γενέσθαι. κῶς ὦν δὴ ταῦτα οἷά τέ ἐστι
“γενέσθαι, καὶ κῶς ἐς σέ τοι τούτων ἀνήκει τῶν
“πρηγμάτων τὸ κῦρος ἔχειν, νῦν ἔρχομαι φράσων.
“ἡμέων τῶν στρατηγῶν ἐόντων δέκα δίχα γίνονται αἱ
“γνῶμαι, τῶν μὲν κελευόντων, τῶν δὲ οὔ συμβάλλειν.
10 “ἢν μέν νυν μὴ συμβάλωμεν, ἔλπομαί τινα στάσιν
“μεγάλην ἐμπεσοῦσαν διασείσειν τὰ Ἀθηναίων φρο-
“νήματα ὥστε μηδίσαι· ἢν δὲ συμβάλωμεν πρίν τι
“καὶ σαθρὸν Ἀθηναίων μετεξετέροισι ἐγγενέσθαι,
“θεῶν τὰ ἴσα νεμόντων οἷοί τέ εἰμεν περιγενέσθαι τῇ
15 “συμβολῇ. ταῦτα ὦν πάντα ἐς σὲ νῦν τείνει καὶ ἐκ
“σέο ἤρτηται· ἢν γὰρ σὺ γνώμῃ τῇ ἐμῇ προσθῇ, ἔστι
“τοι πατρίς τε ἐλευθέρη καὶ πόλις πρώτη τῶν ἐν τῇ
“Ἑλλάδι, ἢν δὲ τὴν τῶν ἀποσπευδόντων τὴν συμ-
“βολὴν ἕλῃ, ὑπάρξει τοι τῶν ἐγὼ κατέλεξα ἀγαθῶν
20 “τὰ ἐναντία.”

The Polemarch Kallimachos is convinced, and four of the Strategi surrender their days of command to Miltiades.

CX. Ταῦτα λέγων ὁ Μιλτιάδης προσκτᾶται τὸν Καλλίμαχον. προσγενομένης δὲ τοῦ πολεμάρχου τῆς γνώμης ἐκεκύρωτο συμβάλλειν. μετὰ δὲ οἱ στρατηγοί, τῶν ἡ γνώμη ἔφερε συμβάλλειν, ὡς
25 ἑκάστου αὐτῶν ἐγίνετο πρυτανηΐη τῆς ἡμέρης, Μιλτιάδῃ παρεδίδοσαν· ὁ δὲ δεκόμενος οὔτι κω συμβολὴν ἐποιέετο, πρίν γε δὴ αὐτοῦ πρυτανηΐη ἐγένετο.

Miltiades waits until his right day for command comes round, ard then draws out the men for action.

CXI. Ὡς δὲ ἐς ἐκεῖνον περιῆλθε, ἐνθαῦτα δὴ ἐτάσσοντο ὧδε Ἀθηναῖοι ὡς συμβαλέοντες· τοῦ μὲν δεξιοῦ κέρεος ἡγέετο ὁ πολέμαρχος Καλλίμαχος· ὁ γὰρ νόμος τότε εἶχε οὕτω τοῖσι Ἀθηναίοισι, τὸν πολέμαρχον ἔχειν κέρας τὸ δεξιόν. ἡγεομένου δὲ 5 τούτου ἐξεδέκοντο ὡς ἠριθμέοντο αἱ φυλαί, ἐχόμεναι ἀλλήλων· τελευταῖοι δὲ ἐτάσσοντο, ἔχοντες τὸ εὐώνυμον κέρας, Πλαταιέες. ἀπὸ ταύτης γάρ σφι τῆς μάχης Ἀθηναίων θυσίας ἀναγόντων ἐς τὰς πανηγύριας τὰς ἐν τῇσι πεντετηρίσι γινομένας κατεύ- 10 χεται ὁ κῆρυξ ὁ Ἀθηναῖος, ἅμα τε Ἀθηναίοισι λέγων γίνεσθαι τὰ ἀγαθὰ καὶ Πλαταιεῦσι. τότε δὲ τασσαμένων τῶν Ἀθηναίων ἐν τῷ Μαραθῶνι ἐγίνετο τοιόνδε τι· τὸ στρατόπεδον ἐξισούμενον τῷ Μηδικῷ στρατοπέδῳ, τὸ μὲν αὐτοῦ μέσον ἐγίνετο ἐπὶ τάξιας 15 ὀλίγας, καὶ ταύτῃ ἦν ἀσθενέστατον τὸ στρατόπεδον, τὸ δὲ κέρας ἑκάτερον ἔρρωτο πλήθεϊ.

The charge.

CXII. Ὡς δέ σφι διετέτακτο καὶ τὰ σφάγια ἐγίνετο καλά, ἐνθαῦτα ὡς ἀπείθησαν οἱ Ἀθηναῖοι, δρόμῳ ἵεντο ἐς τοὺς βαρβάρους. ἦσαν δὲ στάδιοι 20 οὐκ ἐλάσσονες τὸ μεταίχμιον αὐτῶν ἢ ὀκτώ. οἱ δὲ Πέρσαι ὁρέοντες δρόμῳ ἐπιόντας παρεσκευάζοντο ὡς δεξόμενοι, μανίην τε τοῖσι Ἀθηναίοισι ἐπέφερον καὶ πάγχυ ὀλεθρίην, ὁρέοντες αὐτοὺς ὀλίγους, καὶ τούτους δρόμῳ ἐπειγομένους, οὔτε ἵππου ὑπαρχούσης σφι 25 οὔτε τοξευμάτων. ταῦτα μέν νυν οἱ βάρβαροι κατεί-

καζον, Ἀθηναῖοι δὲ ἐπείτε ἀθρόοι προσέμιξαν τοῖσι βαρβάροισι, ἐμάχοντο ἀξίως λόγου. πρῶτοι μὲν γὰρ Ἑλλήνων πάντων τῶν ἡμεῖς ἴδμεν δρόμῳ ἐς πολεμίους ἐχρήσαντο, πρῶτοι δὲ ἀνέσχοντο ἐσθῆτά
5 τε Μηδικὴν ὁρέοντες καὶ τοὺς ἄνδρας ταύτην ἐσθημένους· τέως δὲ ἦν τοῖσι Ἕλλησι καὶ τὸ οὔνομα τὸ Μήδων φόβος ἀκοῦσαι.

The Athenian centre is repulsed; but their two wings turn the enemy, and then close up and engage and beat the forces that had repulsed their centre, and follow them with slaughter to their ships.

CXIII. Μαχομένων δὲ ἐν τῷ Μαραθῶνι χρόνος ἐγίνετο πολλός. καὶ τὸ μὲν μέσον τοῦ στρατοπέδου
10 ἐνίκων οἱ βάρβαροι, τῇ Πέρσαι τε αὐτοὶ καὶ Σάκαι ἐτετάχατο· κατὰ τοῦτο μὲν δὴ ἐνίκων οἱ βάρβαροι, καὶ ῥήξαντες ἐδίωκον ἐς τὴν μεσόγαιαν, τὸ δὲ κέρας ἑκάτερον ἐνίκων Ἀθηναῖοί τε καὶ Πλαταιέες. νικῶντες δὲ τὸ μὲν τετραμμένον τῶν βαρβάρων φεύγειν ἔων,
15 τοῖσι δὲ τὸ μέσον ῥήξασι αὐτῶν συναγαγόντες τὰ κέρεα ἀμφότερα ἐμάχοντο καὶ ἐνίκων Ἀθηναῖοι. φεύγουσι δὲ τοῖσι Πέρσῃσι εἵποντο κόπτοντες, ἐς ὃ ἐπὶ τὴν θάλασσαν ἀπικόμενοι πῦρ τε αἴτεον καὶ ἐπελαμβάνοντο τῶν νεῶν.

Kallimachos and Stesileos fall. Kynegeiros loses his right hand in the struggle at the ships.

20 CXIV. Καὶ τοῦτο μὲν ἐν τούτῳ τῷ πόνῳ ὁ πολέμαρχος Καλλίμαχος διαφθείρεται, ἀνὴρ γενόμενος ἀγαθός, ἀπὸ δ' ἔθανε τῶν στρατηγῶν Στησίλεως ὁ Θρασύλεω· τοῦτο δὲ Κυνέγειρος ὁ Εὐφορίωνος

ἐνθαῦτα ἐπιλαβόμενος τῶν ἀφλάστων νεὸς τὴν χεῖρα ἀποκοπεὶς πελέκεϊ πίπτει, τοῦτο δὲ ἄλλοι Ἀθηναίων πολλοί τε καὶ ὀνομαστοί.

Seven of the Persian ships are taken: the rest sail towards Sunium. A treasonable signal.

CXV. Ἑπτὰ μὲν δὴ τῶν νεῶν ἐπεκράτησαν τρόπῳ τοιούτῳ Ἀθηναῖοι, τῇσι δὲ λοιπῇσι οἱ βάρ- 5
βαροι ἐξανακρουσάμενοι, καὶ ἀναλαβόντες ἐκ τῆς νήσου, ἐν τῇ ἔλιπον, τὰ ἐξ Ἐρετρίης ἀνδράποδα, περιέπλωον Σούνιον, βουλόμενοι φθῆναι τοὺς Ἀθηναίους ἀπικόμενοι ἐς τὸ ἄστυ. αἰτίη δὲ ἔσχε ἐν Ἀθηναίοισι ἐξ Ἀλκμαιωνιδέων μηχανῆς αὐτοὺς ταῦτα 10
ἐπινοηθῆναι· τούτους γὰρ συνθεμένους τοῖσι Πέρσῃσι ἀναδέξαι ἀσπίδα ἐοῦσι ἤδη ἐν τῇσι νηυσί.

The Athenian army returns to Athens in time to meet the Persian fleet, which, after waiting a short time off Phalerum, returned to Asia.

CXVI. Οὗτοι μὲν δὴ περιέπλωον Σούνιον· Ἀθηναῖοι δέ, ὡς ποδῶν εἶχον, τάχιστα ἐβοήθεον ἐς τὸ ἄστυ, καὶ ἔφθησάν τε ἀπικόμενοι πρὶν ἢ τοὺς βαρ- 15
βάρους ἥκειν, καὶ ἐστρατοπεδεύσαντο ἀπιγμένοι ἐξ Ἡρακλείου τοῦ ἐν Μαραθῶνι ἐν ἄλλῳ Ἡρακλείῳ τῷ ἐν Κυνοσάργεϊ. οἱ δὲ βάρβαροι τῇσι νηυσὶ ὑπεραιωρηθέντες Φαλήρου (τοῦτο γὰρ ἦν ἐπίνειον τότε τῶν Ἀθηναίων), ὑπὲρ τούτου ἀνακωχεύσαντες τὰς νέας 20
ἀπέπλωον ὀπίσω ἐς τὴν Ἀσίην.

Numbers of the slain, 6400 Persians, 192 Athenians. How Epizelos lost his sight.

CXVII. Ἐν ταύτῃ τῇ ἐν Μαραθῶνι μάχῃ ἀπέθα-

νον τῶν βαρβάρων κατὰ ἑξακισχιλίους καὶ τετρακοσίους ἄνδρας, Ἀθηναίων δὲ ἑκατὸν ἐνενήκοντα καὶ δύο· ἔπεσον μὲν ἀμφοτέρων τοσοῦτοι, συνήνεικε δὲ αὐτόθι θῶυμα γενέσθαι τοιόνδε, Ἀθηναῖον ἄνδρα
5 Ἐπίζηλον τὸν Κουφαγόρεω ἐν τῇ συστάσι μαχόμενόν τε καὶ ἄνδρα γινόμενον ἀγαθὸν τῶν ὀμμάτων στερηθῆναι, οὔτε πληγέντα οὐδὲν τοῦ σώματος οὔτε βληθέντα, καὶ τὸ λοιπὸν τῆς ζόης διατελέειν ἀπὸ τούτου τοῦ χρόνου ἐόντα τυφλόν. λέγειν δὲ αὐτὸν
10 ἤκουσα περὶ τοῦ πάθεος τοιόνδε τινὰ λόγον, ἄνδρα οἱ δοκέειν ὁπλίτην ἀντιστῆναι μέγαν, τοῦ τὸ γένειον τὴν ἀσπίδα πᾶσαν σκιάζειν· τὸ δὲ φάσμα τοῦτο ἑωυτὸν μὲν παρεξελθεῖν, τὸν δὲ ἑωυτοῦ παραστάτην ἀποκτεῖναι. ταῦτα μὲν δὴ Ἐπίζηλον ἐπυθόμην
15 λέγειν.

Warned in a dream Datis restores an image of Apollo.

CXVIII. Δᾶτις δὲ πορευόμενος ἅμα τῷ στρατῷ ἐς τὴν Ἀσίην, ἐπεί τε ἐγένετο ἐν Μυκόνῳ, εἶδε ὄψιν ἐν τῷ ὕπνῳ. καὶ ἥτις μὲν ἦν ἡ ὄψις, οὐ λέγεται, ὁ δέ, ὡς ἡμέρη τάχιστα ἐπέλαμψε, ζήτησιν ἐποιέετο
20 τῶν νεῶν, εὑρὼν δὲ ἐν Φοινίσσῃ νηῒ ἄγαλμα Ἀπόλλωνος κεχρυσωμένον ἐπυνθάνετο ὁκόθεν σεσυλημένον εἴη· πυθόμενος δὲ ἐξ οὗ ἦν ἱροῦ, ἔπλωε τῇ ἑωυτοῦ νηῒ ἐς Δῆλον. καὶ ἀπίκατο γὰρ τηνικαῦτα οἱ Δήλιοι ὀπίσω ἐς τὴν νῆσον, κατατίθεταί τε ἐς τὸ ἱρὸν
25 τὸ ἄγαλμα, καὶ ἐντέλλεται τοῖσι Δηλίοισι ἀπαγαγεῖν τὸ ἄγαλμα ἐς Δήλιον τὸ Θηβαίων· τὸ δ' ἐστὶ ἐπὶ θαλάσσῃ Χαλκίδος καταντίον. Δᾶτις μὲν δὴ ταῦτα ἐντειλάμενος ἀπέπλεε, τὸν δὲ ἀνδριάντα τοῦτον Δήλιοι οὐκ ἀπήγαγον, ἀλλά μιν δι' ἐτέων εἴκοσι

Θηβαῖοι αὐτοὶ ἐκ θεοπροπίου ἐκομίσαντο ἐπὶ Δήλιον.

The captured Eretrians are treated kindly by Darius and assigned lands in Kissia.

CXIX. Τοὺς δὲ τῶν Ἐρετριέων ἠνδραποδισμένους Δᾶτίς τε καὶ Ἀρταφέρνης, ὡς προσέσχον ἐς τὴν Ἀσίην πλέοντες, ἀνήγαγον ἐς Σοῦσα. βασιλεὺς 5 δὲ Δαρεῖος πρὶν μὲν αἰχμαλώτους γενέσθαι τοὺς Ἐρετριέας ἐνεῖχέ σφι δεινὸν χόλον οἷα ἀρξάντων ἀδικίης προτέρων τῶν Ἐρετριέων, ἐπείτε δὲ εἶδέ σφεας ἀπαχθέντας παρ' ἑωυτὸν καὶ ὑποχειρίους ἑωυτῷ ἐόντας, ἐποίησε κακὸν ἄλλο οὐδὲν, ἀλλά σφεας 10 τῆς Κισσίης χώρης κατοίκισε ἐν σταθμῷ ἑωυτοῦ, τῷ οὔνομά ἐστι Ἀρδέρικκα, ἀπὸ μὲν Σούσων δέκα καὶ διηκοσίους σταδίους ἀπέχοντι, τεσσεράκοντα δὲ ἀπὸ τοῦ φρέατος, τὸ παρέχεται τριφασίας ἰδέας· καὶ γὰρ ἄσφαλτον καὶ ἅλας καὶ ἔλαιον ἀρύσσονται ἐξ αὐτοῦ 15 τρόπῳ τοιῷδε· ἀντλέεται μὲν κηλωνηΐῳ, ἀντὶ δὲ γαυλοῦ ἥμισυ ἀσκοῦ οἱ προσδέδεται· ὑποτύψας δὲ τούτῳ ἀντλέει καὶ ἔπειτα ἐγχέει ἐς δεξαμενήν· ἐκ δὲ ταύτης ἐς ἄλλο διαχεόμενον τράπεται τριφασίας ὁδούς. καὶ ἡ μὲν ἄσφαλτος καὶ οἱ ἅλες πήγνυνται 20 παραυτίκα, τὸ δὲ ἔλαιον συνάγουσι ἐν ἀγγηΐοισι, τὸ οἱ Πέρσαι καλέουσι ῥαδινάκην· ἔστι δὲ μέλαν καὶ ὀδμὴν παρεχόμενον βαρέαν. ἐνθαῦτα τοὺς Ἐρετριέας κατοίκισε βασιλεὺς Δαρεῖος, οἳ καὶ μέχρι ἐμέο εἶχον τὴν χώρην ταύτην φυλάσσοντες τὴν ἀρχαίην γλῶσ- 25 σαν. τὰ μὲν δὴ περὶ Ἐρετριέας ἔσχε οὕτω.

Three days after the full moon, 2000 Spartans arrive at Athens. Curiosity to see the slaughtered Medes induces them to march to Marathon.

CXX. Λακεδαιμονίων δὲ ἧκον ἐς τὰς Ἀθήνας δισχίλιοι μετὰ τὴν πανσέληνον, ἔχοντες σπουδὴν πολλὴν καταλαβεῖν οὕτω, ὥστε τριταῖοι ἐκ Σπάρτης ἐγένοντο ἐν τῇ Ἀττικῇ. ὕστεροι δὲ ἀπικόμενοι τῆς
5 συμβολῆς ἱμείροντο ὅμως θηήσασθαι τοὺς Μήδους, ἐλθόντες δὲ ἐς τὸν Μαραθῶνα ἐθηήσαντο. μετὰ δὲ αἰνέοντες Ἀθηναίους καὶ τὸ ἔργον αὐτῶν ἀπαλλάσσοντο ὀπίσω.

Were the Alkmaeonidae guilty of the treasonable signal to the Persians? Their antecedents are against it.

CXXI. Θῶυμα δέ μοι, καὶ οὐκ ἐνδέκομαι τὸν
10 λόγον, Ἀλκμαιωνίδας ἄν κοτε ἀναδέξαι Πέρσῃσι ἐκ συνθήματος ἀσπίδα, βουλομένους ὑπὸ βαρβάροισί τε εἶναι Ἀθηναίους καὶ ὑπὸ Ἱππίῃ, οἵτινες μᾶλλον ἢ ὁμοίως Καλλίῃ τῷ Φαινίππου, Ἱππονίκου δὲ πατρὶ, φαίνονται μισοτύραννοι ἐόντες. Καλλίης τε γὰρ
15 μοῦνος Ἀθηναίων ἁπάντων ἐτόλμα, ὅκως Πεισίστρατος ἐκπέσοι ἐκ τῶν Ἀθηνέων, τὰ χρήματα αὐτοῦ κηρυσσόμενα ὑπὸ τοῦ δημοσίου ὠνέεσθαι, καὶ τὰ ἄλλα τὰ ἔχθιστα ἐς αὐτὸν πάντα ἐμηχανᾶτο.

[*Account of Kallias.*]

CXXII. [Καλλίεω δὲ τούτου ἄξιον πολλαχοῦ
20 μνήμην ἐστὶ πάντα τινὰ ἔχειν. τοῦτο μὲν γὰρ τὰ προλελεγμένα, ὡς ἀνὴρ ἄκρος ἐλευθερῶν τὴν πατρίδα, τοῦτο δὲ τὰ ἐν Ὀλυμπίῃ ἐποίησε, ἵππῳ νικήσας,

τεθρίππῳ δὲ δεύτερος γενόμενος, Πύθια δὲ πρότερον ἀνελόμενος, ἐφανερώθη ἐς τοὺς Ἕλληνας πάντας μεγίστῃσι δαπάνῃσι, τοῦτο δὲ κατὰ τὰς ἑωυτοῦ θυγατέρας ἐούσας τρεῖς οἷός τις ἀνὴρ ἐγένετο. ἐπειδὴ γὰρ ἐγένοντο γάμου ὡραῖαι, ἔδωκέ σφι δωρεὴν μεγα- 5 λοπρεπεστάτην ἐκείνῃσί τε ἐχαρίσατο. ἐκ γὰρ πάντων τῶν Ἀθηναίων τὸν ἑκάστη ἐθέλοι ἄνδρα ἑωυτῇ ἐκλέξασθαι, ἔδωκε τούτῳ τῷ ἀνδρί.]

The Alkmaeonidae could have had no wish to enslave Athens, and for my part I acquit them.

CXXIII. Καὶ οἱ Ἀλκμαιωνίδαι ὁμοίως ἢ οὐδὲν ἔσσον τούτου ἦσαν μισοτύραννοι. θῶυμα ὦν μοι, 10 καὶ οὐ προσίεμαι τὴν διαβολήν, τούτους γε ἀναδέξαι ἀσπίδα, οἵτινες ἔφευγόν τε τὸν πάντα χρόνον τοὺς τυράννους, ἐκ μηχανῆς τε τῆς τούτων ἐξέλιπον οἱ Πεισιστρατίδαι τὴν τυραννίδα. καὶ οὕτω τὰς Ἀθήνας οὗτοι ἦσαν οἱ ἐλευθερώσαντες πολλῷ μᾶλλον ἤπερ 15 Ἁρμόδιός τε καὶ Ἀριστογείτων, ὡς ἐγὼ κρίνω. οἱ μὲν γὰρ ἐξηγρίωσαν τοὺς ὑπολοίπους Πεισιστρατιδέων Ἵππαρχον ἀποκτείναντες, οὐδέ τι μᾶλλον ἔπαυσαν τοὺς λοιποὺς τυραννεύοντας, Ἀλκμαιωνίδαι δὲ ἐμφανέως ἠλευθέρωσαν, εἰ δὴ οὗτοί γε ἀληθέως 20 ἦσαν οἱ τὴν Πυθίην ἀναπείσαντες προσημαίνειν Λακεδαιμονίοισι ἐλευθεροῦν τὰς Ἀθήνας, ὥς μοι πρότερον δεδήλωται. CXXIV. Ἀλλὰ γὰρ ἴσως τι ἐπιμεμφόμενοι Ἀθηναίων τῷ δήμῳ προεδίδοσαν τὴν πατρίδα. οὐ μὲν ὦν ἦσάν σφεων ἄλλοι δοκιμώτεροι 25 ἔν γε Ἀθηναίοισι ἄνδρες, οὐδ' οἳ μᾶλλον ἐτετιμέατο. οὕτω οὐδὲ λόγος αἱρέει ἀναδεχθῆναι ἔκ γε ἂν τούτων ἀσπίδα ἐπὶ τοιούτῳ λόγῳ. ἀνεδέχθη μὲν γὰρ ἀσπίς, καὶ τοῦτο οὐκ ἔστι ἄλλως εἰπεῖν· ἐγένετο γάρ· ὃς μέντοι ἦν ὁ ἀναδέξας, οὐκ ἔχω προσωτέρω εἰπεῖν τούτων.

Origin of the wealth of the Alkmaeonidae. Kroesos taken at his word.

CXXV. Οἱ δὲ Ἀλκμαιωνίδαι ἦσαν μὲν καὶ τὰ
5 ἀνέκαθεν λαμπροὶ ἐν τῇσι Ἀθήνῃσι, ἀπὸ δὲ Ἀλκμαίωνος καὶ αὖτις Μεγακλέος ἐγένοντο καὶ κάρτα λαμπροί. τοῦτο μὲν γὰρ Ἀλκμαίων ὁ Μεγακλέος τοῖσι ἐκ Σαρδίων Λυδοῖσι παρὰ Κροίσου ἀπικνεομένοισι ἐπὶ τὸ χρηστήριον τὸ ἐν Δελφοῖσι συμπρήκτωρ τε
10 ἐγίνετο καὶ συνελάμβανε προθύμως, καί μιν Κροῖσος πυθόμενος τῶν Λυδῶν τῶν ἐς τὰ χρηστήρια φοιτεόντων ἑωυτὸν εὖ ποιέειν μεταπέμπεται ἐς Σάρδις, ἀπικόμενον δὲ δωρέεται χρυσῷ, τὸν ἂν δύνηται τῷ ἑωυτοῦ σώματι ἐξενείκασθαι ἐσάπαξ. ὁ δὲ Ἀλκμαίων
15 πρὸς τὴν δωρεὴν ἐοῦσαν τοιαύτην τοιάδε ἐπιτηδεύσας προσέφερε· ἐνδὺς κιθῶνα μέγαν καὶ κόλπον πολλὸν καταλιπόμενος τοῦ κιθῶνος, κοθόρνους τοὺς εὕρισκε εὐρυτάτους ἐόντας ὑποδησάμενος ἤϊε ἐς τὸν θησαυρὸν, ἐς τόν οἱ κατηγέοντο, ἐσπεσὼν δὲ ἐς σωρὸν ψήγματος,
20 πρῶτα μὲν παρέσαξε παρὰ τὰς κνήμας τοῦ χρυσοῦ ὅσον ἐχώρεον οἱ κόθορνοι, μετὰ δὲ τὸν κόλπον πάντα πλησάμενος χρυσοῦ καὶ ἐς τὰς τρίχας τῆς κεφαλῆς διαπάσας τοῦ ψήγματος καὶ ἄλλο λαβὼν ἐς τὸ στόμα ἐξήϊε ἐκ τοῦ θησαυροῦ, ἕλκων μὲν μόγις τοὺς
25 κοθόρνους, παντὶ δέ τεῳ οἰκὼς μᾶλλον ἢ ἀνθρώπῳ, τοῦ τό τε στόμα ἐβέβυστο καὶ πάντα ἐξώγκωτο. ἰδόντα δὲ τὸν Κροῖσον γέλως ἐσῆλθε, καί οἱ πάντα τε ἐκεῖνα διδοῖ καὶ πρὸς ἕτερα δωρέεται οὐκ ἐλάσσω

ἐκείνων. οὕτω μὲν ἐπλούτησε ἡ οἰκίη αὕτη μεγάλως, καὶ ὁ Ἀλκμαίων οὗτος οὕτω τεθριπποτροφήσας Ὀλυμπιάδα ἀναιρέεται.

Kleisthenes, tyrant of Sikyon, invites candidates for the hand of his daughter.

CXXVI. Μετὰ δὲ, γενεῇ δευτέρῃ ὕστερον, Κλεισθένης μιν ὁ Σικυῶνος τύραννος ἐξήειρε ὥστε πολλῷ ὀνομαστοτέρην γενέσθαι ἐν τοῖσι Ἕλλησι, ἢ πρότερον ἦν. Κλεισθένεϊ γὰρ τῷ Ἀριστωνύμου τοῦ Μύρωνος τοῦ Ἀνδρέω γίνεται θυγάτηρ, τῇ οὔνομα ἦν Ἀγαρίστη. ταύτην ἠθέλησε Ἑλλήνων πάντων ἐξευρὼν τὸν ἄριστον τούτῳ γυναῖκα προσθεῖναι. Ὀλυμπίων ὦν ἐόντων καὶ νικῶν ἐν αὐτοῖσι τεθρίππῳ ὁ Κλεισθένης κήρυγμα ἐποιήσατο, ὅστις Ἑλλήνων ἑωυτὸν ἀξιοῖ Κλεισθένεος γαμβρὸν γενέσθαι, ἥκειν ἐς ἑξηκοστὴν ἡμέρην ἢ καὶ πρότερον ἐς Σικυῶνα ὡς κυρώσοντος Κλεισθένεος τὸν γάμον ἐν ἐνιαυτῷ, ἀπὸ τῆς ἑξηκοστῆς ἀρξαμένου ἡμέρης. ἐνθαῦτα Ἑλλήνων ὅσοι σφίσι τε αὐτοῖσι ἦσαν καὶ πάτρῃ ἐξωγκωμένοι, ἐφοίτεον μνηστῆρες, τοῖσι Κλεισθένης καὶ δρόμον καὶ παλαίστρην ποιησάμενος ἐπ' αὐτῷ τούτῳ εἶχε.

The suitors.

CXXVII. Ἀπὸ μὲν δὴ Ἰταλίης ἦλθε Σμινδυρίδης ὁ Ἱπποκράτεος Συβαρίτης, ὃς ἐπὶ πλεῖστον δὴ χλιδῆς εἷς ἀνὴρ ἀπίκετο (ἡ δὲ Σύβαρις ἤκμαζε τοῦτον τὸν χρόνον μάλιστα), καὶ Σιρίτης Δάμασος Ἀμύριος τοῦ σοφοῦ λεγομένου παῖς. οὗτοι μὲν ἀπὸ Ἰταλίης ἦλθον, ἐκ δὲ τοῦ κόλπου τοῦ Ἰονίου Ἀμφί-

μνηστος Ἐπιστρόφου Ἐπιδάμνιος· οὗτος δὲ ἐκ τοῦ Ἰονίου κόλπου. Αἰτωλὸς δὲ ἦλθε Τιτόρμου τοῦ ὑπερφύντος τε Ἕλληνας ἰσχύϊ καὶ φυγόντος ἀνθρώπους ἐς τὰς ἐσχατιὰς τῆς Αἰτωλίδος χώρης, τούτου
5 τοῦ Τιτόρμου ἀδελφεὸς Μάλης. ἀπὸ δὲ Πελοποννήσου Φείδωνος τοῦ Ἀργείων τυράννου παῖς Λεωκήδης, Φείδωνος δὲ τοῦ τὰ μέτρα ποιήσαντος Πελοποννησίοισι καὶ ὑβρίσαντος μέγιστα δὴ Ἑλλήνων ἁπάντων, ὃς ἐξαναστήσας τοὺς Ἠλείων ἀγωνοθέτας
10 αὐτὸς τὸν ἐν Ὀλυμπίῃ ἀγῶνα ἔθηκε, τούτου τε δὴ παῖς, καὶ Ἀμίαντος Λυκούργου Ἀρκὰς ἐκ Τραπεζοῦντος, καὶ Ἀζὴν ἐκ Παίου πόλιος Λαφάνης Εὐφορίωνος τοῦ δεξαμένου τε, ὡς λόγος ἐν Ἀρκαδίῃ λέγεται, τοὺς Διοσκούρους οἰκίοισι καὶ ἀπὸ τούτου
15 ξεινοδοκέοντος πάντας ἀνθρώπους, καὶ Ἠλεῖος Ὀνομαστὸς Ἀγαίου. οὗτοι μὲν δὴ ἐξ αὐτῆς Πελοποννήσου ἦλθον, ἐκ δὲ Ἀθηνέων ἀπίκοντο Μεγακλέης τε ὁ Ἀλκμαίωνος τούτου τοῦ παρὰ Κροῖσον ἀπικομένου, καὶ ἄλλος Ἱπποκλείδης Τισάνδρου, πλούτῳ καὶ εἴδεϊ
20 προφέρων Ἀθηναίων. ἀπὸ δὲ Ἐρετρίης ἀνθεύσης τοῦτον τὸν χρόνον Λυσανίης, οὗτος δὲ ἀπ' Εὐβοίης μοῦνος. ἐκ δὲ Θεσσαλίης ἦλθε τῶν Σκοπαδέων Διακτορίδης Κραννώνιος, ἐκ δὲ Μολοσσῶν Ἄλκων.

He tests their courage and temper for a year.

CXXVIII. Τοσοῦτοι μὲν ἐγένοντο οἱ μνηστῆρες.
25 ἀπικομένων δὲ τούτων ἐς τὴν προειρημένην ἡμέρην ὁ Κλεισθένης πρῶτα μὲν τὰς πάτρας τε αὐτῶν ἀνεπύθετο καὶ γένος ἑκάστου, μετὰ δὲ κατέχων ἐνιαυτὸν διεπειρᾶτο αὐτῶν τῆς τε ἀνδραγαθίης καὶ τῆς ὀργῆς καὶ παιδεύσιός τε καὶ τρόπου, καὶ ἑνὶ

ἑκάστῳ ἰὼν ἐς συνουσίην καὶ συνάπασι, καὶ ἐς γυμνάσιά τε ἐξαγινέων ὅσοι ἦσαν αὐτῶν νεώτεροι, καὶ τό γε μέγιστον, ἐν τῇ συνιστίῃ διεπειρᾶτο· ὅσον γὰρ κατεῖχε χρόνον αὐτούς, τοῦτον πάντα ἐποίεε καὶ ἅμα ἐξείνιζε μεγαλοπρεπέως. καὶ δή κου μάλιστα 5 τῶν μνηστήρων ἠρέσκοντό οἱ οἱ ἀπ' Ἀθηνέων ἀπιγμένοι, καὶ τούτων μᾶλλον Ἱπποκλείδης ὁ Τισάνδρου καὶ κατ' ἀνδραγαθίην ἐκρίνετο, καὶ ὅτι τὸ ἀνέκαθεν τοῖσι ἐν Κορίνθῳ Κυψελίδῃσι ἦν προσήκων.

Hippokleides 'doesn't care'.

CXXIX. Ὡς δὲ ἡ κυρίη ἐγένετο τῶν ἡμερέων 10 τῆς τε κατακλίσιος τοῦ γάμου καὶ ἐκφάσιος αὐτοῦ Κλεισθένεος, τὸν κρίνοι ἐκ πάντων, θύσας βοῦς ἑκατὸν ὁ Κλεισθένης εὐώχεε αὐτούς τε τοὺς μνηστῆρας καὶ τοὺς Σικυωνίους πάντας. ὡς δὲ ἀπὸ δείπνου ἐγένοντο, οἱ μνηστῆρες ἔριν εἶχον ἀμφί τε μουσικῇ 15 καὶ τῷ λεγομένῳ ἐς τὸ μέσον. προϊούσης δὲ τῆς πόσιος κατέχων πολλὸν τοὺς ἄλλους ὁ Ἱπποκλείδης ἐκέλευσε τὸν αὐλητὴν αὐλῆσαί οἱ ἐμμέλειαν, πειθομένου δὲ τοῦ αὐλητέω ὠρχήσατο. καί κως ἑωυτῷ μὲν ἀρεστῶς ὠρχέετο, ὁ δὲ Κλεισθένης ὁρέων ὅλον τὸ 20 πρῆγμα ὑπώπτευε. μετὰ δὲ ἐπισχὼν ὁ Ἱπποκλείδης χρόνον ἐκέλευσέ οἵ τινα τράπεζαν ἐσενεῖκαι, ἐσελθούσης δὲ τῆς τραπέζης πρῶτα μὲν ἐπ' αὐτῆς ὠρχήσατο Λακωνικὰ σχημάτια, μετὰ δὲ ἄλλα Ἀττικά, τὸ τρίτον δὲ τὴν κεφαλὴν ἐρείσας ἐπὶ τὴν τράπεζαν 25 τοῖσι σκέλεσι ἐχειρονόμησε. Κλεισθένης δὲ τὰ μὲν πρῶτα καὶ τὰ δεύτερα ὀρχεομένου ἀποστυγέων γαμβρὸν ἄν οἱ ἔτι γενέσθαι Ἱπποκλείδην διά τήν τε ὄρχησιν καὶ τὴν ἀναιδείην κατεῖχε ἑωυτόν, οὐ βουλόμενος

ἐκραγῆναι ἐς αὐτὸν, ὡς δὲ εἶδε τοῖσι σκέλεσι χειρονομήσαντα, οὐκέτι κατέχειν δυνάμενος εἶπε· "Ὦ παῖ "Τισάνδρου, ἀπωρχήσαό γε μὴν τὸν γάμον." ὁ δὲ Ἱπποκλείδης ὑπολαβὼν εἶπε· "Οὐ φροντὶς Ἱππο-
5 "κλείδῃ."

Kleisthenes chooses Megakles as his daughter's husband, and consoles the other suitors by a present of a talent.

CXXX. Ἀπὸ τούτου μὲν τοῦτο ὀνομάζεται. Κλεισθένης δὲ σιγὴν ποιησάμενος ἔλεξε ἐς μέσον τάδε· "Ἄνδρες παιδὸς τῆς ἐμῆς μνηστῆρες, ἐγὼ καὶ πάντας "ὑμέας ἐπαινέω, καὶ πᾶσιν ὑμῖν, εἰ οἷόν τε εἴη, χαρι-
10 "ζοίμην ἄν, μήτ' ἕνα ὑμέων ἐξαίρετον ἀποκρίνων μήτε "τοὺς λοιποὺς ἀποδοκιμάζων· ἀλλ' οὐ γὰρ οἷά τέ ἐστι "μιῆς πέρι παρθένου βουλεύοντα πᾶσι κατὰ νόον "ποιέειν, τοῖσι μὲν ὑμέων ἀπελαυνομένοισι τοῦδε τοῦ "γάμου τάλαντον ἀργυρίου ἑκάστῳ δωρεὴν δίδωμι
15 "τῆς ἀξιώσιος εἵνεκεν τῆς ἐξ ἐμεῦ γῆμαι καὶ τῆς "ἐξ οἴκου ἀποδημίης, τῷ δὲ Ἀλκμαίωνος Μεγακλέϊ "ἐγγυῶ παῖδα τὴν ἐμὴν Ἀγαρίστην νόμοισι τοῖσι "Ἀθηναίων." Φαμένου δὲ ἐγγυᾶσθαι Μεγακλέος ἐκεκύρωτο ὁ γάμος Κλεισθένεϊ.

Kleisthenes the Reformer.

20 CXXXI. Ἀμφὶ μὲν κρίσιος τῶν μνηστήρων τοσαῦτα ἐγένετο, καὶ οὕτω Ἀλκμαιωνίδαι ἐβώσθησαν ἀνὰ τὴν Ἑλλάδα· τούτων δὲ συνοικησάντων γίνεται Κλεισθένης τε ὁ τὰς φυλὰς καὶ τὴν δημοκρατίην Ἀθηναίοισι καταστήσας, ἔχων τὸ οὔνομα ἀπὸ τοῦ
25 μητροπάτορος τοῦ Σικυωνίου· οὗτός τε δὴ γίνεται Μεγακλέϊ καὶ Ἱπποκράτης, ἐκ δὲ Ἱπποκράτεος Μεγακλέης τε ἄλλος καὶ Ἀγαρίστη ἄλλη, ἀπὸ τῆς

Κλεισθένεος Ἀγαρίστης ἔχουσα τὸ οὔνομα, ἣ συνοικήσασά τε Ξανθίππῳ τῷ Ἀρίφρονος καὶ ἔγκυος ἐοῦσα εἶδε ὄψιν ἐν τῷ ὕπνῳ, ἐδόκεε δὲ λέοντα τεκεῖν· καὶ μετ' ὀλίγας ἡμέρας τίκτει Περικλέα Ξανθίππῳ.

The fall of Miltiades B.C. 489. *He asks for* 70 *ships and some soldiers.*

CXXXII. Μετὰ δὲ τὸ ἐν Μαραθῶνι τρῶμα 5 γενόμενον Μιλτιάδης, καὶ πρότερον εὐδοκιμέων παρὰ Ἀθηναίοισι, τότε μᾶλλον αὔξετο. αἰτήσας δὲ νέας ἑβδομήκοντα καὶ στρατιήν τε καὶ χρήματα Ἀθηναίους, οὐ φράσας σφι, ἐπ' ἣν ἐπιστρατεύσεται χώρην, ἀλλὰ φὰς αὐτοὺς καταπλουτιεῖν, ἤν οἱ 10 ἕπωνται, ἐπὶ γὰρ χώρην τοιαύτην δή τινα ἄξειν, ὅθεν χρυσὸν εὐπετέως ἄφθονον οἴσονται, λέγων τοιαῦτα αἴτεε τὰς νέας. Ἀθηναῖοι δὲ τούτοισι ἐπαερθέντες παρέδοσαν.

He sails to Paros.

CXXXIII. Παραλαβὼν δὲ ὁ Μιλτιάδης τὴν 15 στρατιὴν ἔπλεε ἐπὶ Πάρον, πρόφασιν ἔχων, ὡς οἱ Πάριοι ὑπῆρξαν πρότεροι στρατευόμενοι τριήρεϊ ἐς Μαραθῶνα ἅμα τῷ Πέρσῃ. τοῦτο μὲν δὴ πρόσχημα λόγου ἦν, ἀτάρ τινα καὶ ἔγκοτον εἶχε τοῖσι Παρίοισι διὰ Λυσαγόρην τὸν Τισίεω, ἐόντα γένος Πάριον, 20 διαβαλόντα μιν πρὸς Ὑδάρνεα τὸν Πέρσην. ἀπικόμενος δὲ ἐς τὴν ἔπλεε ὁ Μιλτιάδης τῇ στρατιῇ ἐπολιόρκεε Παρίους κατειλημένους ἐντὸς τείχεος, καὶ ἐσπέμπων κήρυκα αἴτεε ἑκατὸν τάλαντα, φὰς, ἤν μή οἱ δῶσι, οὐκ ἀπαναστήσειν τὴν στρατιὴν, πρὶν ἢ 25 ἐξέλῃ σφέας. οἱ δὲ Πάριοι, ὅκως μέν τι δώσουσι

Μιλτιάδῃ ἀργυρίου, οὐδὲ διενοεῦντο, οἱ δέ, ὅκως διαφυλάξουσι τὴν πόλιν, τοῦτο ἐμηχανῶντο, ἄλλα τε ἐπιφραζόμενοι, καὶ τῇ μάλιστα ἔσκε ἑκάστοτε ἐπίμαχον τοῦ τείχεος, τοῦτο ἅμα νυκτὶ ἐξηείρετο
5 διπλήσιον τοῦ ἀρχαίου.

The priestess Timo admits him to the temple of Demeter. He is seized with a panic and in retreating injures his thigh.

CXXXIV. Ἐς μὲν δὴ τοσοῦτο τοῦ λόγου οἱ πάντες Ἕλληνες λέγουσι, τὸ ἐνθεῦτεν δὲ αὐτοὶ Πάριοι γενέσθαι ὧδε λέγουσι· Μιλτιάδῃ ἀπορέοντι ἐλθεῖν ἐς λόγους αἰχμάλωτον γυναῖκα, ἐοῦσαν μὲν
10 Παρίην γένος, οὔνομα δέ οἱ εἶναι Τιμοῦν, εἶναι δὲ ὑποζάκορον τῶν χθονίων θεῶν. ταύτην ἐλθοῦσαν ἐς ὄψιν Μιλτιάδεω συμβουλεῦσαι, εἰ περὶ πολλοῦ ποιέεται Πάρον ἑλεῖν, τὰ ἂν αὐτὴ ὑπόθηται, ταῦτα ποιέειν. μετὰ δὲ τὴν μὲν ὑποθέσθαι, τὸν δὲ ἀπικό-
15 μενον ἐπὶ τὸν κολωνὸν τὸν πρὸ τῆς πόλιος ἐόντα τὸ ἕρκος θεσμοφόρου Δήμητρος ὑπερθορεῖν, οὐ δυνάμενον τὰς θύρας ἀνοῖξαι, ὑπερθορόντα δὲ ἰέναι ἐπὶ τὸ μέγαρον ὅ τι δὴ ποιήσοντα ἐντός, εἴτε κινήσοντά τι τῶν ἀκινήτων εἴτε ὅ τι δή κοτε πρήξοντα· πρὸς τῇσι
20 θύρῃσί τε γενέσθαι, καὶ πρόκατε φρίκης αὐτὸν ὑπελθούσης ὀπίσω τὴν αὐτὴν ὁδὸν ἵεσθαι, καταθρώσκοντα δὲ τὴν αἱμασιὴν τὸν μηρὸν σπασθῆναι. οἱ δὲ αὐτὸν τὸ γόνυ προσπταῖσαι λέγουσι.

The Oracle at Delos forbids the punishment of Timo.

CXXXV. Μιλτιάδης μέν νυν φλαύρως ἔχων
25 ἀπέπλεε ὀπίσω, οὔτε χρήματα Ἀθηναίοισι ἄγων

οὔτε Πάρον προσκτησάμενος, ἀλλὰ πολιορκήσας τε ἓξ καὶ εἴκοσι ἡμέρας καὶ δηϊώσας τὴν νῆσον. Πάριοι δὲ πυθόμενοι, ὡς ἡ ὑποζάκορος τῶν θεῶν Τιμὼ Μιλτιάδῃ κατηγήσατο, βουλόμενοί μιν ἀντὶ τούτων τιμωρήσασθαι, θεοπρόπους πέμπουσι ἐς 5 Δελφούς, ὥς σφεας ἡσυχίη τῆς πολιορκίης ἔσχε, ἔπεμπον δὲ ἐπειρησομένους, εἰ καταχρήσονται τὴν ὑποζάκορον τῶν θεῶν ὡς ἐξηγησαμένην τοῖσι ἐχθροῖσι τῆς πατρίδος ἅλωσιν καὶ τὰ ἐς ἔρσενα γόνον ἄρρητα ἱρὰ ἐκφήνασαν Μιλτιάδῃ. ἡ δὲ Πυθίη οὐκ ἔα, φᾶσα 10 οὐ Τιμοῦν εἶναι τὴν αἰτίην τούτων, ἀλλὰ δέειν γὰρ Μιλτιάδεα τελευτᾶν μὴ εὖ, φανῆναί οἱ τῶν κακῶν κατηγεμόνα.

Miltiades is impeached by Xanthippos. He is fined 50 talents, and soon afterwards dies of a mortification of his thigh. His son pays the fine.

CXXXVI. Παρίοισι μὲν δὴ ταῦτα ἡ Πυθίη ἔχρησε. Ἀθηναῖοι δὲ ἐκ Πάρου Μιλτιάδεα ἀπο- 15 νοστήσαντα ἔσχον ἐν στόμασι, οἵ τε ἄλλοι καὶ μάλιστα Ξάνθιππος ὁ Ἀρίφρονος, ὃς θανάτου ὑπαγαγὼν ὑπὸ τὸν δῆμον Μιλτιάδεα ἐδίωκε τῆς Ἀθηναίων ἀπάτης εἵνεκεν. Μιλτιάδης δὲ αὐτὸς μὲν παρεὼν οὐκ ἀπελογέετο (ἦν γὰρ ἀδύνατος ὥστε σηπομένου 20 τοῦ μηροῦ), προκειμένου δὲ αὐτοῦ ἐν κλίνῃ ὑπεραπελογέοντο οἱ φίλοι, τῆς μάχης τε τῆς ἐν Μαραθῶνι πολλὰ ἐπιμεμνημένοι καὶ τὴν Λήμνου αἵρεσιν, ὡς ἑλὼν Λῆμνόν τε καὶ τισάμενος τοὺς Πελασγοὺς παρέδωκε Ἀθηναίοισι. προσγενομένου δὲ τοῦ δήμου 25 αὐτῷ κατὰ τὴν ἀπόλυσιν τοῦ θανάτου, ζημιώσαντος δὲ κατὰ τὴν ἀδικίην πεντήκοντα ταλάντοισι, Μιλτιάδης μὲν μετὰ ταῦτα σφακελίσαντός τε τοῦ μηροῦ καὶ σαπέντος τελευτᾷ, τὰ δὲ πεντήκοντα τάλαντα ἐξέτισε ὁ παῖς αὐτοῦ Κίμων.

How Miltiades took Lemnos.
The Pelasgic builders, driven out of Attica, settle in Lemnos and elsewhere.

CXXXVII. Λῆμνον δὲ Μιλτιάδης ὁ Κίμωνος
5 ὧδε ἔσχε· Πελασγοὶ ἐπεί τε ἐκ τῆς Ἀττικῆς ὑπὸ Ἀθηναίων ἐξεβλήθησαν, εἴτε ὦν δὴ δικαίως εἴτε ἀδίκως· τοῦτο γὰρ οὐκ ἔχω φράσαι, πλὴν τὰ λεγόμενα, ὅτι Ἑκαταῖος μὲν ὁ Ἡγησάνδρου ἔφησε ἐν τοῖσι λόγοισι λέγων ἀδίκως· ἐπείτε γὰρ ἰδεῖν τοὺς
10 Ἀθηναίους τὴν χώρην, τὴν σφίσι αὐτοῖσι ὑπὸ τὸν Ὑμησσὸν ἐοῦσαν ἔδοσαν οἰκῆσαι μισθὸν τοῦ τείχεος τοῦ περὶ τὴν ἀκρόπολίν κοτε ἐληλαμένου, ταύτην ὡς ἰδεῖν τοὺς Ἀθηναίους ἐξεργασμένην εὖ, τὴν πρότερον εἶναι κακήν τε καὶ τοῦ μηδενὸς ἀξίην, λαβεῖν φθόνον
15 τε καὶ ἵμερον τῆς γῆς, καὶ οὕτω ἐξελαύνειν αὐτοὺς οὐδεμίαν ἄλλην πρόφασιν προϊσχομένους τοὺς Ἀθηναίους· ὡς δὲ αὐτοὶ Ἀθηναῖοι λέγουσι, δικαίως ἐξελάσαι. κατοικημένους γὰρ τοὺς Πελασγοὺς ὑπὸ τῷ Ὑμησσῷ ἐνθεῦτεν ὁρμεομένους ἀδικέειν τάδε·
20 φοιτᾶν γὰρ αἰεὶ τὰς σφετέρας θυγατέρας τε καὶ τοὺς παῖδας ἐπ' ὕδωρ ἐπὶ τὴν Ἐννεάκρουνον (οὐ γὰρ εἶναι τοῦτον τὸν χρόνον σφίσι κω οὐδὲ τοῖσι ἄλλοισι Ἕλλησι οἰκέτας), ὅκως δὲ ἔλθοιεν αὗται, τοὺς Πελασγοὺς ὑπὸ ὕβριός τε καὶ ὀλιγωρίης βιᾶσθαί σφεας.
25 καὶ ταῦτα μέντοι σφι οὐκ ἀποχρᾶν ποιέειν, ἀλλὰ τέλος καὶ ἐπιβουλεύοντας ἐπιχειρήσειν ἐπ' αὐτοφώρῳ φανῆναι. ἑωυτοὺς δὲ γενέσθαι τοσούτῳ ἐκεί-

νων ἄνδρας ἀμείνονας, ὅσῳ παρεὸν αὐτοῖσι ἀποκτεῖναι τοὺς Πελασγούς, ἐπεί σφεας ἔλαβον ἐπιβουλεύοντας, οὐκ ἐθελῆσαι, ἀλλά σφι προειπεῖν ἐκ τῆς γῆς ἐξιέναι. τοὺς δὲ οὕτω δὴ ἐκχωρήσαντας ἄλλα τε σχεῖν χωρία καὶ δὴ καὶ Λῆμνον. ἐκεῖνα μὲν δὴ Ἑκαταῖος ἔλεξε, 5 ταῦτα δὲ Ἀθηναῖοι λέγουσι.

'The Lemnian deeds.' The Pelasgians carry off Attic women to Lemnos, whose children giving them alarm they kill both them and their mothers.

CXXXVIII. Οἱ δὲ Πελασγοὶ οὗτοι Λῆμνον τότε νεμόμενοι καὶ βουλόμενοι τοὺς Ἀθηναίους τιμωρήσασθαι, εὖ τε ἐξεπιστάμενοι τὰς Ἀθηναίων ὁρτάς, πεντηκοντέρους κτησάμενοι ἐλόχησαν Ἀρτέ- 10 μιδι ἐν Βραυρῶνι ἀγούσας ὁρτὴν τὰς τῶν Ἀθηναίων γυναῖκας, ἐνθεῦτεν δὲ ἁρπάσαντες τούτων πολλὰς οἴχοντο ἀποπλέοντες, καί σφεας ἐς Λῆμνον ἀγαγόντες παλλακὰς εἶχον. ὡς δὲ τέκνων αὗται αἱ γυναῖκες ὑπεπλήσθησαν, γλῶσσάν τε τὴν Ἀττικὴν 15 καὶ τρόπους τοὺς Ἀθηναίων ἐδίδασκον τοὺς παῖδας. οἱ δὲ οὔτε συμμίσγεσθαι τοῖσι ἐκ τῶν Πελασγίδων γυναικῶν παισὶ ἤθελον, εἴ τε τύπτοιτό τις αὐτῶν ὑπ' ἐκείνων τινός, ἐβοήθεόν τε πάντες καὶ ἐτιμώρεον ἀλλήλοισι· καὶ δὴ καὶ ἄρχειν τε τῶν παίδων οἱ 20 παῖδες ἐδικαίευν καὶ πολλὸν ἐπεκράτεον. μαθόντες δὲ ταῦτα οἱ Πελασγοὶ ἑωυτοῖσι λόγους ἐδίδοσαν· καί σφι βουλευομένοισι δεινόν τι ἐσέδυνε, εἰ δὴ διαγινώσκοιεν σφίσι τε βοηθέειν οἱ παῖδες πρὸς τῶν κουριδιέων γυναικῶν τοὺς παῖδας καὶ τούτων αὐτίκα ἄρχειν 25 πειρῴατο, τί δὴ ἀνδρωθέντες δῆθεν ποιήσουσι. ἐνθαῦτα ἔδοξέ σφι κτείνειν τοὺς παῖδας τοὺς ἐκ τῶν Ἀττικέων γυναικῶν. ποιεῦσι δὴ ταῦτα, προσαπολλύουσι δέ σφεων καὶ τὰς μητέρας. ἀπὸ τούτου δὲ τοῦ ἔργου καὶ τοῦ προτέρου τούτων, τὸ ἐργάσαντο αἱ γυναῖκες τοὺς ἅμα Θόαντι ἄνδρας σφετέρους ἀποκτείνασαι, 5 νενόμισται ἀνὰ τὴν Ἑλλάδα τὰ σχέτλια ἔργα πάντα Λήμνια καλέεσθαι.

This crime was followed by a dearth; and the Delphic Oracle orders the Pelasgians to give the Athenians satisfaction. The Pelasgians will only comply under impossible conditions.

CXXXIX. Ἀποκτείνασι δὲ τοῖσι Πελασγοῖσι τοὺς σφετέρους παῖδάς τε καὶ γυναῖκας οὔτε γῆ καρπὸν ἔφερε οὔτε γυναῖκές τε καὶ ποῖμναι ὁμοίως 10 ἔτικτον καὶ πρὸ τοῦ. πιεζόμενοι δὲ λιμῷ τε καὶ ἀπαιδίῃ ἐς Δελφοὺς ἔπεμπον, λύσιν τινὰ αἰτησόμενοι τῶν παρεόντων κακῶν. ἡ δὲ Πυθίη σφέας ἐκέλευε Ἀθηναίοισι δίκας διδόναι ταύτας, τὰς ἂν αὐτοὶ Ἀθηναῖοι δικάσωσι. ἦλθόν τε δὴ ἐς τὰς Ἀθήνας οἱ 15 Πελασγοί, καὶ δίκας ἐπηγγέλλοντο βουλόμενοι διδόναι παντὸς τοῦ ἀδικήματος. Ἀθηναῖοι δὲ ἐν τῷ πρυτανηΐῳ κλίνην στρώσαντες ὡς εἶχον κάλλιστα καὶ τράπεζαν ἐπιπλέην ἀγαθῶν πάντων παραθέντες ἐκέλευον τοὺς Πελασγοὺς τὴν χώρην σφίσι παραδι-20 δόναι οὕτω ἔχουσαν. οἱ δὲ Πελασγοὶ ὑπολαβόντες εἶπαν· "Ἐπεὰν βορέῃ ἀνέμῳ αὐτημερὸν νηῦς ἐξανύσῃ "ἐκ τῆς ὑμετέρης ἐς τὴν ἡμετέρην, τότε παραδώσο-"μεν." Τοῦτο εἶπαν ἐπιστάμενοι τοῦτο εἶναι ἀδύνατον γενέσθαι· ἡ γὰρ Ἀττικὴ πρὸς νότον κέεται πολλὸν 25 τῆς Λήμνου.

The impossible made possible by Miltiades.

CXL. Τότε μὲν τοσαῦτα· ἔτεσι δὲ κάρτα πολλοῖσι ὕστερον τούτων, ὡς ἡ Χερσόνησος ἡ ἐν Ἑλλησπόντῳ ἐγένετο ὑπ' Ἀθηναίοισι, Μιλτιάδης ὁ Κίμωνος ἐτησίων ἀνέμων κατεστηκότων νηῒ κατανύσας ἐξ Ἐλαιοῦντος τοῦ ἐν Χερσονήσῳ ἐς τὴν Λῆμνον 5 προηγόρευε ἐξιέναι ἐκ τῆς νήσου τοῖσι Πελασγοῖσι, ἀναμιμνήσκων σφέας τὸ χρηστήριον, τὸ οὐδαμὰ ἤλπισαν σφίσι οἱ Πελασγοὶ ἐπιτελέεσθαι. Ἡφαιστιέες μέν νυν ἐπείθοντο, Μυριναῖοι δὲ οὐ συγγινωσκόμενοι εἶναι τὴν Χερσόνησον Ἀττικὴν ἐπολιορ- 10 κέοντο, ἐς ὃ καὶ αὐτοὶ παρέστησαν. οὕτω δὴ τὴν Λῆμνον ἔσχον Ἀθηναῖοί τε καὶ Μιλτιάδης.

NOTES.

[*For names of persons and places see Historical and Geographical Index. G. stands for Goodwin's Greek Grammar. App. for Appendix on the Ionic dialect in Book IX. Clyde for Clyde's Greek Syntax. Madvig for Madvig's Greek Syntax, Eng. Transl.*]

CHAPTER I.

2. **οὕτω** 'as I have described', see 5, 126. The death of Aristagoras would seem to have been in B.C. 497—6. 1

τύραννος. Histiaios was still nominally tyrant of Miletos, though he had resigned the actual government to his son-in-law Aristagoras [5, 30], who had in his turn committed it to Pythagoras [5, 126].

3. **μεμετιμένος** 'having been allowed to depart', i.e. from Susa, where Darius had been retaining him, on the pretext of requiring his advice, but really as a State prisoner. The Attic form of the word is *μεθειμένος*. Cp. 5, 108; 7, 229.

παρῆν ἐς 'arrived at', a common brachylogy, see p. 13, l. 4 *παρῆν ἐς τὴν Ἀσίην*, p. 4, l. 10 *συλλέγεσθαι ἐς*.

5. **κατὰ κοῖόν τι** 'on what ground', cp. p. 2, l. 9.

7. **δῆθεν** 'as he pretended'. Both *δῆθεν* and *δή* are used to express the insincerity of a pretext, or the writer's belief of such insincerity, cp. 9, 5 *ὡς παρ' ἑωυτοῦ δῆθεν* 'as though from his own pocket as he pretended'.

9. **ἀτρεκείην τῆς ἀποστάσιος** 'the exact truth about the revolt', cp. 4, 152 *τῶν ἀτρεκείην ἴδμεν* 'of whom we have a trustworthy account'.

10. **τοι** 'let me tell you'. This particle is frequently used to introduce short sententious remarks.

11. **σὺ** made still more emphatic by its position. For the part taken by Histiaios in beginning the Ionian revolt, see 5, 30, 35, and the Index.

CHAPTER II.

13. **ἐς τὴν ἀπόστασιν ἔχοντα** 'referring to the revolt', cp. p. 10, l. 10; p. 41, l. 10.

15. **ὑπὸ...νύκτα** 'under cover of the next night', cp. 9, 58 ὑπὸ τὴν παροιχομένην νύκτα...διαδράντας.

2 1. **Σαρδώ.** Sardinia was known to the Ionian navigators, and its great fertility made its possession a favourite scheme of the time; just as that of Sicily was afterwards. See 1, 170, where we are told that Bias advised the Ionians to move there in a body when Kyros' general Harpagos was everywhere enslaving them. For Histiaios' promise see 5, 106.

ὑποδεξάμενος κατεργάσεσθαι 'by promising to subdue'. Far the larger number of MSS. have κατεργάσασθαι, and the aorist infinitive in this and many other cases is defended as an acknowledged Greek idiom. Stein quotes 5, 106; 7, 134 (R τίσειν) as other instances. But on the whole I am inclined to accept the arguments of Madvig (*Adversaria*, pp. 156—182), who would change these aorist infinitives after words of promising and the like into futures; and the dictum of Cobet that 'the aorist infinitive in indirect discourse is only admissible when the aorist indicative would have been used in the direct' (*Variae Lect.* p. 97 sq.): especially as in this instance the Vatican MS. (R) has the future. [See Goodwin, *Moods and Tenses*, § 23.]

2. **ὑπέδυνε** 'endeavoured to insinuate himself into', the imperfect indicates his failure. ὑποδύνειν 'to undertake', cp. 4, 120 ὑπέδυσαν τὸν πόλεμον.

4, 5. **νεώτερα πρήσσειν** 'to be engaged in introducing some innovations', meaning 'some severity'. **ἐκ Δαρείου** 'as agent of Darius'. Cp. p. 11, l. 20; p. 22, l. 13.

6. **τὸν πάντα λόγον** 'the truth', cp. 9, 13 πυθόμενος τὸν πάντα λόγον. Also 4, 152.

CHAPTER III.

9. **κατ' ὅ τι.** See p. 1, l. 5.

9—11. **ἐπέστειλε...εἴη ἐξεργασμένος.** The Greek idiom admits of the use of the same tense and mood in an oblique question as would be used in the direct. Thus ἐπέστειλε is a *dramatic* indicative, the direct question would have been τί ἐπέστειλας; But the optative is also correct:

and when there are two clauses in dependent sentences, especially when the latter contains a consequence of, or a more remote contingency than, the former, Herodotos often uses the two moods, perhaps for variety, cp. p. 45, l. 22; 5, 97 ἔλεγε...ὡς οὔτε ἀσπίδα οὔτε δόρυ νομίζουσι, εὐπετέες τε χειρωθῆναι εἴησαν: and the next sentence, ἐβουλεύσατο...ἐπιστείλειε, where however there is a change of subject. Just as in final sentences containing two clauses he will use the subjunctive in one and the optative in the other. See on 8, 6, 7; 9, 51, and cp. the next sentence.

11. **τὴν γενομένην αἰτίην** 'what had really been his motive', used as the past tense of τὴν οὖσαν. His real motive is stated in 5, 35 to be discontent with his honourable detention at Susa, and a hope that in case of an Ionian disturbance he should be sent down to quell it. **αὐτοῖσι** belongs to ἐξέφαινε.

12. **οὐ μάλα** 'not at all'. **ἐξέφαινε** 'allowed to appear'.

13. **Φοίνικας ἐξαναστήσας.** Though Herodotos says that there was no foundation for ascribing this intention to the king, it would not have appeared altogether improbable to the Chians, remembering the transference of the Paeonians in a body from Europe to Asia [5, 13—15], and seeing the good service which the Phoenikian ships did the king.

15. **ἐπιστείλειε** sc. Histiaios. **οὐδέν τι κ.τ.λ.** These words are a comment of the historian, and are not connected by any particle with the previous sentence: compare a similar case in c. 21 οὐδὲν ὁμοίως οἱ Ἀθηναῖοι.

16. **ἐδειμάτου** 'he was trying to frighten', with the sense of false alarm. Thus in Aristophanes, *Ranae* 145, Dionysios says in reply to the description of the monsters he will meet in Hades,

μή μ' ἔκπληττε μηδὲ δειμάτου·
οὐ γάρ μ' ἀποτρέψεις.

CHAPTER IV.

18. **μετὰ** adverbial, p. 39, l. 14. **δι' ἀγγέλου ποιεύμενος** 'acting by means of a messenger'.

20. **βιβλία** 'a letter', written on paper made from the byblus, which was now common in Ionia, though a kind of parchment made from skins had formerly been used there. See 5, 58. [The MSS. are mostly in favour of βυβλία, Inscriptions of βιβλία.]

ὡς προλελεσχηνευμένων 'which might imply that they had conversed with him before on the subject of a revolt'. This and the other compound περιλεσχήνευτος (2, 135) are peculiar to Herodotos, though the simple deponent λεσχηνεύομαι occurs in other writers. For λέσχη (1) 'club-house', see Hom. *Odyss.* 18, 329, (2) 'discussion', see 2, 32; 9, 71.

22. διδοῖ...ἐνεχείρισε. The former is an historic present, the latter an aorist; and the change of tense seems only referable to the same tendency to introduce variety noticed above (c. 3) in regard to the mood.

25. τὰ ἀμοιβαῖα sc. βιβλία 'the answers', an unusual and probably poetical expression.

3 1. τούτων...γενομένων 'these persons being detected'.

CHAPTER V.

4. δὴ 'accordingly' or 'of course', p. 13, l. 22.

5. κατῆγον 'conducted him home'; used, like κατιέναι and κάτοδος, especially of a return from exile or foreign residence, see p. 13, l. 13; p. 42, l. 2, and 5, 30. Though in this instance the Chians did not 'restore' Histiaios; they merely shipped him to Miletos. But κατάγειν also is especially applicable to return by sea, cp. νέας...καταχθείσας ἐς τὰς 'Αφέτας 8, 4.

9. οἷα...γευσάμενοι 'seeing that they had tasted freedom', or, 'as was natural after their taste of freedom'. νυκτὸς γὰρ p. 6, l. 11.

10. ἐπειρᾶτο κατιὼν 'tried to effect his return'. Herodotos often constructs πειρᾶσθαι with a participle, cp. 9, 26 ἐπειρῶντο κατιόντες: 1, 84 ἐπειρᾶτο προσβαίνων. But he also sometimes uses the infinitive, cp. p. 81, l. 25; 9, 31 ἐπειρῶντο...ποιέεσθαι. Cp. 8, 142.

12. ἀπωστὸς...γίνεται = ἀπωθέεται 'is banished' or 'expelled'. Again a poetical expression, cp. Soph. *Aj.* 1020.

13. οὐ γὰρ ἔπειθε...ὥστε δοῦναι 'for he could not prevail upon them to give him ships', though they had helped him in other ways.

15. ἔπεισε, notice the aorist used of the one successful act, whereas the imperfect in the last sentence expressed his unsuccessful attempt at persuading. οἱ δὲ, the Lesbians.

17. ἰζόμενοι 'taking up a position', an odd word to use in reference to a naval squadron. It is common of land forces, see 9, 2, 17, 26; p. 43, l. 17.

τὰς ἐκ τοῦ Πόντου ἐκπλωούσας the merchant vessels engaged in the corn trade from the shores of the Black Sea, then, as now, a great wheat-producing district. See 4, 14—18, 108 where several semi-Greek 'staple towns' for this trade are mentioned, especially Olbia and Gelonus. ἐλάμβανον, imperfect, 'they made a practice of seizing'.

18. **πλὴν ἢ ὅσοι αὐτῶν**, using a masculine for the crews or ship-masters, where grammatically the ships should be meant, is common in Herodotos, cp. p. 4, l. 25. For **πλὴν ἢ** cp. 2, 111. πλὴν = πλέον and may naturally be followed by ἤ. See Arist. *Nubes* 361. Sometimes it is used like *amplius* in Latin without affecting the case, cp. Soph. *Aj.* 1238 οὐκ ἄρ' Ἀχαιοῖς ἄνδρες εἰσὶ πλὴν ὅδε; Xen. *Oecon.* 4, 6; Arist. *Plut.* 106; Eurip. *Hippol.* 599.

CHAPTER VI.

20. **Ἱστιαῖος μὲν...ἐπὶ δὲ Μίλητον** 'while Histiaios and the Mytileneans were thus employed, against Miletos itself a large naval and military host was in hourly expectation'. For this method of expressing simultaneousness cp. p. 4, l. 3.

25. **περὶ ἐλάσσονος ποιησάμενοι** 'regarding as of less importance'; cp. περὶ πλείστου ποιεύμενος 8, 40.

1. **νεωστί**, see 5, 116. The Cyprians were active mariners, and they and the Kilikians, Phoenikians, and Egyptians formed the best part of the Persian fleet. 4

CHAPTER VII.

3. **μὲν...δὲ**, p. 3, l. 20.

5. **προβούλους σφέων αὐτῶν** 'some deputies to consult in their behalf' or 'to represent them', thus taking σφέων αὐτῶν as an objective genitive. Some have explained it to mean 'some of their own citizens as deputies'. But see 7, 172 where the deputies assembled at the Isthmus are πρόβουλοι τῆς Ἑλλάδος. The assembly of Ionian deputies at the Panionium does not appear to have been a periodical one, or to have amounted to a League government. The reunion at this place was primarily for religious purposes, and to celebrate a festival [ὁρτή 1, 143]: and in accordance with Greek ideas the essential condition for admission to joint religious worship was community of blood. Still on occasions of special necessity such meetings were held as almost amounted to a federal government: but they fell short of it in the fact that they were occasional and not regular; and in having no system of appointing permanent or annual officials, in fact a government. Somewhat similar were the meetings of deputies in Thessaly, Boeotia, and those in the Isthmus, as in B.C. 480, for the whole of Greece. The elaboration of a system of permanent federal government was reserved for the Achaians (and in some degree to the Aetolians) two centuries later.

6. **ἐς τοῦτον τὸν χῶρον**, that is to the sacred inclosure of the Temple of Poseidon on Mykale, which went by the name of Panionium.

8. **ἀντίξοον**=*ἐνάντιον*, a word apparently characteristic of the Ionic dialect.

τὰ τείχεα...Μιλησίους 'that the Milesians should protect their walls by themselves'. **ῥύεσθαι** a poetical word which Herodotos often uses, cp. 4, 135 *στρατόπεδον ῥ.*, 7, 154 *τὴν σφετέρην χώρην ῥ.*

10. **συλλέγεσθαι ἐς**. Cp. *παρῆν ἐς* p. 1, l. 3 and p. 6, l. 8.

12. **ἐπὶ τῇ πόλι** 'over against the city of Miletos'.

CHAPTER VIII.

14. **πεπληρωμένῃσι τῇσι νηυσί**. The dative of accompanying circumstances, 'with their ships manned'.

18. **εἴχοντο τούτων** 'came next to them'. For *ἔχεσθαι* with gen. = 'to hold on to', 'keep up a continuous line with', cp. 9, 28. The datives **δυώδεκα νηυσὶ** *κ.τ.λ.* may be regarded as partly instrumental, partly defining quantity. To this combined Ionian fleet four important cities are not mentioned as sending contingents, Ephesos, Kolophon, Lebedos, Klazomenae. The two first seem always to have held aloof from the Ionian confederacy as much as possible [1, 147; infr. c. 16], Lebedos probably followed the lead of Ephesos, and Klazomenae was in the hands of the Persians [5, 123].

21. **πρὸς δὲ τούτοισι** 'and near these were drawn up'; or it might mean 'and besides these were drawn up'.

25. **πάντων δὲ τούτων**: for the masculine cp. p. 3, l. 18.

CHAPTER IX.

5 3. **ἀπίκατο**=*ἀφιγμέναι ἦσαν* See App. D. II. (a). **τὴν Μιλησίην** 'the territory of Miletos'.

5. **καταρρώδησαν** [App. A. II. (10)] 'were thoroughly alarmed'.

μὴ οὐ...γένωνται. After words of fearing *μὴ οὐ*=*ut* (*ne non*), 'lest they should prove not to be able'. After a leading verb in the past tense the ordinary rule is to have the optative, but by the *dramatic* principle so common in Greek the subjunctive is often employed, as that which the persons fearing would actually have used, cp. 1, 165 *οἱ Φωκαίες τὰς νήσους οὐκ ἐβούλοντο πωλέειν δειμαίνοντες μὴ ἐμπόριον γένωνται.* Here the Persian leaders would have said *δεδοίκαμεν μὴ οὐ γενώμεθα* 'we fear we shall not be able'. On the other hand the historical mood is also used, as οὐκέτι ἐπετίθεντο δεδοικότες μὴ ἀποτμηθείησαν Xen. *An.* 3, 4, 29. Goodwin, *M. and T.*, § 46.

6. **ὑπερβαλέσθαι** 'to overcome'; cp. 7, 163. Herodotos also uses it as = 'to postpone' 9, 45, 'to be dilatory' 9, 51. **καὶ οὕτω** 'and in that case' i.e. if they failed to beat the Greeks at sea.

7. **οἷοί τε** 'able'. The suffix τε is a survival of a use much more common in earlier Greek. 'Its force is that of an undeclined τις', Monro, *Homeric Grammar*, § 108, and it survived in certain other connexions, cp. ὅσον τε (9, 23), ἐπ' ᾧ τε (9, 4), ἅτε, ὥστε.

μὴ οὐκ ἐόντες 'if they were not', 'unless they were'. μὴ οὐ is used with participles depending on *negative* expressions, cp. p. 61, l. 5; see also 2, 110, οὐκ ὢν δίκαιον εἶναι ἱστάναι ἔμπροσθε τῶν ἐκείνου ἀναθημάτων μὴ οὐκ ὑπερβαλλόμενον τοῖσι ἔργοισι.

7, 8. **πρός τε Δαρείου** 'and should also run the risk of some severity at the hands of Darius'. τε answers to οὔτε in l. 6. p. 9, l. 15.

8. **κινδυνεύσωσι λαβεῖν**. Herodotos constructs κινδυνεύειν in three ways, (1) with the infinitive, as here: (2) with the dative of the thing risked τῇ ψυχῇ 7, 209: Ἑλλάδι 8, 60: (3) with the preposition περί, cp. 9, 74 περὶ ἐκείνης [Πελοποννήσου] κινδυνεύειν. Lastly it is used as = δοκεῖν, cp. 4, 105 κινδυνεύουσι οἱ ἄνθρωποι γόητες εἶναι.

10. **οἳ ὑπ' Ἀρισταγόρεω...τῶν ἀρχέων** 'who had been deposed by Aristagoras'. The first measure of Aristagoras, when he resolved to raise the Ionian revolt, was to lay down his own authority at Miletos, and to secure the expulsion of the tyrants from the other Ionian towns, who were all likely to stand up for the Persian rule as supporting their own, see 5, 37.

11. **ἔφευγον ἐς Μήδους** 'were in exile and had taken refuge with the Persians'. For this pregnant meaning of ἔφευγον, cp. παρῆν ἐς, p. 1, l. 3: συλλέγεσθαι ἐς, p. 4, l. 10. Cp. 2, 152 ὅς οἱ τὸν πατέρα Νεκὼν ἀπέκτεινε τοῦτον φεύγοντα τότε ἐς Συρίην.

14. **τις ὑμέων** = ἕκαστος ὑμέων, cp. 8, 118. **εὖ ποιήσας φανήτω** 'show himself a benefactor (εὐεργέτης) of the king'. A register of such men was kept. See 8, 86. Thucyd. 1, 129, 137.

16. **πειράσθω ἀποσχίζων** 'let each try to detach', cp. p. 3, l. 10.

20. **βιαιότερον ἕξουσι οὐδὲν** 'nor shall they have any harsher rule to submit to than they had before', cf. 3, 15 ἔνθα τοῦ λοιποῦ διαιτᾶτο ἔχων οὐδὲν βίαιον 'under no restraint, or harsh treatment'. He means 'they shall be treated as mildly as before', for he would not acknowledge that they had had anything βίαιον to complain of.

20, 21. **εἰ δὲ ταῦτα οὐ ποιήσουσι.** The use of *οὐ* after *εἰ* is justified by regarding *οὐ-ποιήσουσι* as a negative verb, 'but in case they refuse-to-do this'. It is more naturally used with *ἐᾶν*, cp. Soph. *Aj.* 1131 *εἰ τοὺς θανόντας οὐκ ἐᾷς θάπτειν*: but with other verbs also, cp. Homer *Il.* 15, 162 *εἰ δέ μοι οὐκ ἐπέεσσι ἐπιπείσεται.* So some MSS. have *εἰ δὲ ταῦτα οὐ ποιήσεις* in 1, 112. **διὰ μάχης ἐλεύσονται** 'they will fight', cp. *δι' ἡσυχίας εἶναι* 1, 206; and Xenoph. *An.* 3, 2, 8 *διὰ πολέμου ἰέναι.*

22—26. For the fulfilment of these threats see c. 32.

22. **ἐπηρεάζοντες** 'using threatening language'; the word is used in Attic Greek to indicate insolent and insulting language or action, but not with this special meaning of 'threatening'. **τά πέρ σφεας κατέξει** 'which shall actually befall them'.

23. **ἐξανδραποδιεῦνται** [App. D. III. note 2] future passive, though in 1, 66 it is active. Cp. p. 9, l. 23.

25. **ἐς Βάκτρα.** To send to Baktra was to send to the farthest province of the Empire.

CHAPTER X.

6 3. **ἀγνωμοσύνῃ διεχρέοντο** 'persisted in an attitude of obstinate defiance'. So *τρόπῳ διαχρᾶσθαι* 7, 9; *τῇ ἀληθηΐῃ διαχρᾶσθαι* 3, 72; 7, 102; *ἀβουλίῃ διαχρᾶσθαι* 7, 210. For *ἀγνωμοσύνῃ* cp. 9, 3. Herodotos seems to regard the action of the Ionians as at least ill-advised.

4. **προσίεντο** 'accept', p. 71, l. 11. **τὴν προδοσίην** = *προδοῦναι ἑαυτούς.*

6. **ἰθέως...Περσέων** 'immediately after the arrival of the Persians'.

CHAPTER XI.

8. **μετὰ δὲ**, see p. 2, l. 18. **ἐς** 'at', cp. p. 4, l. 10.

9. **ἀγοραί** 'public assemblies', 'meetings for discussion', cp. p. 31, l. 29. *ἀγορά* (Ionic -*η*) was the designation of an 'assembly of the people', as opposed to the *βουλή* or council of elders. See Hom. *Il.* 2, 51—53. This was differently named in different states, as in Athens *ἐκκλησία*, and in Doric states *ἡλία* or *ἀλία.* Herodotos also uses *ἀγορή* in the more common meaning of 'market place', see 5, 101 etc.

καὶ δή κου 'and indeed on one occasion or another'; the particle *κου* (*που*) gives the sense of indefiniteness, properly of place (9, 18), which in this instance amounts almost to time, i.e. at one or other of the meetings (*ἀγοραί*).

10. **ἠγορεύοντο.** As *βουλεύεσθαι* means to take part in the proceedings of a *βουλή*, so *ἠγορεύεσθαι* means to take part in the proceedings of an *ἀγορή*, i.e. 'to make an oration': it is the word used in Homer, see *Il.* 4, 1.

11. **ἐπὶ ξυροῦ ἀκμῆς** a proverbial expression for a position of imminent peril: cp. *Il.* 10, 173 *νῦν γὰρ δὴ πάντεσσιν ἐπὶ ξυροῦ ἵσταται ἀκμῆς.* For a speech introduced by **γὰρ** giving the ground for the statements which follow, cp. p. 29, l. 20; p. 33, l. 1; p. 39, l. 14.

13. **καὶ τούτοισι** 'and that too'.

15. **οἷοί τε** p. 5, l. 7. **ὑπερβαλόμενοι** p. 5, l. 6.

17. **διαχρήσεσθε**, cp. l. 3.

18. **μὴ οὐ δώσειν**: for *μὴ οὐ* with infinitives and participles after a negative verb, see p. 5, l. 7.

20. **θεῶν τὰ ἴσα νεμόντων** 'if only the gods are impartial', i.e. do not favour your enemies more than you. Cp. p. 64, l. 14; and the epigram of Euripides over the Athenians who fell at Syracuse (Plut. *Nikias* c. 18):

οἵδε Συρακοσίους ὀκτὼ νίκας ἐκράτησαν
ἄνδρες, ὅτ' ἦν τὰ θεῶν ἐξ ἴσου ἀμφοτέροις.

23. **ἐπιτράπουσι σφέας αὐτούς**, that is, they appointed him to be their commander. But how little hold this gave him over them was soon made apparent.

CHAPTER XII.

25. **ἐπὶ κέρας** 'in column'. The ships seem to have been rowed out in two columns, which at a certain distance swung round and kept up a mimic war with each other, practising the *diekplus*, the manoeuvre, that is, by which ships were rowed swiftly through the enemy's line and then turned and charged. Thucydides (6, 32, 50 etc.) always has *ἐπὶ κέρως* in this phrase, which means with the ships 'following one behind the other': cp. Thucyd. 2, 90 *κατὰ μίαν ἐπὶ κέρως παραπλέοντας.*

27. **τῇσι νηυσὶ** 'with his ships', instrumental dative, cp. p. 21, l. 14. **δι' ἀλληλέων** is to be taken closely with *διέκπλοον*. The **ἐπιβάται** are armed men fighting on the decks, when the ships grapple.

ὁπλίσειε 'get them under full armour', implying a drill on the decks to teach them what to do in case of a fight. Ordinarily they would not go on board until the day of battle.

28. **ἔχεσκε.** App. D. 1. f.

7 1. **δι' ἡμέρης** 'the whole day long'.

5. **ἡλίῳ**: the ἐπιβάται, being on deck, would be exposed to the sun, the rowers would feel the fatigue (ταλαιπωρίῃσι) especially.

ἑωυτοὺς=ἀλλήλους, cp. p. 81, l. 22.

6. **παραβάντες** 'having offended'. It is not the ordinary word to use with a person. The idea of transgressing law suggests that of sinning against a god.

ἀναπίμπλαμεν 'are we enduring'. Cp. 5, 4 ὅσα μιν δέει ἀναπλῆσαι κακά, 9, 87 γῆ ἡ Βοιωτίη πλέω μὴ ἀναπλήσῃ.

7. **ἐκπλώσαντες ἐκ τοῦ νόου** 'having gone out of our senses', cp. 3, 155 ἐξέπλωσας τῶν φρενῶν. A metaphor from the course of a ship, Aeschyl. *P. V.* 902 ἔξω δὲ δρόμου φέρομαι λύσσης πνεύματι μάργῳ, as in a like sense the race-course is used, cp. Soph. *Aj.* 182 φρενόθεν ἐπ' ἀρίστερα ἔβας.

8. **νέας τρεῖς** 'only three ships'. The Phokaeans had once been among the most powerful naval cities of Ionia; but a large number of their most enterprising citizens had removed to Sardinia and Italy rather than submit to Kyros' general Harpagos [1, 163—7], and the town seems not to have recovered the loss.

ἐπιτρέψαντες...ἔχομεν hardly distinguishable from a perfect: 'we have committed and still are committing ourselves'. Cp. 1, 37 ἀμφοτέρων με τουτέων ἀποκληΐσας ἔχεις: 1, 27 τοὺς δουλώσας ἔχεις. See p. 12, l. 30; p. 17, l. 4; p. 73, l. 20.

10. **καὶ δὴ** 'and already', cp. 9, 18 καὶ δὴ διετείνοντο τὰ βέλεα, 9, 6 ὁ δὲ ἐπιὼν καὶ δὴ ἐν Βοιωτίῃ ἐλέγετο εἶναι.

13. **καὶ ὁτιῶν ἄλλο** 'anything else in the world'.

15. **τοῦ λοίπου** gen. of time, 'for the future'.

16. **πειθώμεθα αὐτοῦ**. The genitive is on the analogy of ὑπακούειν 'to listen to and obey'. So in 1, 59 οὐκ ὦν ταῦτα παραινέσαντος Χίλωνος πείθεσθαι ἐθέλειν τὸν Ἱπποκράτεα, though in this last the genitive might be called absolute. Thucyd. 7, 72, 2 σφῶν πείθεσθαι.

17. **οἷα στρατιὴ** 'as though they were a land army'.

18. **ἐσκιητροφέοντο** 'they made themselves comfortable in the shade', as opposed to the work in the sun l. 5, cp. Eur. *Bacch.* 458 οὐχ ἡλίου βολαῖσιν ἀλλ' ὑπὸ σκιᾶς.

19. **ἐθέλεσκον** App. D. 1. f.

20. **ἀναπειρᾶσθαι** 'to practise', used especially of naval drill, cp. Thucyd. 7, 7, 4 οἱ δὲ Συρακόσιοι ναυτικὸν ἐπλήρουν καὶ ἀνεπειρῶντο.

CHAPTER XIII.

21. **τὰ γινόμενα ἐκ τῶν Ἰώνων** 'what was being done by the Ionians', p. 11, l. 20; cp. 9, 16 ὅ τι δέει γενέσθαι ἐκ τοῦ θεοῦ: 5, 21 ζήτησις μεγάλη ἐκ τῶν Περσέων ἐγίνετο. 8, 114, 120, 140.

22. **ἐνθαῦτα δή**—like οὕτω δή—recapitulates and makes definite the point arrived at by the previous sentence. 'It was then, I say, that '(they received) the proposals from Aeakes son of Syloson which he 'formerly sent them at the instance of the Persians, begging them to 'abandon their alliance with the Ionians,—the Samians, I say, seeing 'that the want of discipline on the part of the Ionians was serious 'received these proposals'. There is no verb for the subject οἱ στρατηγοὶ τῶν Σαμίων, or to govern the object ἐκείνους λόγους. For a similar break in a sentence, cp. 9, 84.

25. **ὦν** marks the resumption of the sentence, the *epanalepsis*, as it is called.

28. **τὰ βασιλέος πρήγματα** 'the power of the king', cp. 3, 137 καταρρωδέοντες τὰ Περσικὰ πρήγματα. **ὑπερβαλέσθαι**, cp. p. 5, l. 6.

1. **πενταπλήσιον** not to be taken literally. It is only a strong way of saying 'much greater', cp. the use in Latin of *sexcentiens*. 8

2. **προφάσιος...ἐπιλαβόμενοι**, cp. the construction of ἔχεσθαι p. 47, l. 18, 'having therefore got hold of a pretext', cp. 3, 36.

3. **ἐν κέρδεϊ ἐποιεῦντο** 'they regarded it as a lucky opportunity'. One of the numerous phrases made up with ποιεῖσθαι: cp. 2, 121, 4 τὸν ἐκκεχυμένον οἶνον συγκομίζειν ἐν κέρδεϊ ποιευμένους 'regarding it as a stroke of good luck to get the spilt wine': 9, 42 ἐν ἀδείῃ ποιεύμενοι 'considering it safe': 1, 131 ἐν νόμῳ ποιεῖσθαι 'to think it lawful'.

5. **παρ' ὅτευ** 'from whom it was that'.

8. **ἀπεστέρητο.** See on p. 5, l. 10. **τὴν ἀρχὴν** accus. of remoter object after a verb taking two accusatives, retained with the passive verb. G. § 197, note 2.

CHAPTER XIV.

10. **ὦν** resuming the narrative from c. 6.

11. **ἐπὶ κέρας**, see on p. 6, l. 25. Rawlinson translates the phrase in c. 12 'in column', and in this chapter 'in line'. But the movement is the same in both cases: the ships start one behind the other, and turn into line to face the enemy.

13. **συγγράψαι** 'to record in my history'.

17. **ἀειράμενοι τὰ ἱστία** 'having hoisted their sails'. The ships while in action would be rowed: and indeed the larger sails were often removed on shore during the battle for the sake of clearing the decks. See 8, 94; Xen. *Hellen.* 1, 1, 13; 6, 2, 27.

19. **ἀνηκουστήσαντες** 'having declined to obey', on the analogy of *ἀπείθειν*: for an opposite change see p. 7, l. 16.

20, 21. **ἔδωκε...ἀναγραφῆναι** 'granted them the privilege of having their names inscribed'. **πατρόθεν** 'with the names of their fathers', cp. 8, 90 *οἱ γραμματισταὶ ἀνέγραφον πατρόθεν τὸν τριήραρχον καὶ τὴν πόλιν.*

22. **ἀγαθοῖσι** 'brave' as in l. 14.

23. **ἐν τῇ ἀγορῇ**, that is, in the market-place of the Capital of Samos, which was called by the same name as the Island.

τοὺς προσεχέας, see p. 4, ll. 23—5.

25. **ὡς δὲ**=*οὕτω δὲ* 'and in these circumstances'. Cp. 9, 35; for *ὡς* cp. p. 43, l. 9; p. 50, l. 19.

CHAPTER XV.

27. **περιέφθησαν τρηχύτατα** 'suffered most severely', cp. p. 23, l. 22.

28. **ἐθελοκακέοντες** 'running away without waiting to be beaten', it differs from *κακοὶ γενόμενοι* (l. 14) in that the terror or cowardice is deliberately assumed. Cp. 8, 22, 69.

9 2. **ἐπιβατεύοντας** 'serving as marines', *ἐπιβάται* p. 6, l. 27.

3. **ἐδικαίευν**, App. D. III. (3). 'They determined not to be like the cowards among them', lit. 'they did not think they ought'. See p. 41, l. 1 and cp. 3, 79 *ἐδικαίευν καὶ αὐτοὶ ἕτερα τοιαῦτα ποιέειν. δικαιόω*=(1) I make right, (2) I hold to be right: it then passes to a sense not distinguishable from *ἀξιόω*, the more common word in Attic in this sense.

5. **διεκπλώοντες** 'sailing through the enemy's line': practising the *diekplus*, see p. 6, l. 26.

ἐναυμάχεον, ἐς ὃ 'they kept up the fight until'.

CHAPTER XVI.

7. **τῇσι λοιπῇσι.** For the case see on p. 4, l. 18.

8. **ἀποφεύγουσι** 'effect their escape'. **ἐς τὴν ἑωυτῶν sc. γῆν.**

9. **οὗτοι δὲ.** The main subject when repeated, or represented by a pronoun following a relative sentence, is often introduced by *δέ* repeated. Madvig § 188, Rem. 4.

11. **νέας μὲν δή.** The *δὴ* emphasises the *νέας*: 'as for their ships, they ran them ashore on Mykale and abandoned them'. **αὐτοῦ** i.e. on the nearest shore, without attempting to go elsewhere.

13. **κομιζόμενοι** 'in the course of their march'.

νυκτός τε...καὶ ἐόντων...θεσμοφορίων 'they arrived after dark, and when the women were celebrating the thesmophoria there'. Both the genitives express time and circumstance: and the two explain why the suspicions of the Ephesians were aroused. The carrying off of women at a time when they were engaged in a celebration from which men were excluded was no uncommon thing. We have an instance in c. 138 of such a transaction. The Thesmophoria was a festival of Demeter Thesmophoros, the goddess of law and civilisation; and was celebrated in various parts of Greece: as at Athens, Sparta, Thebes, Miletos, Syracuse, and many other places; see 2, 171, where they are spoken of as secret mysteries introduced from Egypt. The Ephesian Thesmophoria were, according to Strabo [xiv. 1, § 3], presided over by members of the family of the founder of the Ionian colony at Ephesos, Androklos son of Kodros of Athens.

15. **ἐνθαῦτα δὴ** 'it was in these circumstances', p. 7, l. 22.

οὔτε answered by *τε* in l. 16. Cp. p. 5, l. 7. **ὡς εἶχε** 'what the true state of the case was'.

17. **πάγχυ καταδόξαντες** 'having quite made up their minds'. *καταδοκέω* is only a strengthened *δοκέω*, but seems generally used with an idea of suspicion, cp. 3, 27 *πάγχυ σφέας καταδόξας ἑαυτοῦ κακῶς πρήξαντος χαρμόσυνα ταῦτα ποιεῖν.* Cp. 8, 69.

18. **ἐπὶ τὰς γυναῖκας** 'to carry off the women'. Thus of foragers 9, 51 *τοὺς ἐπὶ τὰ σιτία οἰχομένους.* Arist. *Ranae* 111 *ἦλθες ἐπὶ τὸν Κέρβερον* 'to carry off Kerberos'. **πανδημεὶ** 'with a levy *en masse*'. Cp. p. 62, l. 4. It is the natural word in such cases, whereas *πανστρατιῇ* refers to the formal levies of all arms. **ἔκτεινον** 'began slaughtering'.

CHAPTER XVII.

21. **τὰ πρήγματα,** see p. 7, l. 28. **διεφθαρμένα** 'ruined', p. 51, l. 16.

23. **ἀνδραποδιεῦνται** passive future, see p. 5, l. 23.

24. **ἰθέως ὡς εἶχε** 'exactly as he was', i.e. without waiting to take on board anything from home.

γαύλους trading vessels of a structure peculiar to the Phoenikians. Cp. 3, 136—7; 8, 97. It properly means a 'bucket', see p. 69, l. 17; though the accent differs, *γαῦλος* and *γαυλός*.

25. **καταδύσας** 'having disabled' by making them water-logged. See 8, 90; and Thucyd. 1, 50, 1 *οἱ Κορίνθιοι τὰ σκάφη οὐχ εἷλκον ἀναδούμενοι τῶν νεῶν ἃς καταδύσειαν, πρὸς δὲ τοὺς ἀνθρώπους φονεύειν ἐτράποντο.* It is not therefore meant that the ships were 'sunk', or no goods would have been got off them.

27. **Καρχηδονίων καὶ Τυρσηνῶν.** The two great naval powers in the Western Mediterranean. See Historical Index.

CHAPTER XVIII.

10 1. **ὑπορύσσοντες τὰ τείχεα** 'undermining the walls'. The method was to dig a mine under the walls, put wooden pins or props in to support them, and, when the time came, to burn the props, so that the wall might come down and a breach be effected. Polyb. 21, 28.

2. **κατ' ἄκρης** 'entirely': lit. from the summit down: for unless the citadel was taken the besieged might still hold out. Stein aptly quotes Thucyd. 4, 112 *ἄνω καὶ ἐπὶ τὰ μετέωρα τῆς πόλεως ἐτράπετο βουλόμενος κατ' ἄκρας καὶ βεβαίως ἑλεῖν αὐτήν.*

3. **ἕκτῳ ἔτεϊ** in the sixth year from the revolt of Aristagoras. Reckoning that as being between July, B.C. 500 and July, B.C. 499, the sixth year would be completed in July, 494. The siege of Miletos must have lasted some months, and we may date the battle of Lade in the Autumn of B.C. 495.

4. **συμπεσεῖν** 'agreed'.

5. **ἐς Μίλητον** 'in reference to Miletos', cp. p. 1, l. 13.

CHAPTER XIX.

7. **περὶ σωτηρίης** 'as to how their city was to be preserved', i.e. against the Lakedaemonians and Kleomenes, see cc. 76—81. This unfavourable answer perhaps induced them to bribe Kleomenes. See c. 82.

8. **ἐπίκοινον** 'joint', referring to both cities in one answer.

9. **παρενθήκην** 'an addition' or 'appendix', cp. 1, 156. **ἔχρησε,** sc. *ἡ Πυθίη.*

10. ἐπεὰν γένωμαι. The first part of the oracle referring to the Argives is given in c. 77.

13—16. And it is then, oh Miletos, contriver of deeds that are evil,
Many shall have in thee a feast and a glorious booty.
Oft shall thy matrons wash the feet of their long-haired masters.
My temple at Didyma then shall become a charge unto others.

The Delphic oracle was nearly always Dorian in its sympathies; which perhaps accounts for its severity of tone to the Ionian Miletos.

15. κομήταις, sc. Persians, who wore long hair. Stein quotes the epitaph on Aeschylos, see Anthol. *Appendix* 3, where the Mede is called *βαθυχαιτήεις*. The priest who drew up the verses from the mouth of the Pythia may have thought of the word on this occasion, as having just hinted that the Argives were to suffer from the Lakedaemonians, who were also *κομῆται* [7, 208; Plut. *Nik.* 19]. πόδας νίψουσι, i.e. as slaves. Homer *Odyss.* 19, 358—391.

16. Διδύμοις at Didyma or -mi, near Miletos where there was a temple dedicated to Zeus and Apollo. It was also called *Branchidae* from the name of the priests who had the care of it. 'Then others shall have the care of our temple at Didyma', i.e. the present Milesians shall be removed.

19. ἐν ἀνδραπόδων λόγῳ ἐγίνοντο 'became classed as slaves', 'were treated as slaves', cp. p. 12, l. 30.

21. ἐνεπίμπρατο 'was set fire to'. It was soon afterwards restored, if Strabo is right in saying that it was again burnt by Xerxes [14, 1, 5]. Its subsequent restoration was attempted on such a large scale that it was never finished [Strabo l.c.], though Pausanias speaks of it as standing in his time [7, 5, 4].

22. πολλάκις...ἐποιησάμην. See 1, 92; 5, 36. The most valuable treasures in the temple were the gifts of Kroesos.

CHAPTER XX.

26. κατοίκισε, compare Darius' treatment of the Eretrians, c. 119.

2. τὰ ὑπεράκρια the hill country in the district of Miletos. **11** Καρσὶ Πηδασεῦσι 'to Karians of the town of Pedasa'.

CHAPTER XXI.

5. οὐκ ἀπέδοσαν τὴν ὁμοίην, sc. *χάριν*, 'did not return their kindness'.

8. **ἡβηδὸν** 'old and young alike', 'from youth upwards'. **ἀπεκείραντο** 'cut their hair', as a sign of mourning. This was a custom of the Greeks, as well as of the East generally. See 2, 36; 9, 24.

9. **προσεθήκαντο** 'assumed', 'entertained'. The first aorist middle *ἐθηκάμην* is not used in Attic.

10. **τῶν...ἴδμεν**, attraction of relative into case of an antecedent not expressed. G. § 153 note. **ἐξεινώθησαν** 'were united by ties of friendship'. Cp. Aeschyl. *Choeph.* 689 *γνωστὸς γενέσθαι καὶ ξενωθῆναι.*

οὐδὲν ὁμοίως, see on p. 2, l. 15.

12. **ὑπεραχθεσθέντες** 'exceedingly grieved'. The nominative participle after *δῆλον ἐποίησαν*, on the analogy of verbs of declaring and showing, when the participle refers to the subject of the verb. G. § 280.

τῇ τε ἄλλῃ πολλαχῇ, καὶ δὴ 'in the many other ways and particularly when'. For **πολλαχῇ** cp. 1, 42 *πολλαχῇ ἂν ἴσχον ἐμωυτὸν* 'for many considerations', 'in many points of view'. The definite article seems to imply that the other things which the Athenians did were notorious.

13. **καὶ δὴ καὶ**, cp. p. 26, l. 6.

ποιήσαντι...θέητρον 'when he composed and brought out a drama "The Capture of Miletos" the audience was moved to tears'. The datives express the circumstances, and therefore the time, at which the event took place: cp. p. 14, l. 11; 5, 97 *νομίζουσι δὴ (τοῖς Ἀθηναίοις) ταῦτα καὶ διαβεβλημένοισι ἐς τοὺς Πέρσας, ἐν τούτῳ τῷ καιρῷ ὁ Μιλήσιος... ἀπίκετο.* See Madv. § 38. Or it may be explained, with Stein, as a kind of dative of advantage, as though the tears of the spectators were a tribute to the talents of the poet: cp. p. 25, l. 16.

14. **διδάξαντι.** The author of the play took part in the training of the actors and chorus, and was therefore said to 'teach' it. Cp. 1, 23, where Arion is said to have been the first *διδάσκειν διθύραμβον* So Aeschylos is made to say of himself (Arist. *Ran.* 1026) *εἶτα διδάξας Πέρσας μετὰ τοῦτ' ἐξεδίδαξα νικᾶν.*

15. **τὸ θέητρον** = *οἱ θεώμενοι.* Cp. Arist. *Eq.* 233 *τὸ γὰρ θέατρον δεξιόν.*

16. **χιλίῃσι δραχμῇσι** about £40. After *ζημιόω* the penalty is expressed by the dative, cp. Thucyd. 2, 65, 3 *αὐτὸν ἐζημίωσαν χρήμασιν.* 8, 21 *τετρακοσίους φυγῇ ζημιώσαντες.*

17. **χρᾶσθαι** 'bring on the stage'. Such reproductions were not

uncommon. Thus Iophon, the son of Sophokles, reproduced his father's plays, Arist. *Ran.* 73—9.

CHAPTER XXII.

18. **μέν νυν**, see on p. 24, l. 13.

19. **τοῖσί τι ἔχουσι** 'who were possessed of any property'; thus *οἱ ἔχοντες*='the rich', Eurip. *Alc.* 57.

τὸ μὲν ἐς τοὺς Μήδους...ποιηθὲν 'what their generals had done in regard to the Medes', see c. 13. For **ἐς** cp. p. 1, l. 13; p. 10, l. 10.

20. **ἐκ τῶν στρατηγῶν**, see p. 7, l. 21.

21—3. **ἐδόκεε βουλευομένοισι...μένοντας** 'they resolved on consultation that they would migrate, and that they would not remain to be slaves to the Medes and Aeakes'. The change of case is owing to the influence of the infinitive. Cp. p. 59, l. 3. 5, 109; 8, 111.

24. **οἱ ἀπὸ Σικελίης** 'who live *in* Sicily'. In speaking of persons or things at a distance the Greeks could speak of them either as *in* such and such a place, or *from* it, according to the point from which they are regarded. Thus 9, 76 *τὸ ἀπ' ἑσπέρης κέρας* 'the west wing'.

26. **ἐπεκαλέοντο τοὺς Ἴωνας** 'invited the Ionians to come over'. The people of Zankle (Messina), who were colonists from the Ionian Chalkis in Euboea, naturally turned to the men of the same blood, when they wished to strengthen their position in Sicily; probably as against the incursions of Etruscan mariners and adventurers.

12

2. **Σικελῶν** 'of the native Sikels' as opposed to the Σικελιῶται or Greek inhabitants of Sicily.

3. **τῆς Σικελίης** a topographical genitive (partitive), cp. p. 14, l. 3: 9, 13, 30.

5. **οἱ ἐκπεφευγότες**, that is, when Miletos was captured, c. 18.

CHAPTER XXIII.

6. **ἐν ᾧ** 'in course of which transaction'.

9. **πόλιν τῶν Σικελῶν** 'a city of the native Sikels', see l. 2. Stein thinks that its name has probably been lost from the text: but it is not needed for the purposes of the narrative.

13. **εἴη** the optative in *oratio obliqua*. **ἐπ' ἣν ἔπλωον** 'which was the object of their voyage'.

14. **ἐᾶν χαίρειν** 'to give up all idea of', lit. 'to say good-bye to', cp. 9, 41, 44.

18. **ἐπεκαλέοντο**, p. 11, l. 26, 'invited to come to their aid'.

19. **δή** shows that Herodotos is giving, not his own statement of the reason, but the motive of the Zankleans.

21. **μούναρχον**. He is called *βασιλεύς* in l. 8, but not *τύραννος*, and he was therefore probably a legal and constitutional sovereign.

26. **εἰρημένος** 'settled', 'agreed upon'.

27, 28. **ἐπίπλων** 'moveable property'. **λαβεῖν** 'that he should receive'. The infinitive is in apposition to *μισθός*.

30. **ἐν ἀνδραπόδων λόγῳ** 'as slaves', cp. p. 10, l. 19. **εἶχε δήσας** 'put into prison and retained them there', cp. p. 7, l. 8; p. 73, l. 20.

31. **κορυφαίους** (*κορυφή* 'top') 'leading men', cf. p. 56, l. 12; 3, 82, 159.

CHAPTER XXIV.

13 4. **παρῆν ἐς** 'came to', cp. p. 1, l. 3. **ἀνέβη** 'went up the country', probably to Susa.

7. **παραιτησάμενος** 'having obtained the king's leave': that is, he did not fly, or get off on some feigned pretext, as Histiaios did, see c. 2.

9. **ἐς ὃ γήραϊ...Πέρσῃσι** 'and there remained in very wealthy circumstances until his death, which happened in Persia at an advanced age'. Enriched no doubt by presents from the King, as Demaratos, Themistokles, and others were.

11. **περιεβεβλέατο** 'had possessed themselves of'. Cp. 3, 71 *ἰδίῃ περιβαλλόμενοι κέρδος*, 8, 8 *πολλά τε καὶ αὐτὸς περιεβάλετο*, 9, 39 *τὰ λοιπὰ ἤλαυνον περιβαλόμενοι.*

CHAPTER XXV.

13. **κατῆγον** 'restored', cp. p. 3, l. 5.

17. **τὴν ἔκλειψιν τῶν νεῶν**. Subjective genitive 'the fact of their ships having left the line', p. 8, l. 17.

20. **ὑποκυψάσας...προσηγάγοντο**: the construction is changed in the two clauses: *τὰς μὲν...ὑποκυψάσας* is in opposition to **Καρίην**, while *τὰς δὲ...προσηγάγοντο* is a fresh sentence: 'The Persians forthwith took possession of Karia,—some of its cities having voluntarily submitted, while others they reduced by force'. Cp. p. 15, l. 23.

CHAPTER XXVI.

23. **ἐόντι περὶ Βυζάντιον** 'cruising about in the neighbourhood of Byzantium'. His special object was the corn ships coming from the Pontus. See c. 5. **συλλαμβάνοντι** 'stopping and seizing', cp. p. 26, l. 15.

25. **δὴ** 'accordingly', p. 3, l. 4.

26. **ἐπιτράπει** 'committed', historical present. For the form see App. A. 2, 5, cp. p. 6, l. 23.

3. **τῆς Χίης χώρης** a topographical genitive, see on p. 12, l. 3. 14

4. **δή**, p. 13, l. 25.

5. **οἷα δὴ—ναυμαχίης** 'as was only natural in their weakened state from damage received in the sea-fight'. For **ἐκ** of agent or instrument, see p. 7, l. 21; p. 11, l. 20.

CHAPTER XXVII.

8. **φιλέει**, sc. *ὁ θεός*, 'there are wont to be forewarnings'. Cp. *ὕει*, *νίφει* and other verbs expressing natural processes. **κως** 'it seems', *nescio quomodo.*

9. **καὶ γὰρ** 'for in fact'.

10—14. **τοῦτο μὲν...τοῦτο δὲ** 'in the first place'...'in the second place', p. 23, l. 15; p. 38, l. 14.

11. **σφι πέμψασι** 'when they sent', not governed by *ἀπενόστησαν*, but like the datives in p. 11, l. 13.

χορὸν νεηνιέων probably at the Pythian festival. For the choruses sent by the various cities on such occasions, see the account of the *θεωρία* of Nikias at Delos in Plutarch *Nikias* c. 3.

14. **ὀλίγῳ**, sc. *χρόνῳ* dative of the time how long before, p. 51, l. 23.

13—18. **ὑπολαβὼν...ὑπολαβοῦσα** 'catching' or 'intercepting' them.

18. **ἐς γόνυ...ἔβαλε** 'brought the state down', a metaphor from wrestling, for which Stein quotes Simonides fr. 158 **Μίλωνος τόδ'** *ἄγαλμα...ὃς ποτὶ Πίσῃ ἑπτάκι νικήσας ἐς γόνατ' οὐκ ἔπεσεν.*

19. **ἐπὶ δὲ τῇ ναυμαχίῃ ἐπεγένετο** 'and to crown the misfortune of the sea-fight came Histiaios'.

21. **καταστροφὴν ἐποιήσατο** = *κατεστρέψατο.*

CHAPTER XXVIII.

24. **οἱ Φοίνικες**, that is, the Phoenikian fleet serving Darius.

15 2. **αὐτὸς δὲ** emphatic for *ὁ δὲ* to bring out the contrast with *Θάσον μέν*, 'Thasos indeed he left unplundered, but himself and his army he took in all haste to Lesbos'.

4. **πέρην διαβαίνει** 'crossed to the continent opposite', p. 55, l. 15. **ἐκ τοῦ Ἀταρνέος.** As Atarneus had belonged to Chios, Histiaios would regard himself as having now rights over it, owing to his victory over the Chians, see c. 26.

7.. **οὐκ ὀλίγης** common *litotes* for *μεγάλης*.

CHAPTER XXIX.

12. **τῆς Ἀταρνείτιδος...χώρης**, topographical genitive, see p. 14, l. 3.

συνέστασαν 'held out', 'maintained the fight'. Herodotos uses *συνίστασθαι* as = (1) 'to be opposed', 'to clash', 1, 208, 218; 7, 79, 142; (2) 'to be involved in', 'to endure', e.g. *λιμῷ, καμάτῳ, πόνῳ*, 7, 170; 8, 74; 9, 89. The meaning in the present case is an extension of the first of these. Cp. also p. 62, l. 19.

14. **τό τε δὴ ἔργον** 'and so the credit of the victory belonged entirely to the cavalry'—cp. 9, 102 the Athenians at Mykale acted *ὅκως ἑωυτῶν γένηται τὸ ἔργον καὶ μὴ Λακεδαιμονίων*. 8, 102. Plutarch *Nikias* c. 19 *καὶ ἐκείνου τὸ πᾶν ἔργον γεγονέναι φησὶ Θουκυδίδης*.

17. **ἁμαρτάδα.** The Ionic *ἁμαρτάς* = *ἁμαρτία*, cp. 8, 140.

φιλοψυχίην 'love of life' of an unworthy sort, amounting to cowardice, Plato *Apol.* 37 C. **ἀναιρέεται** historic present. It is difficult to be quite certain of its exact significance here. Stein translates 'fasste, gab sich hin', *concepit*, p. 38, l. 21. But it seems to be somewhat more than that, 'he conceived and ventured to practise'; there is a notion of voluntary assumption conveyed by it as *ἀναιρέεσθαι πόνους* p. 62, l. 6. Cp. 7, 16 *ἀναιρέεσθαι γνώμην* 'to deliberately adopt an opinion'.

19. **συγκεντηθήσεσθαι** 'to be stabbed to death', cp. 3, 77. For the regular future infin. after *ἔμελλον* see G. § 202, 3. And for exceptions to the rule see Rutherford, *New Phrynichus*, p. 420 sqq. *σὺν* is intensive, cp. the Latin *confixus, confossus*.

20. **γλῶσσαν μετεὶς** 'speaking in Persian', which he had learnt during his residence in the Persian court. Cp. 9, 16 Ἑλλάδα γλῶσσαν ἱέντα...μετιέναι πολλὰ τῶν δακρύων: p. 19, l. 22 μετιέναι βλαστόν. In connexion with the voice it is common in poétry, cp. τῷ μεθέντι τὸν λόγον Soph. *O. T.* 784 (St.).

CHAPTER XXX.

22. **εἰ ἀνήχθη ἀγόμενος** 'had he been taken up country to the King'. This is an emendation for ἄχθη, and though ἄχθη ἀγόμενος may be defended by such expressions as φεύγων ἐκφεύγει 5, 95, yet the constant use of ἀνάγειν and ἀναβαίνειν [p. 13, l. 4] with reference to journeys into central Asia makes ἀνήχθη nearly certain. Cp. p. 16, l. 1; p. 21, l. 22; ἀνήνεικαν in l. 29. **ὁ δὲ** for δὲ in apodosis, cp. p. 31, l. 27.

23. **ἀπῆκέ τ' ἄν**, sc. the King: the subject is changed in the two clauses, cp. p. 13, l. 20. **οὔτε...τε**, p. 9, l. 15.

24. **νῦν δὲ** 'but as it was'.

25. **ἵνα μὴ...γένηται**, for the *dramatic* subjunctive after a verb in the historic tense, see Goodwin, *Moods and Tenses*, p. 70.

28. **ἀνεσταύρωσαν**, cp. the treatment of the body of Leonidas 7, 238; and the proposal as to that of Mardonios 9, 78.

29. **ταριχεύσαντες** 'having salted', or 'embalmed', cp. 9, 120 καὶ τεθνεὼς καὶ τάριχος ἐών.

31. **ἐπαιτιησάμενος** 'having blamed', sometimes with accus. of the crime for which blame is given; μέζονα ἐπαιτιώμενος (αὐτοῖς) 1, 26.

2. **περιστείλαντας** 'having adorned', or 'swathed', cp. 2, 90. 16

3. **μεγάλως ἑωυτῷ τε καὶ Πέρσῃσι εὐεργέτεω** 'of a man who had been in an eminent degree a benefactor to himself and the Persians'. The word εὐεργέτης is here an adjective taking the dative, on the analogy of the phrase πολλοῦ ἄξιος Ἑλλάδι or the like. Cp. Eur. *Herc. F.* 117 βροτοῖσιν εὐεργέτης. The εὐεργεσία of Histiaios was the exertion of his influence to prevent the breaking of the bridge over the Danube during Darius' Skythian expedition, 4, 137. For the formal roll of such 'benefactors' of the King kept at the Persian court, see 8, 85.

CHAPTER XXXI.

7. **τῷ δευτέρῳ ἔτεϊ** 'in the next year', that is the spring of B.C. 493. For δεύτερος='next', cp. p. 73, l. 4.

ἀνέπλωσε 'put to sea again'.

9. **ὅκως δὲ λάβοι** 'and whenever it (the army) took'. The optative with a temporal adverb of indefinite frequency, p. 33, l. 15, G. § 213. 3. The use of *ὅκως = ὁπότε* seems peculiar to Herodotos; cp. 1, 17 *ὅκως μὲν εἴη ἐν τῇ γῇ καρπὸς ἁδρὸς, τηνικαῦτα ἐσέβαλλε τὴν στρατιήν*. 5, 63; 8, 91.

10. **ἐσαγήνευον**. This hunting down the men in an island by joining hands, and thus forming a line right across it, could only have been possible in a very small island without a town, in which perhaps some men may have taken refuge. To suppose it applied to anything larger is absurd. It looks like one of those tales, so common after a campaign, founded on one knows not what single incident, and eagerly repeated, exaggerated, and multiplied by heated and terrified imaginations. It seems unlikely that anything like general depopulation of the islands or cities took place; for the Ionians were still numerous and active in the Persian wars 20 years later. The same story was told of the proceedings of Datis in Eretria before Marathon, Plato *Menex.* 240 A, B.

16. **κατὰ ταὐτὰ** 'in the same systematic way'.

17. **οἷά τ'**, see on p. 5, l. 7.

CHAPTER XXXII.

18. **ἐνθαῦτα** 'in these circumstances', 'in the course of these proceedings'. **οὐκ ἐψεύσαντο** 'did not belie', 'did not leave unfulfilled'.

19. **τὰς ἐπηπείλησαν**: the relative represents a cognate accusative. For the threats see p. 5, l. 23 sq.

23. **καλλιστεύειν** 'to be most beautiful' as *ἀριστεύειν* 'to be bravest'. For the force of *ἀνά* in *ἀνασπάστους* cp. p. 16, l. 1.

24. **δή**, p. 3, l. 4; p. 13, l. 25.

25. **αὐτοῖσι τοῖς ἱροῖσι** 'temples and all', a dative absolute, or, of accompanying circumstances, as in the common *αὐτοῖσι ἀνδράσι* (p. 53, l. 19), G. § 188, 5. The burning of the temples was an act especially shocking to Greek sentiment; and when, nearly two centuries later, Alexander the Great professed to be avenging the wrongs inflicted on the Greeks by the Persians, Polybios (5, 10) declares that he was careful to abstain from similar sacrilege. The Persians however affected to act in revenge for temples destroyed at Sardis (c. 101).

26. **τὸ τρίτον**. The three subjugations of the Ionians to which Herodotos refers were (1) that by Alyattes of Lydia and his son Kroesos,

B.C. 600—550 [1, 14—19, 51—3], (2) by Kyros and his generals, B.C. 546—5 [1, 141—160], (3) the suppression of the Ionian revolt, B.C. 500—494.

CHAPTER XXXIII.

2. **τὰ ἐπ' ἀριστερὰ ἐσπλώοντι**, that is, the Thrakian cities on the western shore of the Hellespont and Propontis. 17

3. **αὐτοῖσι τοῖσι Πέρσῃσι** 'by the Persians themselves', i.e. not, as in the present instance, by their Phoenikian dependants. See 5, 117.

4. **ὑποχείρια ἦν** 'were already subject'. **γεγονότα κατ' ἤπειρον** 'having been so subjected by a land attack'.

7. **τὰ τείχεα** 'forts' or 'fortified towns', thus Sestos is called *ἰσχυρότατον τεῖχος τῶν ταύτῃ* 9, 115. Cp. *Κιμμέρια τείχη* 4, 12.

12. **οἴκησαν** 'settled in'. It was already founded.

κατακαύσαντες 'having utterly wasted with fire', used generally of objects that can be entirely consumed, as a temple (p. 58, l. 12), a city (8, 33), a house (4, 79).

14. **πυρὶ...νείμαντες** 'having committed them to the flames'. So the fire itself is said *ἐπινέμεσθαι* (5, 101).

15. **ἐξαιρήσοντες** 'intending to destroy', cp. p. 77, l. 26, arising perhaps from the idea of removing the inhabitants, like *ἐξανιστάναι*, see 5, 16 *τοὺς ἐν τῇ λίμνῃ κατοικημένους ἐξαιρέειν*. But Herodotos uses it as simply = 'to take' a city, cp. 5, 65, 122; 9, 86.

16. **προσσχόντες** 'when they touched there'.

17. **οὐδὲ ἔπλωσαν ἀρχήν** 'did not sail at all', cp. p. 49, l. 23; 7, 26 *οὐδὲ γὰρ ἀρχὴν ἐς κρίσιν τούτου πέρι ἐλθεῖν οἶδα*. Rarely without a negative, as in 8, 132 *ἐόντες ἀρχὴν ἑπτά*.

19. **ὑπὸ βασιλέϊ** 'under the power of the Great King'.

CHAPTER XXXIV.

7. **ἐπάγεσθαι ἐπὶ τὴν χώρην** 'to invite over into their country'. 18 **οἰκιστὴν** as 'founder', that is, of a new colony, or perhaps only of a new dynasty and government.

9. **ξείνια** 'hospitality' of any sort; hence frequently 'a banquet' as in 9, 89.

τὴν ἱρὴν ὁδὸν 'the sacred way', namely that leading from Delphi through Phokis to Chaeronea. It was called sacred, as was the road from Athens to Eleusis, because of the sacred processions, and the pious enquirers of the Oracle who passed along it under the protection of the God. The road is carefully described by Pausanias 10, 5, 6.

10, 11. **ἤϊσαν...ἐκτράπονται** 'they *were* going along the road... but no one inviting them they *turned* out of the direct road towards Athens'. The historic present *ἐκτράπονται* is for the aorist, p. 2, l. 22.

CHAPTER XXXV.

14. **οἰκίης τεθριπποτρόφου** 'of a family wealthy enough to keep a four-horse racing chariot'. That is, for the races at the Olympic or other games. Cp. p. 73, l. 2. The keeping of such horses was regarded at Athens as especially the duty of the rich, cp. Isokrates *de Big.* § 33 *ἱπποτροφεῖν...ὃ τῶν εὐδαιμονεστάτων ἔργον ἐστί*. In after times we find that citizens contrived to combine the duty of serving in the cavalry with that of training horses for the races, for the sake of economy. Lysias *Orat.* 22 § 63 *ὅτε ἵππευεν, οὐ μόνον ἵππους ἐκτήσατο λαμπροὺς ἀλλὰ καὶ ἀθλητάς, οἷς ἐνίκησεν Ἰσθμοῖ καὶ Νεμέᾳ, ὥστε τὴν πόλιν κηρυχθῆναι καὶ αὐτὸν στεφανωθῆναι.* 'When he was serving in the cavalry he not only purchased splendid chargers, but racing horses also, with which he won prizes at the Isthmian and Nemean games, so that the city was proclaimed and he was himself crowned'. This passage will show the reason which made such men eminent and popular in their states, cp. Dem. *de Cor.* § 320.

18. **τοῖσι προθύροισι** 'in the portico (or verandah) before his house'. This seems to answer to the *αἴθουσα* of the Homeric house, which was much used, even sometimes for sleeping (*Odyss.* 4, 297). It is always plural in Herodotos (see p. 52, l. 15; 3, 35, 140) and must not be confounded with *πρόθυρον* 'the great gate of the court-yard' (*αὐλή*). It is there that Solon is said to have hung up his arms, as a sign that he had done his work and no longer meant to take part in politics (*λαβὼν τὰ ὅπλα καὶ πρὸ τῶν θυρῶν θέμενος εἰς τὸν στενωπόν*, Plutarch *Sol.* 30).

20. **αἰχμάς.** Carrying arms had been discontinued in Greece by this time, and was a sign of *barbarism*. Cp. Thucyd. 1, 6, 1 *πᾶσα γὰρ ἡ Ἑλλὰς ἐσιδηροφόρει διὰ τὰς ἀφράκτους οἰκήσεις καὶ οὐκ ἀσφαλεῖς παρ' ἀλλήλους ἐφόδους, καὶ ξυνήθη τὴν δίαιταν μεθ' ὅπλων ἐποιήσαντο, ὥσπερ οἱ βάρβαροι.*

προσεβώσατο 'he shouted to them to come to him'.

24. **μιν πείθεσθαι** 'that he would obey', *ἐδέοντο αὐτοῦ* being treated as a simple jussive verb. Cp. 1, 141 *Κύρου δεηθέντος...ἀπίστασθαι σφέας ἀπὸ Κροίσου οὐκ ἐπείθοντο.*

25. **ἀχθόμενον τῇ Πεισιστράτου ἀρχῇ.** As usual with Greek tyrannies, the most determined opponents of Peisistratos were the rich and noble.

28. **εἰ ποιέῃ,** the deliberative subjunctive retained in indirect discourse in spite of *εἰ*, on the *dramatic* principle, whereby the mood and tense which would have been used in direct is used in an indirect sentence, cp. p. 49, l. 30; 5, 42, 67; Aesch. *in Ctes.* § 202. Goodwin, *M. & T.* p. 155.

CHAPTER XXXVI.

29. **κελευούσης** 'bidding him go'. **οὕτω δὴ** often used to introduce action taken in circumstances described by the preceding clause. Sometimes *οὕτω* stands by itself, and sometimes *δὴ* in almost the same sense.

30. **Ὀλύμπια ἀναραιρηκὼς** 'having won a victory at the Olympic games'. So *Ὀλυμπιάδα ἀνελέσθαι*, p. 59, l. 2.

2. **ἔσχε** 'took control of', p. 20, l. 22. 19

7. **στάδιοι.** The Stade was 600 Greek or 582 English feet. The distance across the isthmus will thus be somewhat under four English miles; and the length of the peninsula just over 46 miles.

CHAPTER XXXVII.

10. **αὐχένα** 'isthmus'.

14. **ἦν ἐν γνώμῃ γεγονὼς** 'was known to' *or* 'was high in the favour of' (L. and Sc.). For the paraphrastic tense cp. p. 7, l. 20. It is a singular use of *γνώμη*, which never, as far as I am aware, means 'knowledge' or 'acquaintance'. Stein says that there is an idea of esteem contained in it; and Abicht translates it 'had been dear and intimate' (*lieb und vertraut*). But that implies some meaning of *γνώμη* akin to 'inclination', which is equally removed from the common usage. *γνώριμος* 'acquaintance', less than *φίλος* or *ξένος*. Dem. *de Cor.* § 284.

16. **προηγόρευε** 'he publicly warned'.

17. **πίτυος τρόπον** 'after the manner of a pine'. For the more common *δίκην*, G. § 160. 2. Cp. *τρόπον αἰγυπιῶν* Aesch. *Ag.* 49,

Αἰγίδος τρόπον Eurip. *Ion* 1423. It is a poetical expression, though not uncommon in Herodotos, see 1, 197; 3, 98. There is perhaps an allusion to the ancient name of Lampsacus, Pityoessa. But it was, or afterwards became, a common proverb. See Suidas s. v. *δίκην· ὁ μὲν ἐξετρίβη πίτυος δίκην.* Aelian *v. h.* 6, 13 *ἐκτρῖβον τοὺς τυράννους πίτυος δίκην ἢ στερίσκον τῶν παίδων.*

18. **πλανωμένων ἐν τοῖσι λογοισι** 'differing in their explanations'. Thus in 2, 115 of a man 'prevaricating', not sticking to the same story, *πλανωμένου ἐν τῷ λόγῳ.*

19. **τί ἐθέλει...εἶναι** 'what the expression meant'. Cp. 4, 131 *γνῶναι τὸ ἐθέλει τὰ δῶρα λέγειν.*

20. **μόγις κοτὲ μαθὼν** 'having at length with some difficulty understood it'.

22. **μετίει** 'puts forth', cp. p. 15, l. 20.

23. **πανώλεθρος ἐξαπόλλυται** 'perishes utterly and to extinction', or, 'beyond hope of revival'; unlike the olive which would quickly revive even after burning. See 8, 55.

24. **λύσαντες μετῆκαν** 'released Miltiades from prison and allowed him to depart'.

CHAPTER XXXVIII.

25. **διὰ Κροῖσον** 'thanks to the interference of Kroesos'.

20 3. **θύουσι.** The technical word for the worship of heroes is *ἐναγίζειν*, not *θύειν*, 2, 44: but the latter is often used in a general sense to include both. **ὡς νόμος οἰκιστῇ**, see p. 18, l. 7, and compare the conduct of the people of Amphipolis to Brasidas [B.C. 422] who 'put up a fence round his tomb, and continued to sacrifice to him, and honour his memory with yearly games and victories, and dedicated the colony to him as a founder'. Thucyd. 5, 11.

4. **ἐγγίνεται**=*licet.* Cp. 1, 132 *τῷ θύοντι ἰδίῃ μούνῳ οὔ οἱ ἐγγίγνεται ἀρᾶσθαι ἀγαθά.*

6. **κατέλαβε** 'it befell', frequent in Herod., cp. 9, 49, 75, 105 *τοῦτον κατέλαβε κέεσθαι.*

7. **πληγέντα τὴν κεφαλὴν.** See G. § 160.

τῷ πρυτανηΐῳ 'in the court-house', the place of assembly for the magistrates, whether their title was *πρυτάνεις* or not, cp. 7, 197 *λήϊτον καλέουσι τὸ πρυτανήϊον οἱ Ἀχαιοί.*

8. **καὶ ὑποθερμοτέρου** 'and a somewhat bitter one too'.

CHAPTER XXXIX.

10. **ἐνθαῦτα** 'thereupon', p. 9, l. 15.

12. **τὰ πρήγματα** 'the government', cp. p. 9, l. 21. But it does not appear from Herodotos' narrative that Miltiades was an independent tyrannus. Rather he seems to hold the government under Athens, without losing his citizenship, and when he returned was indicted for *tyrannis* as being subject to Athenian law, c. 104.

15. **δῆθεν**, p. 1, l. 7.

16. **ἐν ἄλλῳ λόγῳ** 'in another part of my book', see c. 103.

17. **εἶχε κατ' οἴκους** 'kept within doors', cp. 3, 79; as a sign of grief, cp. Eurip. *Hipp.* 131 *ἐντὸς ἔχειν οἴκων.*

18. **ἐπιτιμέων** 'by way of showing respect to . [This meaning however of *ἐπιτιμᾶν* is exceedingly doubtful. And various emendations have been attempted: *ἔτι τιμέων, ἔτι πενθέων, πενθέων* (Cobet). In the two passages quoted by Stein for *ἐπιτίμιον*, in that from Aeschylos *S. c. Th.* 1024 it may mean 'a penalty', and in that from Sophokles *El.* 915 *τἀπιτίμια* has been altered by most editors to *τἀπιτύμβια*. The only other place in which Herodotos uses *ἐπιτιμᾶν* is 4, 43 *τὴν ἀρχαίην δίκην ἐπιτιμῶν* 'inflicting the ancient punishment'.]

δηλαδὴ 'as he gave out', intimating that it was not the real reason. Cp. 4, 135 *προφάσιος τῆσδε δηλαδή, ὡς αὐτὸς...ἐπιθήσεσθαι μέλλοι τοῖσι Σκύθῃσι.* In this compound *δή = δῆθεν*, cp. 9, 11.

21. **ὡς συλλυπηθησόμενοι** 'as though with the view of joining in his mourning'. **κοινῷ στόλῳ**, cp. 5, 63, 91.

22. **ἴσχει**, p. 19, l. 2.

23. **βόσκων** 'keeping 500 mercenary guards', cp. Thucyd. 7, 48, 5 *καὶ ναυτικὸν πολὺ ἔτι ἐνιαυτὸν ἤδη βόσκοντας.*

CHAPTER XL.

26. **νεωστὶ μὲν ἐληλύθεε.** This must refer to his *return* after his expulsion by the Skythians; for he had been many years in the Chersonese when this happened; having been sent there by the Peisistratidae, whose power in Athens terminated in B.C. 510. The most probable arrangement of dates seems to be

Miltiades comes to the Chersonese	B.C. 512
Darius' Skythian Expedition	B.C. 508
Miltiades expelled by the Skythae, but restored the same year	B.C. 495
Miltiades flies to Athens	B.C. 493

But it seems strange that Herodotos should refer this movement of the Nomad Skythians to the expedition of Darius so many years before.

μὲν...δὲ expressing simultaneousness, p. 3, l. 20, cp. the use of *καὶ...καὶ* 9, 67 and *τε...καὶ* 8, 83; 9, 55.

27. **κατελάμβανε**, *impersonal*, see above, l. 6.

21 1. **ἄλλα...χαλεπώτερα** 'other troubles more severe than those in which he was now involved'. For *πρήγματα* = troubles, cp. 7, 147 *πρήγματα ἔχειν στρατηλέοντας*. Thucyd. 8, 48, 3 *πράγματα ἔχειν...τοῖς Ἀθηναίοις προσθέμενον*. Arist. *Plut.* 652. In the Orators it commonly refers to lawsuits and such troublesome business.

2. **τρίτῳ ἔτεϊ τούτων** 'in the third year *before* these', see l. 8.

3. **ἐρεθισθέντες** 'having been provoked by Darius', i.e. in his Skythian Expedition of B.C. 508 (?). They seem to have been induced by the first successes of the Ionian revolt to try retaliation.

6. **ἔφευγε...ἐς ὃ** 'was in exile from the Chersonese until'.

7. **κατήγαγον**, see p. 3, l. 5.

8. **τότε**, i.e. at the time Herodotos is now speaking of, in B.C. 495.

CHAPTER XLI.

14. **τῇσι νηυσί** 'with their ships', for this dative see p. 6, l. 27.

15. **καταφεύγει** 'effects his escape'.

22. **ἀνήγαγον** 'took him up country to the King', cp. p. 16, l. 1.

χάριτα μεγάλην καταθήσεσθαι 'that they would secure for themselves great gratitude from the King', lit. 'lay up for themselves a store of gratitude'. Cp. 7, 178 *χάριν ἀθάνατον κατέθεντο*. 9, 60 *χάριν θέσθε*. 9, 78 *κλέος καταθέσθαι*. Thucyd. 1, 33 *ὡς ἂν μετ' ἀειμνήστου μαρτυρίου τὴν χάριν καταθεῖσθε*. 4, 87, 4 *ἀΐδιον δόξαν καταθέσθαι*. The passive is *κεῖσθαι*, see Thucyd. 1, 129 *κεῖταί σοι εὐεργεσία ἐν τῷ ἡμετέρῳ οἴκῳ ἀνάγραπτος*.

23. **ὅτι δὴ** 'because, as they reflected', *δὴ* showing that it is the thought, not of the writer, but of the persons of whom he is speaking. In Latin it would be indicated by the subjunctive, *quod sententiam tulisset*. For the circumstances alluded to see 4, 137. It seems strange that Miltiades should have remained so long without being attacked, after thus notoriously advising the breaking of the bridge,—which would have been fatal to the King.

25. **τὴν σχεδίην** 'the temporary bridge of boats' across the Danube.

29. **οἶκον καὶ κτῆσιν**, see p. 15, l. 23.

30. **τὰ ἐς Πέρσας κεκοσμέαται** 'who have been reckoned as Persians'. That is, they had all the rights and privileges of Persians, cp. 3, 91 *ἐς τὸν Αἰγύπτιον νομὸν Κυρήνη τε καὶ Βάρκη ἐκεκοσμέατο*. The meaning comes from the military use of *κοσμεῖσθαι = τάσσεσθαι*. Notice that *τέκνα*, like some other neuter words which mean persons, may have a plural verb.

CHAPTER XLII.

2. **ἐς νεῖκος φέρον** 'hostile', opposed to *εἰρηναῖα* below l. 15. Cp. 4, 90 *ἐς ἄκεσιν φέροντα* 'medicinal'.

3. **τάδε μὲν**. The clause answering to this, indicating the other measure concerning the Ionians, is in l. 8 *καὶ τὰς...*, the construction being broken by the repetition *ταῦτά τε ἠνάγκασε*.

4. **τούτου τοῦ ἔτεος** 'in the course of this year', gen. of the time 'within which'.

ὁ Σαρδίων ὕπαρχος, see 5, 25. This appears to be equivalent to a Satrap, Sardis being the seat of government for the second or Ionian *νομὸς* or Satrapy. The governor (*ὕπαρχος*) of Sardis however seems to have been in a special sense a military officer appointed at Sardis,—owing to its contiguity to the Ionian Greeks and the necessity of keeping a strong hand over them,—rather in the position of the military (*φρούραρχος*), as opposed to the civil, governor of the later organisation of the Persian Empire described by Xenophon in *Oeconomicus* 4, 8—10.

6. **σφίσι αὐτοῖσι = ἀλλήλοις**, cp. p. 7, l. 5; p. 81, l. 22.

7. **δωσίδικοι** 'submitting to arbitration', or, 'legal forms of redress'. It is a rare word, not apparently occurring again until Polybios (4, 4).

φέροιεν καὶ ἄγοιεν 'pillage', by carrying off property and driving off cattle, Lat. *ferre agere*. Cp. 1, 88 *φέρειν καὶ ἄγειν τὰ σά*: 3, 39 *ἔφερε καὶ ἦγε πάντας*: 9, 31 *ἔφερόν τε καὶ ἦγον τὴν Μαρδονίου στρατιήν*. Herodotos seems to hold the opinion, afterwards strongly held by Xenophon, that the institutions of Darius and the organisation of the Persian Empire were what were sorely needed in Greece.

9. **κατὰ παρασάγγας** 'in parasangs', a measure as he explains of 30 stades, or about 3½ miles. It is still called a *farsakh*, and it is supposed to be the distance a mule or camel can walk in an hour.

11. **οἱ κατὰ χώρην διατελέουσι ἔχοντες** 'which remain exactly as they were', i.e. the *φόροι*. For the tribute imposed on the various

Satrapies under Darius, see 3, 89. The tax according to this statement was calculated on the land, so much a square parasang. How such tax was contributed individually by the people living in this tract was a matter for local authorities to determine. Each district thus marked out for taxing purposes was called a *νομός* (3, 89). The *πρῶτος νομὸς* or *νομὸς Ἰωνικὸς* consisted of the Ionians, Asiatic Magnesians, Aeolians, Karians, Lykians, Milyans, Pamphylians, paying altogether 400 talents to the royal treasury. The measure of Artaphernes here described does not appear to have affected the amount of this tribute, but to have been an internal arrangement for levying it on an equal scale.

12. **ἔτι καὶ ἐς ἐμὲ** 'up to my time'. Herodotos does not apparently mean to the time at which he is writing, but to the time within his personal knowledge. For after the battle of Mykale (B.C. 479) most of the towns and Islands of the Asiatic Greeks were really though not professedly free from the Persians. This freedom was formally secured by the 'Peace of Kimon' B.C. 449—8 and lasted till the 'Peace of Antalkidas' B.C. 387. Yet that this freedom did not extend at once to all Greek towns is shown by the fact that the King was, in B.C. 465, able to assign Magnesia, Myus and Lampsakos to the support of Themistokles, Thucyd. 1, 138. Probably the same official rating may have nominally remained, although some of the towns ceased to pay.

13. **ἐξ Ἀρταφέρνεος**: for *ἐκ* = *ὑπὸ* cp. p. 11, l. 20.

14. **πρότερον**, i.e. in the original arrangement of Darius referred to above and described in 3, 89—90.

τὰ εἶχον = *ὡς εἶχον* sc. *οἱ φόροι*.

CHAPTER XLIII.

15. **εἰρηναῖα** opposed to *ἐς νεῖκος φέρον* l. 2.

16. **καταλελυμένων** 'having been superseded in their commands'. So 7, 16 *καταλύειν τὸν στόλον* 'to disband the army'. Cp. *παραλελυμένων* 5, 75.

19. **πολλὸν δὲ ναυτικὸν** 'and a large number of men to serve in the ships', which were to meet him.

22. **ἐν τῇ Κιλικίῃ** where the Phoenikian fleet would meet him, cp. p. 54, l. 14.

24. **ἐπὶ τὸν Ἑλλήσποντον** 'with a view of reaching the Hellespont', where they were to meet the fleet.

23 1. **ἐνθαῦτα** 'at this point', p. 20, l. 10.

2. **μὴ ἀποδεκομένοισι** 'who do not believe', in the same sense *ἐνδέχομαι* 3, 115.

3. **τοῖσι ἑπτά...ἀποδέξασθαι.** The 'Seven' are the conspirators who killed the Magus who pretended to be Smerdis the son of Kyros. See 3, 80.

4. **τυράννους...καταπαύσας.** This measure was apparently prompted, partly by a wish to try a policy of conciliation, and partly by the fact that experience had shown that the tyranni could not be relied upon for loyalty to the Great King, whenever their power was well established and their resources enlarged.

7. **ἐς τὸν Ἑλλήσποντον** '*into* the Hellespont', with his fleet: the land army march *ἐπὶ τὸν Ἑλλήσποντον*.

8. **χρῆμα πολλὸν** 'a large number'. Herodotos uses this word to express anything large, as 'a huge boar' *μέγα χρῆμα συὸς* 1, 36. And especially of number 3, 109 *πολλὸν χρῆμα τῶν τέκνων*, 3, 130 *πολλὸν χρῆμα χρυσοῦ*. Cf. Arist. *Nub.* 1 *ὦ Ζεῦ βασιλεῦ, τὸ χρῆμα τῶν νύκτων ὅσον.*

10. **ἐπορεύοντο δὲ ἐπὶ** 'and the object of their march was to attack Eretria and Athens'.

CHAPTER XLIV.

12. **πρόσχημα** 'ostensible object', cp. p. 77, l. 18. 7, 157 *πρόσχημα ποιεύμενος ὡς ἐπ' Ἀθήνας ἐλαύνει.*

15, 16. **τοῦτο μὲν...τοῦτο δὲ** 'in the first place'...'in the second place', p. 38, l. 14.

15. **οὐδὲ χεῖρας ἀνταειραμένους** 'without their even venturing to resist him', cp. 3, 144 *οὔτε τίς σφι χεῖρας ἀνταείρεται.*

18. **τὰ ἐντὸς Μακεδόνων ἔθνεα** 'the nations east of Makedonia', which to the Persian would be on 'this side' of Makedonia. The Makedonians, whose ruler was now Amyntas, had sent earth and water to Megabazos (5, 17, 18); but the Persian envoys at his court had been murdered by his son Alexander for insolence to some Makedonian women (5, 18—20); and though this matter had been hushed up, it seems that the Persian court considered an expedition against the country necessary.

20. **πέρην** 'to the continental side', p. 15, l. 4. Of an island p. 55, l. 15.

21. **τὸν Ἄθων περιέβαλλον** 'they tried to round the promontory of Athos'.

23. ἄπορος 'impossible to make head against'. So of troops 'impossible to get at' 4, 46 ἄμαχοί τε καὶ ἄποροι προσμίσγειν.

τρηχέως περιέσπε, cp. p. 8, l. 27.

26. ὑπὲρ δὲ δύο μυριάδας ἀνθρώπων, sc. διαφθαρῆναι.

ὥστε = ὡς or ἅτε, cp. 9, 37; 8, 118.

28. θηρίων appears to mean 'sharks', but as there seems a doubt whether any such fish have ever been known in the Archipelago, some have interpreted Herodotos as meaning the wild animals on the peninsula, as Mr Grote seems to do. This is putting a strain upon the language of Herodotos, especially as he has just said that the sea was θηριωδεστάτη. Sharks seem to have at times infested the coast of Greece, see Aeschines *in Ctes.* 130, Plut. *Phoc.* 28, where some *mystae* are said to have been devoured by them while purifying in the sea.

24 1. νέειν. The natives of the Levant were generally famous swimmers (see 8, 9); but these men were many of them from the interior of Asia (c. 43).

CHAPTER XLV.

8. οὐ γὰρ δὴ 'for, as is well known, he did not leave the country until etc.'; for this δή, cp. p. 23, l. 15.

10. μέντοι 'however, it was not till after the subjugation of these that he led off his army'.

11. ἅτε...προσπταίσας 'taking into consideration that he had suffered a disaster'. προσέπταιον μεγάλως 5, 62.

13. οὗτος μέν νυν 'so then this expedition returned to Asia with disgrace'. μέν νυν = μὲν οὖν (nearly), p. 25, l. 15. Cp. below l. 25, μέν γε.

CHAPTER XLVI.

16. δευτέρῳ ἔτεϊ τούτων 'in the second year after this'.

18. μηχανῴατο, App. D. II. c, optative in reported speech depending on διαβληθέντας, 'having been accused to him of being engaged in contriving'.

20. οἱ γὰρ δὴ Θάσιοι· 'for the fact was that the Thasians, on the ground of having been subjected to a siege by Histiaios (c. 28), and being possessed of large revenues, were employing the money in having ships built'.

The gold mines in Thasos and the opposite Thrakian coast afterwards fell into the hands of the Makedonian kings and were the great source of their revenue. See 5, 17; 9, 75.

22. **ναυπηγεύμενοι**, the middle is used of the people who get the ships built for them. The actual artizans are said ναυπηγεῖν, see Aristoph. *Plutus* 513 ἐθελήσει τίς χαλκεύειν ἢ ναυπηγεῖν ἢ ῥάπτειν ἢ τροχοποιεῖν;

23. **περιβαλλόμενοι** 'surrounding themselves with'. Cp. Thucyd. 1, 8, 3 καί τινες καὶ τείχη περιεβάλλοντο.

25. **μέν γε** 'at any rate' = γοῦν, cp. l. 13.

1. **τὸ ἐπίπαν** 'on an average' or 'as a general rule' every year, 25 cp. 8, 60, 3 οἰκότα βουλευομένοισι ὡς τὸ ἐπίπαν ἐθέλει γίνεσθαι.

4. **ἐοῦσι καρπῶν ἀτελέσι** 'being thus free from all imposts upon the products of their lands'.

6. **προσῆλθε**, used as the only aorist available for προσεῖμι, of which the imperfect is used in l. 2 and 4. The imperfect is used in l. 2 and 4 to express the yearly recurrence of the revenue; the aorist in l. 6 to express the more definite notion of the year of the highest point in the revenue. They both are tenses of the verb expressing the notion of the substantive πρόσοδος.

CHAPTER XLVII.

8. **τὰ οἱ Φοίνικες ἀνεῦρον.** The Phoenikians from early times had been the most active commercial people in the Mediterranean, and had discovered and worked gold mines in various parts as far as Spain, and carried the gold to Tyre; where in remembrance of their Thasian settlement there was a temple to 'Herakles Thasios' 2, 44. **κτίσαντες** 'having colonized'.

10. **τοῦ Φοίνικος** 'the son of Phoenix', and therefore brother of Europa. This legend is part of the tradition which traced much of the civilisation of Europe to Phoenikia.

13. **οὖρος μέγα ἀνεστραμμένον** 'a great mountain having been ransacked in the search'. The construction appears to be an accusative absolute. G. § 278, Clyde § 64. See on p. 40, l. 13.

15. **μὲν νύν**, p. 24, l. 13.

CHAPTER XLVIII.

19. **ἀπεπειρᾶτο** 'attempted to ascertain the disposition of the Greeks'. The compound ἀποπειρᾶσθαι means 'to satisfy oneself by

experiment'. Cp. 9, 21 and 2, 73 ᾠὸν πλάσσειν ὅσον τε δυνατός ἐστι φέρειν, μετὰ δὲ πειρᾶσθαι αὐτὸ φορέοντα, ἐπεὰν δὲ ἀποπειρηθῇ κ.τ.λ.

21. **διέπεμπε** 'he sent in different directions'.

22. **ἀνὰ Ἑλλάδα** 'throughout Greece', i.e. European Greece. This local meaning of ἀνά is fairly frequent in Homer and Herodotos, but seems to have had a tendency to drop out of common usage in the Attic period except in composition, not being once found in Aristophanes, or (I believe) in Thucydides, though the tragedians still used it. Cp. p. 82, l. 5.

23. **βασιλέϊ γῆν τε καὶ ὕδωρ** 'earth and water for the King', symbols of authority over an entire country. See 5, 17; 7, 13, 31; 8, 46.

δὴ 'accordingly', p. 3, l. 4; p. 13, l. 25.

26. **νέας μακρὰς** 'ships of war'; as opposed to στρόγγυλαι (1, 163), and ὁλκάδες (c. 26), cp. πλοῖα μακρὰ Thucyd. 1, 14, the *longae naves* of the Romans. But though the term is a well-known one, both Thucydides and Herodotos generally use the more definite name of trireme for a war-ship. Cp. 5, 30.

ποιέεσθαι 'to have made', 'to cause to be made', cp. note on p. 24, l. 22.

CHAPTER XLIX.

26 3. **τὰ προΐσχετο αἰτέων** 'that for which he was making formal demand', i.e. earth and water. Sometimes the construction is reversed, as on p. 5, l. 17 προϊσχόμενοι ἐπαγγέλλεσθε. Cp. the two constructions of the simple ἔχω. It is possible to say κυπτάζεις ἔχων (Arist. *Nub.* 509) or ἀτιμάσας ἔχει 'continues to treat with dishonour' Eurip. *Med.* 33.

4. **νησιῶται** 'islanders', as opposed to the people in the continent, and in this generic sense, is used without the article, cp. 8, 46, and νῆσοι 9, 3.

6. **καὶ δὴ καί**, p. 11, l. 13, 'and among others the Aeginetans'.

7. **ἐπεκέατο**, App. D. II. a, 'attacked them'.

ἐπὶ σφίσι ἔχοντας 'as being hostile to themselves' (σφι = αὐτοῖς, σφίσι = ἑαυτοῖς). For the construction and meaning cp. Hom. *Od.* 22, 73 ἐπὶ δ' αὐτῷ πάντες ἔχωμεν. Thucyd. 8, 105 τὰς ἐπὶ σφίσι ναῦς ἐχούσας (Abicht).

9. **προφάσιος ἐπελάβοντο**, cp. p. 8, l. 2.

10. **φοιτέοντές τε ἐς Σπάρτην**. Athens was at present almost an open town, with few ships of war, and had not begun to think

of supremacy in Greece. This appeal to Sparta in such an international business was a clear acknowledgment of her superiority and commanding position among the Greek States.

κατηγόρεον...τὰ πεποιήκοιεν 'stated as an accusation against them what they alleged they had done'. The optative indicates the substance of the accusation, not the thought of the writer. The vexatious nature of the charge could only be defended by the plea that Aegina was not as open to attack by the Persians, and therefore had no such excuse, as the other islands, against whom the Athenians make no accusation.

CHAPTER L.

13. **πρὸς ταύτην τὴν κατηγορίην** 'in view of this accusation', or 'to investigate this accusation'.

15. **συλλαβεῖν** 'arrest', cp. 2, 114 *ἄνδρα τοῦτον συλλαβόντες ἀπάγετε παρ' ἐμέ*, cp. p. 13, l. 23.

16. **ἐπειρᾶτο συλλαμβάνων**, cp. p. 3, l. 10; p. 5, l. 16.

17. **ἐν δὲ δὴ καὶ** 'and among the rest Krios'. A combination of particles used to mark the opposition to *ἄλλοι* as in 5, 90. The *δὴ* can hardly be translated; it serves, as often, to mark the introduction of a particular instance or anecdote. For *ἐν* cp. p. 6, l. 10; p. 49, l. 3.

18. **οὐκ ἔφη** *negavit.*

19. **χαίροντα** 'without a struggle', or 'with impunity', p. 43, l. 9.

20. **τοῦ κοινοῦ**, cp. p. 8, l. 20. **ἄνευ Σπαρτιητέων τοῦ κοινοῦ** 'without the consent of the Spartan body public', opposed as the ruling class to the other inhabitants of Lakedaemon. See p. 31, l. 20.

21. **ἀναγνωσθέντα χρήμασι** 'bribed', in the sense in which *πεισθέντα* is frequently used.

ἅμα γὰρ ἄν...συλλαμβάνειν 'for otherwise he would have come with the other king to arrest them'. The aorist or imperfect indicative with *ἄν*, expressing an unfulfilled condition, becomes a participle or infinitive in oratio obliqua; the imperfect or present infinitive or participle representing the present or imperfect indicative, and the aorist participle or infinitive the aorist indicative. Here *ἂν συλλαμβάνειν* represents what in direct speech would have been *συνελάμβανες ἄν* 'you would have been arresting them'. G. § 211.

23. **ἐξ ἐπιστολῆς τῆς Δημαρήτου** 'in accordance with a message from Demaratos'. There was a quarrel of some standing between the

two kings, from the fact that Demaratos had quitted Kleomenes in his expedition against Attica, when he went to revenge his failure to effect a revolution there by the expulsion of the Alkmaeonidae [5, 72—5].

27 3. **συνοισόμενος** 'about to meet with'. This is an unusual meaning for *συμφέρεσθαι* which generally stands for (1) 'to come together', (2) with adv. 'to turn out well or ill, (3) 'to accord with'. Abicht and Stein quote Homer *Iliad* 11, 736 *συμφερόμεσθα μάχῃ*.

CHAPTER LI.

5. **διέβαλλε** 'was engaged in trying to discredit'.

7. **οἰκίης τῆς ὑποδεεστέρης** 'of the junior royal house'. See Index s.v. Eurysthenes.

8. **κατὰ πρεσβυγενείην κως** 'but to a certain extent on the ground of being the elder'. For **κατὰ** cp. p. 1, l. 5.

CHAPTER LII.

10. **ὁμολογέοντες οὐδενὶ ποιητῇ** 'agreeing with no poet'. No works of any poet dealing with the return of the Herakleidae into the Peloponnese remain: but the story told by Apollodoros 2, 8, 2 is probably founded on some poet; and, according to him, Aristodemos was killed by lightning at Naupaktos when preparing to invade the Peloponnese; and his two sons Eurysthenes and Prokles, who drew lots with the other Herakleidae, Temenos and Kresphontes, obtained Lakonia.

12. **τοῦ Ὕλλου**, son of Herakles.

13. **ἐς ταύτην τὴν χώρην**, sc. Lakonia.

18. **ἐπιδόντα** 'having just lived to see'. Cp. the meaning of **ἐπιζῆν** in 1, 120 *εἰ ἐπέζωσε καὶ μὴ ἀπέθανε πρότερον*.

20. **κατὰ νόμον** 'according to custom'. Herodotos no doubt attributes to these heroic times the rule of primogeniture which existed as long as we know the history of the Peloponnese, and may be said perhaps to be founded on natural order.

21. **ποιήσασθαι**, notice the middle, 'to make *their* king'.

22. **οὐκ ὦν δή σφεας ἔχειν** 'they were not however able to decide which of the two they should choose'. The infinitives here and in l. 26 are negatived by *οὐ* according to the rule in indirect discourse, see p. 37, l. 8. **ὦν δή** = *δ' οὖν*, cp. l. 27. **ὥστε** = *ἅτε*, cp. 8, 118; 9, 37.

23. **γνῶναι** 'to ascertain' which was the elder.

ἢ καὶ πρὸ τούτου 'even before this occasion arose', **ἢ καὶ** *vel etiam.*

25. **καὶ τὸ κάρτα** 'quite well'. Cp. 4, 181 *καὶ τὸ κάρτα γίνεται ψυχρόν.*

26. **βουλομένην δὲ εἴ κως...** 'but wishing, if it were in any way possible, that both should become kings'. Cp. 9, 14 *θέλων εἴ κως τούτους πρῶτον ἕλοι.* Stein also illustrates the aorist in this phrase from 7, 128 *ἐδίδου τὰ βιβλία ὁ Βαγαῖος εἴ οἱ ἐνδεξαίατο ἀπόστασιν ἀπὸ Ὀροίτεω.*

27. **ὧν δή**, l. 22.

29. **πέμπειν ἐς Δελφούς.** Thus we find the Oracle consulted in a somewhat similar difficulty in regard to Demaratos c. 66. It is interesting to observe again and again the indications of the Oracle of Delphi being the general referee of all Greece, the one source to which all look for impartial advice and information in all difficulties great and small. When the belief in its fairness faded, it quickly fell into disrepute.

28

1. **ὅ τι χρήσωνται τῷ πρήγματι** 'how they were to manage the affair'. So 8, 20 the direct *οὐδὲν χρησάμενοι τοῖσι ἔπεσι* 'not being able to make anything of the verses'.

4. **δή** 'as I say', often used in summing up the result of previous sentences. Cp. p. 25, l. 23.

6. **ὑποθέσθαι** 'suggested an explanation', cp. p. 78, l. 13.

7. **εἶναι.** The verb in the relative sentence in an oratio obliqua may follow the mood of the main verbs, or be in the indicative for its own clause. Here the former construction is adopted. Cp. p. 80, l. 9.

9. **φυλάξαι** 'to watch and see'. The aorist is used because it is not meant to set a general watch upon the mother; but only to observe one particular and definite action. Cp. 5, 12 *φυλάξαι ὅτι χρήσεται τῷ ἵππῳ ἡ γυνή.*

10, 11. **ἢν κατὰ ταὐτὰ φαίνηται αἰεὶ ποιεῦσα** 'if she prove to invariably follow the same course'.

11. **τοὺς δὲ** 'they, on their part'. For *δὲ* in apodosis, cp. p. 15, l. 22; p. 31, l. 27. G. § 227, 2. **ἕξειν** 'ascertain', p. 38, l. 18.

12. **πλανᾶται καὶ ἐκείνη** 'but if she too varies in her practice'. For **πλανᾶται**, cp. p. 19, l. 18.

13. **ἐναλλάξ** 'first in one order and then in the other'.

14, 15. **σφεας τράπεσθαι** 'they must have recourse to some other plan'. The infin. depends on *ὑποθέσθαι*, l. 8.

15. **δὴ** 'accordingly', p. 3, l. 4.

18. **τὸν πρότερον** 'the elder'.

19. **ἐφυλάσσετο** 'she was being watched'.

21. **ἐν τῷ δημοσίῳ.** This may mean 'in the public state-house', or merely 'publicly', 'at the public cost and under the public protection'. This part of the legend may be founded on the institution at Sparta, attributed to Lykurgos, whereby all boys were taken from their mothers at 7 years old to be educated in common [Plutarch, *Lycurg.* 16]; while the king, when grown up, was supported and guarded by the state; living with the Polemarchs, and with three of the peers (*ὅμοιοι*) in constant attendance on him. [Xen. *Rep. Lac.* 13.]

24. **διαφόρους**, p. 12, l. 11.

25. **ὡσαύτως**, sc. *διαφόρους*.

CHAPTER LIII.

27, 28. **ταῦτα** 'what I have just stated'. **τάδε** 'what I am going to state'. **ὑπ' Ἑλλήνων**, by Greeks generally, as opposed to Lakedaemonians alone.

29. **γράφω** 'I will now write', present for immediate future; p. 48, l. 16. **δὴ**, p. 28, l. 4.

29 1. **τοῦ θεοῦ ἀπεόντος** 'omitting the God', i.e. Zeus father of Perseus. He means that if the pedigree is only carried as far as Perseus, these Dorian kings are true-born Greeks: beyond that all is doubtful.

2. **καταλεγομένους καὶ ἀποδεικνυμένους** 'correctly traced and declared to belong to'. The participle for infinitive after a verb *declarandi* understood from *λεγόμενα* or *γράφω*.

3. **ἤδη γὰρ τηνικαῦτα** 'for already by that time'. Cf. Plutarch *Nik.* 9 *ἤδη δέ που καὶ Ἀλκιβιάδης ἐνεφύετο τηνικαῦτα τοῖς Ἀθηναίοις.* But *τηνικαῦτα ἤδη* 'then for the first time' Arist. *Eccles.* 789.

ἐς Ἕλληνας ἐτέλεον 'began to be counted as Greeks', cp. p. 63, l. 3. **οὗτοι** Perseus and his generation, who settled in Argos. The descent of the Dorian kings beyond Herakles is

Perseus
|
Alkaios
|
Amphitryon
|
Herakles.

Herodotos wishes to show that the family were reckoned Greeks before Herakles.

5. **ἀνέκαθεν** 'farther back in the pedigree'.

οὐκ ἔπεστι...θνητοῦ 'no second name belonging to a mortal father is assigned to Perseus', i.e. no second (mortal) father is named as being the father of Perseus, as Amphitryon of Herakles.

8, 10. **μοι** dat. of agent with perfect passive, p. 17, l. 3. **ἀπὸ Δανάης ...καταλέγοντι** 'if you trace the pedigree from Danaë', i.e. by the mother's side. **φαινοίατο ἂν ἐόντες** 'would be shown to be'.

11. **Αἰγύπτιοι ἰθαγενέες** 'Egyptian in origin' 'true-born Egyptians', cp. 2, 17 *οὐκ ἰθαγενέα στόματα* (Νίλου) *ἀλλ' ὀρυκτά.*

CHAPTER LIV.

11. **μέν νυν**, p. 24, l. 13. **ταῦτα γεγενεηλόγηται** 'such is the genealogy made out in accordance with the account given by Greeks'. *γενεηλογέω* is a transitive verb. Cp. 2, 143 *Εκαταίῳ...γενεηλογήσαντι ἑωυτόν.*

13. **ὁ Περσέων λόγος.** Xerxes was said afterwards to have used the similarity of the name of Perseus to claim kindred with the Argives, and induce them to side with him against the Lakedaemonians: see 7, 61, 150.

14. **ἐγένετο Ἕλλην** 'became a naturalised Greek'.

15. **πατέρας** 'ancestors'. **ὁμολογέοντας...οὐδέν** 'had no connexion with Perseus in regard to kinsmanship', 'were not in any way connected by blood with Perseus', cp. 1, 142 *αὗται αἱ πόλεις τῇσι προτέρῃσι λεχθείσῃσι ὁμολογέουσι κατὰ γλῶσσαν οὐδέν.*

CHAPTER LV.

18. **καὶ...εἰρήσθω**, a formula for dismissing a subject, 'so much on that point'.

19. **ὅ τι**=*καθ' ὅτι* 'on what grounds' or 'how it came about', and in virtue of what achievements, 'that though they were Egyptians, they obtained the kingships over Dorians'. He speaks of 'kingships', because he is thinking not only of Sparta, but of Argos, Elis, and Messenia, which according to the legend were all first held by different members of the Herakleidae.

ἀποδεξάμενοι (*δείκνυμι*) 'having performed', a favourite word with Herodotos. Cp. 9, 71 *ἔργα ἀποδέξασθαι μεγάλα*.

20. **ἄλλοισι**, for the case, see l. 8. He is referring especially to Hekataeos, Pherekydes, and other λογοποιοί. For the position of **γάρ** see p. 6, l. 11.

21. **οὐ κατελάβοντο** 'did not embrace in their accounts', lit. 'did not occupy' or 'take possession of', as in p. 20, l. 12.

CHAPTER LVI.

24. **ἱρωσύνας δύο...οὐρανίου.** Lakedaemon, the eponymous hero of Lakedaemonia, was said to be a son of Zeus [Pausan. 3, 1, 1]; *Zeus Lakedaemonius* is therefore the representative of Lakonian nationality as Zeus Hellenius (9, 7) of Hellenic. *Zeus Uranius* is on the other hand the supreme God, and as such is approached by the head of the nation. The best illustration of the 'two priesthoods' will be the following passages from Xenophon: *Rep. Lac.* 13, 2, 'When the king is going out on an expedition, he and his staff first offer sacrifice at home to Zeus Agetor. If the sacrifices then are favourable, the fire-bearer (8, 8) takes fire from the altar and leads the way to the frontier; and there the king sacrifices to Zeus and Athenè'. *Id.* 15 'Lykurgos ordained that the king should offer all public sacrifices in behalf of the city, as holding direct from a God'. These however were special sacrifices offered on going out to war; for their regular sacrifices see next chapter.

25. **καὶ πόλεμόν γε...χώρην** 'and moreover should direct war into whatever territory they chose'. This does not appear to mean that they had power of declaring war; but that war being declared they had the entire control and direction of it. The checks upon their conduct of foreign affairs were (1) the necessity of the two acting together; (2) and later, at any rate, the necessity of the previous vote or consent of the Spartan Peers; (3) and later still, the rule that ten Councillors should attend a king in the camp, as well as the Ephors. Thucyd. 5, 63.

καὶ...γε 'and even'.

27. **εἰ δὲ μή** 'otherwise'. The Greek idiom retains the negative in this phrase, even though the idea requires no negation. Here he means 'if he does hinder him'. **ἐν τῷ ἄγεϊ** 'under the curse', i.e. he was outlawed. Cp. p. 52, l. 9. **διακωλυτήν.** A remarkable case of resistance to a movement ordered by Pausanias at Plataea is recorded

in 9, 53 sq. But then Pausanias was only a king's guardian and not a king himself.

28. **πρώτους.** So Xenophon [*R. L.* 13, 6] 'When the king is leading, if no one appears to oppose him, none marches in front of him except the Skiritae and the cavalry videttes'. [The Skiritae refer to the *Σκιρίτης λόχος* composed of 600 men of Skiritis, a district on the northern Lakonian frontier, who generally formed the 'forlorn hope' of the Lakonian armies. Thucyd. 5, 67, 68.]

29. **ἑκατὸν...λογάδας.** These hundred picked men were a part of 300 picked cavalry who had this special duty, see 7, 124. Thucyd. 5, 72, 4. A scheme for the yearly retirement of a certain number of them is mentioned in 1, 67.

1. **ἐπὶ στρατιῆς** 'when with the army', i.e. 'on service'. Others 30 read *στρατητης* 'expedition'.

προβάτοισι 'sheep' for offerings; as at home he had one out of every brood of pigs 'that he might never be at a loss for victims'. Xen. *R. L.* 15.

2. **ἐξοδίῃσι,** see Notes on Text.

3. **τὰ νῶτα** 'the chines', which had always been the share of the kings, see Hom. *Il.* 7, 321 where Agamemnon presents it as a special mark of favour to Ajax, *νώτοισιν δ' Αἴαντα διηνεκέεσσι γέραιρεν Ἥρως Ἀτρείδης.* Cp. *Odyss.* 4, 65.

CHAPTER LVII.

6, 7. **ἢν θυσίη...ποιέηται.** The king represented the whole people at a public sacrifice, cp. Xenoph. *Rep. Laced.* 15 *ἔθηκε γὰρ θύειν βασιλέα πρὸ τῆς πόλεως τὰ δημόσια ἅπαντα.* The word **δημοτελὴς** means 'to which all contribute', or 'in which all share', cp. Thucyd. 2, 15, 3 *ἑορτὴ δημοτελής.*

7. **πρώτους ἵζειν** 'that they (the kings) should sit in the place of honour'. The infinitive is dependent on *δέδοται.* With **ἄρχεσθαι** there is a change of subject, 'They (the people, or the servants) begin with them', in serving the food,—still however dependent on *δέδοται.*

9. **διπλήσια τὰ πάντα** 'double of everything'. So 9, 81 *πάντα δέκα* 'tenfold of everything'. Xen. *R. Lac.* 15 *καὶ διμοιρίᾳ γε ἐπὶ τῷ δείπνῳ ἐτίμησεν οὐχ ἵνα διπλάσια καταφάγοιεν, ἀλλ' ἵνα καὶ ἀπὸ τοῦδε τιμῆσαι ἔχοιεν εἴ τινα βούλοιντο* 'that they might from it send complimentary presents to any one'. See a case of a similar present by Themistokles, Plut. *Them.* 7. It does not seem to differ from *διπλήσια πάντα* in l. 23.

10. **σπονδαρχίας** 'right of pouring the first libation'. At banquets, such as followed a sacrifice, the *σπονδαί* were poured out at the end of the meal when the drinking began. **εἶναι τούτων** 'should be the privilege of these kings'.

11. **νεομηνίας ἀνὰ πάσας** 'every new moon'. **καὶ ἑβδόμας ἱσταμένου τοῦ μηνὸς** 'and the seventh day of every month'. The months were divided into three decades. In the first the days are reckoned as such and such a day *μηνὸς ἱσταμένου*, in the second as such and such *μηνὸς μεσοῦντος*, and in the last ten as such and such *μηνὸς φθίνοντος*. According to Hesiod (*Op.* 768) the 1st, 4th and 7th are holy days: the last as the birthday of Apollo.

13. **ἱρήϊον τέλεον** either (1) full grown, or (2) 'perfect', 'without blemish'. Cp. 1, 183; Thucyd. 5, 47, 8; Homer *Il.* 1, 65 *αἴ κέν πως ἀρνῶν κνίσης αἰγῶν τε τελείων βούλεται ἀντιάσας ἡμῖν ἀπὸ λοιγὸν ἀμῦναι.* From Xen. *R. L.* 15 it seems that the victims were often swine.

ἐς Ἀπόλλωνος, sc. *ἱρόν*, 'to take to the temple of Apollo'. For the pregnant use of *ἐς* see p. 1, l. 3.

14. **μέδιμνον...Λακωνικήν.** The Lakonian *μέδιμνος* = 1½ Attic, which latter = 11 gals. 4 pints English. Therefore the Lakonian medimnos = 17 gals. 2 pints. The **τετάρτη**, or quart, is a fourth part of some measure, but it seems uncertain what its capacity was. We find that the allowance for the king of barley-meal per month is the same as the amount contributed by each citizen to the *συσσίτια*, i.e. a *μέδιμνος*, Plutarch *Lycurg.* 12. It is reasonable therefore to suppose that the amount of wine was also the same, which Plutarch states at 8 *χόες*, which is two thirds of an Attic *μετρητής*. The Lakonian measures of capacity were therefore probably different, but we have not sufficient information to determine what they were.

The following explanation is founded on a comparison of Herodotos and Xenophon, but is offered only as a conjecture. We find in line 22, that the allowance to the king when absent from a banquet was 2 *χοίνικες* = $\frac{1}{24}$ of the monthly allowance of a *μέδιμνος*. Supposing then that the allowance of a *κοτύλη* of wine is the same proportion, and starting with the supposition that his monthly allowance of wine is the same also as the monthly contribution of each citizen (8 *χόες*), we may construct the following table:

1 *μετρητής* = 32 *χόες* (96 *κοτύλαι*),
1 *χοῦς* = 3 *κοτύλαι*.

Then the monthly allowance of 8 *χόες* will be a ¼ of a *μετρητής* (*τετάρτη*); and the allowance to the king in his absence from the banquet would be also $\frac{1}{24}$ of 8 *χόες* = ⅓ *χοῦς* = 1 *κοτύλη*. The allowance granted to the Spartans in Sphakteria was 2 *χοίνικες* of barley meal and 2 κοτύλαι of wine, Attic measure, which was probably smaller than the Lakonian, but we are not told the period for which this allowance was made. Thucyd. 4, 16.

15. **προεδρίας** 'place of honour', which was also one of the few marks of honour bestowed by the Spartans on strangers, cp. 9, 73. **ἐξαιρέτους** 'special'.

16. **προξείνους ἀποδεικνύναι** 'to appoint men to receive foreigners'. That is, to act in the name of the state in the reception of citizens of another state, with which Sparta had the relations known as προξενία. Sometimes such states chose their own πρόξενοι without entering into the formal relation with the other state as such. Individuals also had their ξένοι in various states. See on 9, 85.

τούτοισι προσκεῖσθαι 'is the function of these' (kings), cp. 1, 119 παρέφερον τοῖσι προσέκειτο τὴν κεφαλὴν τοῦ παιδός.

18. **Πυθίους.** See Index. The Pythii, besides their regular duty of visiting the Oracle at Delphi, acted, according to Müller (*Dorians* 2, p. 15) as assessors to the kings in the Gerusia. Cp. Xen. *R. L.* 15 ἔδωκε δ' αὖ καὶ συσκήνους δύο ἑκατέρῳ προσελέσθαι, οἳ δὴ καὶ Πύθιοι καλοῦνται.

19. **σιτεόμενοι** with acc., cp. 3, 98 σιτέονται ἰχθῦν.

20. **ἐπὶ τὸ δεῖπνον,** i.e. to the syssitia.

22. **δύο χοίνικας,** that is, $\frac{1}{24}$ of the month's allowance to the kings. See on l. 14.

23. **διπλήσια πάντα** does not seem to differ from τὰ πάντα in l. 9. **τὠυτὸ δὲ τοῦτο** 'and on this same principle', i.e. by having a double share, and occupying the place of honour. For omission of preposition cp. Plat. *Protag.* ἀλλ' αὐτὰ ταῦτα νῦν ἥκω παρά σε.

27. **πατρούχου...ἔχειν** 'in the case of a virgin who has succeeded to her father's property, as to whom it belongs to have her to wife'. Thus in Athens the ἄρχων ἐπώνυμος, as representative of the judicial character of the ancient kings, had the guardianship of orphan heiresses. The object in the case of Sparta was to prevent the accumulation of property in single families by free marriages of heiresses. The principle of the equal division of land in Lakonia however gradually disappeared, in spite of this and other precautions, so that by the middle of the 3rd century B.C. the number of landowners was said to be only 100. See Plutarch *Agis* 5. Thirlwall's *History of Greece*, vol. 8, p. 132.

28. For ἰκνέεται = προσήκει cp. 9, 26; p. 35, l. 24; p. 47, l. 8.

ἢν μή περ....ἐγγυήσῃ 'unless her father shall have betrothed her', i.e. in his lifetime, or by will.

29. **θετὸν...ποιέεσθαι** 'to adopt'. Such adoptions would be under the authority of the king also, on account of the rights of property which they would involve; just as in Rome they were performed before the *curiae*.

31 2. **τοῖσι γέρουσι.** The members of the Gerusia were appointed for life (Polyb. 6, 45), and were not less than 60 years old (Plut. *Lycurg.* 10). They were elected by all the full citizens, and formed a high court in criminal cases involving capital sentences. Their power, however, was limited by the subsequent enlargement of the functions and influence of the Ephors.

3. **τοὺς μάλιστα...προσήκοντας** 'those of the Gerusia most nearly related to them'. Cp. p. 75, l. 9.

5. **δύο ψήφους τιθεμένους** 'giving two votes for the kings'. The words are ambiguous, leaving it doubtful whether Herodotos means that each king was represented by a member of the Gerusia, who gave two votes in his name and one for himself, or whether he means that such member or members gave two votes for the two kings, and then their own. It was the false impression perhaps caused by Herodotos' words, taken in the former sense, that Thucydides contradicts in 1, 20; as he does also the notion as to the Pitanetan Lochos mentioned by Herodotos in 9, 53.

CHAPTER LVIII.

8. **ἐκ τοῦ κοινοῦ**, p. 8, l. 20.

13. **καταμιαίνεσθαι** 'to assume signs of deep mourning'. A solitary instance of such a use of the word, which means 'to defile', 'to deface', and it must be taken to include the cutting of the hair (p. 11, l. 8), and perhaps tearing of the clothes (3, 66) and defiling the head with dust, as well as the wearing of black clothes [Paus. 4, 14, 3 *προείρητο...ἐπὶ τὰς ἐκφορὰς τῶν βασιλέων...ἄνδρας ἐκ τῆς Μεσσηνίας καὶ τὰς γυναῖκας ἐν ἐσθῆτι ἥκειν μελάνῃ καὶ τοῖσι παράβασιν ἀπεκεῖτο ποινή*]. Such signs of mourning were forbidden to the Spartans for the loss of private friends or relations: they marked a kind of consecration of the kings to heroship (Xen. *R. L.* 15).

17. **γὰρ ὦν** 'for, as is well known', 'for of course'.

20. **χωρὶς Σπαρτιητέων** 'over and above Spartans', 'not counting Spartans', who had of course to go too.

ἀριθμῷ 'in a fixed number', cp. Thucyd. 2, 72, 6 *παράδοτε...δένδρα ἀριθμῷ τὰ ὑμέτερα καὶ ἄλλο εἴ τι δυνατὸν ἐς ἀριθμὸν ἐλθεῖν.*

21. **τῶν περιοίκων** 'of the Perioeci', the free but unenfranchised inhabitants of Lakonia.

24. **κόπτονται** 'they beat their breasts'.

25. **φάμενοι.** This middle participle of *φημί* is rare in Attic. **αἰεὶ** 'from time to time'.

26. **τοῦτον δὴ** 'that he of course'. Here *δὴ* like *adeo* emphasises the word which it follows.

27. **τούτῳ δὲ** 'in his honour', see on p. 11, l. 13. For **δὲ** in apodosis, see p. 15, l. 22.

28. **εὖ ἐστρωμένη** 'with rich cushions and coverings'.

29. **ἀγορὴ** 'public business' such as is transacted in the *ἀγορή*, of whatever kind, cp. p. 6, l. 9. **δέκα ἡμερέων**, gen. of the time within which. See p. 7, l. 15; p. 9, l. 13.

1. **ἀρχαιρεσίη** 'a meeting for electing to offices', such as Ephors, or to fill up vacancies in the Gerusia. 32

CHAPTER LIX.

Stein thinks that this and the following chapter were added at a later time to the text.

4. **ἐνίστηται** 'is entering upon his reign'. Cp. 3, 67 *ἠπιστέατο Σμέρδιν τὸν Κύρου βασιλέα ἐνεστεῶτα* 'they understood that Smerdis son of Kyros had become king'.

6. **τῷ δημοσίῳ** 'the treasury', p. 30, l. 13.

8. **φόρον μετίει** 'remits the tribute due'. Thus, on the death of Cambyses, the Magus Smerdis sent round to every nation granting a remission of military service and tribute for three years (3, 67).

CHAPTER LX.

9. **καὶ τάδε** 'in the following respects also with the Egyptians'. **καὶ** refers to the whole clause, not to *τάδε* alone.

11. **τὰς πατρωΐας τέχνας.** On the various *γένεα* or 'castes' of the Egyptians, devoted to various trades or callings, see 2, 164, where seven are mentioned. Other writers have stated the number as three,

counting the priestly and warlike as two, and all the others under the general head of artificers; others count five, and others six. Herodotos here seems to mean that the system was carried on in regard to all callings in life, of which he instances three, which happen to correspond with some hereditary trades in Sparta; where the αὐλητής as giving time to war dances and marches; the cooks as furnishing the syssitia; and the heralds as being closely connected with military expeditions, and authorities on matters of international etiquette or law, would be important persons. The family of heralds were called Talthybiadae, and were believed to be descended from Talthybios, herald of Agamemnon, 7, 134.

13, 14. **οὐ κατὰ...παρακληΐουσι** (for the absence of a conjunction cp. p. 2, l. 15) 'others cannot on the score of possessing a loud voice take up this calling and shut them out'. **ἐπιτιθέμενοι.** (1) 'applying oneself to', sc. *τῇ κηρυκηΐῃ*. Cp. 1, 1 *ναυτιλίῃσι μακρῇσι ἐπιθέσθαι*. 1, 96 *δικαιοσύνην ἐπιθέμενος ἤσκεε*. (2) 'attacking', cp. p. 63, l. 5.

15. **ἐπιτελέουσι**, sc. *τὰς τέχνας*, 'they practise these arts'.

CHAPTER LXI.

16. **μὲν δή**, p. 13, l. 22.

18. **προεργαζόμενον** 'working *for* Greece', as in 2, 158 fin. *τῷ βαρβάρῳ αὐτὸν* (*ὀρύσσοντα*) *προεργάζεσθαι*. It is the verb expressing the meaning of *προὔργου* 'serviceable'.

διέβαλε, sc. *πρὸς Αἰγινήτας* 'he decried Kleomenes to the Aeginetans', by this message or letter, see c. 50.

19. **φθόνῳ καὶ ἄγῃ** 'jealousy and envy'. Cp. 8, 69 *ἀγαιόμενοι καὶ φθονέοντες αὐτῇ*.

21. **ἐβούλευε** 'began to entertain the design', p. 27, l. 20.

διὰ πρῆγμα τοιόνδε 'by the help of the following circumstance', 'using the following circumstance as a basis of his attack'.

22. **ἐπίβασιν ἐς αὐτὸν ποιεύμενος** = *ἐπιβαίνων αὐτῷ* 'attacking him'. Nearly any verb can be represented by *ποιεῖσθαι* and the cognate substantive. See Index. The use of *ἐς* with a verb expressing hostility is peculiar, and must be accounted for by considering the literal meaning of *ἐπίβασις* 'a stepping upon', and 'approaching', *aggrediens*.

33 1. **καὶ οὐ γὰρ** 'and because he did not think etc.'. For the position of *γὰρ* introducing the reason before the clause stating the resulting action see p. 6, l. 11; and Index.

4. **τῷ προσεκέετο** 'of whom he was fond', cp. 1, 123 Κύρῳ...προσέκειτο ὁ Ἅρπαγος. Thucyd. 7, 50, 4 θειασμῷ τε καὶ τῷ τοιούτῳ προσκείμενος. For a different meaning see p. 30, l. 17.

7. **καὶ ταῦτα** 'and that too'. **ἐξ αἰσχίστης** 'after being the ugliest possible'.

8. **ἐοῦσάν μιν.** We may regard this as accusative after ἐφόρεε, the construction being interrupted by the parenthetical clause οἷα...τοιάδε, or after ὁρέουσα understood from the next clause.

9. **οἷα** 'in view of the fact that she was'.

θυγατέρα ἐοῦσαν follows the case of ἐοῦσάν μιν in l. 8. **τε...καὶ** (l. 10) connect the clauses which give the reasons for the nurse's action.

10. **πρὸς δὲ** 'and besides', cp. μετὰ p. 2, l. 18.

11, 12. **συμφορὴν...ποιευμένους** 'regarding her appearance in the light of a misfortune', 'being greatly vexed at her appearance', p. 49, l. 26. Cp. 9, 77 συμφορὴν ἐποιεῦντο, 5, 90 συμφορὴν ἐποιεῦντο διπλόην.

ταῦτα ἕκαστα, sc. (1) that she was an ugly child of rich parents, (2) that her parents were distressed at it. **μαθοῦσα** 'understanding'. It almost amounts to 'for these two reasons': cp. τί μαθὼν 'why?'

13. **ἐς τὸ τῆς Ἑλένης ἱρόν**, as a heroine renowned for perfect beauty, Helen might be expected to bestow beauty on her worshippers.

15. **ὅκως δὲ ἐνείκειε** 'and whenever she had brought the babe there', cp. 1, 17 ὅκως τὸν καρπὸν διαφθείρειε. The optative of indefinite frequency with ὅκως = ὁπότε, p. 16, l. 9, Madv. § 133.

16. **ἵστα...ἐλίσσετο** 'was accustomed to place', 'was accustomed to pray'.

17. **καὶ δή κοτε** 'and so once', for καὶ δή in continuance of a story cp. p. 6, l. 9; p. 11, l. 13.

19. **φέρει** 'what she is carrying'; for the use of this form, so soon after the other form ἐφόρεε (l. 12) and with φορέει in l. 20, Stein quotes other examples from 1, 133; 5, 25; 9, 11. The difference between the two is perhaps slight, but φορέω seems generally to be used with a frequentative sense, such as 'to wear' armour etc. Cp. also 2, 73 κομίζειν δὲ οὕτω· πρῶτον τῆς σμύρνης ᾠὸν πλάσσειν, ὅσον τε δυνατός ἐστι φέρειν, μετὰ δὲ πειρᾶσθαι αὐτὸ φορέοντα..., where the three verbs are used in a slightly different sense 'to convey' to a place, 'to support the weight of', and, 'to carry about'. Here the woman asks 'what have you got in your arms?' The nurse replies that she is 'carrying (as her habitual duty) a child'.

21, 22. **οὐ φάναι** 'refused', 'said 'no''. **ἀπειρῆσθαι...μηδενὶ** 'had been forbidden to show it to anyone'. For *μὴ* after negative verbs, see Madvig § 210. **ἐκ τῶν γειναμένων** 'by the parents', cp. p. 11, l. 20: p. 22, l. 13.

22. **πάντως** 'in spite of everything', 'at all hazards'.

24. **περὶ πολλοῦ ποιευμένην** 'making a great point of seeing the child', cp. p. 3, l. 25.

ἰδέσθαι = *ἰδεῖν*, rare in Attic prose.

οὕτω δὴ 'in these circumstances', cp. p. 8, l. 25.

26. **εἶπαι** for *εἰπεῖν* is confined to Herodotos and late Attic.

27. **δή**, see l. 17.

28—30. **δὴ...δὴ** refer back to line 3, *ὧδε γαμέει* and *ἦν οἱ φίλος*. They may be translated, 'and so, as I said': and **οὗτος δή...** 'this was the man I mentioned as the friend of Ariston'.

CHAPTER LXII.

30, 31. **τὸν δὲ...ὁ ἔρως** 'now it appears (*ἄρα*, cp. 8, 8) that a passion for this woman was exciting Ariston'. **ἄρα** like *γὰρ* in l. 1 gives the reason for what follows. **γυναικὸς** is an objective genitive. **ἔκνιζε.** Herodotos uses this word three times in the meaning of 'to excite', 'to irritate' (7, 10, 5; 7, 12), but otherwise it seems wholly confined to Poetry in this metaphorical sense. Cp. Theocr. 4, 59 *τὰν κυάνοφρυν ἐρωτίδα τᾶς ποκ' ἐκνίσθη*. 5, 122 *κἠγὼ μὲν κνίσδω τινά*.

34 1. **δὴ** 'accordingly', p. 3, l. 4; p. 13, l. 25.

4. **τὴν ὁμοίην**, sc. *δωτίνην*, but cp. p. 11, l. 5.

5. **ἀμφὶ τῇ γυναικί**, p. 75, l. 15. This use of *ἀμφὶ* is poetical. Madvig § 72. It is nearly synonymous with *περὶ* (see other uses with dat. in 1, 140; 5, 19, 52) and gradually disappeared from use: not, for instance, occurring at all in Polybios.

6. **ἐπὶ τούτοισι** 'upon the above terms'.

7. **ἐπήλασαν**, sc. **ἀλλήλοις**, 'they imposed strong oaths on each other', cp. 1, 146 *σφίσι αὐτῇσι ὅρκους ἐπήλασαν*. It is a forcible word. *ἐλαύνω* and its compounds are used to describe many things connected with the exertion of force: to ride, to row, to lead an army, to march, to hammer, to build, and many other acts of similar nature, may be translated by it. **μετὰ δὲ** adverbial, p. 2, l. 18.

8. **ὅ τι δὴ ἦν** 'whatever it was', cp. Hom. *Il.* 13, 446 *ἦ ἄρα δή τι εἴσκομεν ἄξιον εἶναι Τρεῖς ἑνὸς ἀντὶ πεφάσθαι*; 'in any way', p. 78, l. 18.

9. **τὴν ὁμοίην**, see l. 4 and p. 11, l. 5.

10. **ἐνθαῦτα δὴ**, see p. 7, l. 22.

11. **ἐπειρᾶτο ἀπάγεσθαι** 'tried to take away for himself'. For construction of *πειρᾶσθαι*, see p. 3, l. 10; p. 81, l. 25. Notice the force of the middle *ἀπάγεσθαι* and of *φέρεσθαι* in l. 10.

12. **ἔφη καταινέσαι** 'he said he consented'. This aorist stands where perhaps a future infinitive would be used in English: but the Greek use of the instantaneous aorist must be observed; it is especially appropriate to such words as *κατῄνεσα* or *ἐπῄνεσα* used to express a single and complete act of assent. See Donaldson's *Greek Grammar*, pp. 417, 418.

13. **τῇ παραγωγῇ** 'the deception which the trick had put upon him'. **τῆς ἀπάτης** is subjective, and *παραγωγή* has the meaning of *παράγειν* 'a leading on one side', 'a leading astray'.

CHAPTER LXIII.

15. **οὕτω δὴ**, p. 33, l. 24.

ἐσηγάγετο 'took to himself', cp. 5. 39 **ἄλλην** *ἐσαγαγέσθαι*.

18. **τοῦτον δὴ** 'this very Demaratos'. For *δὴ* with reference to a previous mention, see p. 33, l. 28.

19. **ἐν θώκῳ κατημένῳ** 'sitting in council'. Pausanias who tells the story (3, 7, 7) says *μετὰ τῶν ἐφόρων κατημένῳ τηνικαῦτα ἐν βουλῇ ἦλθεν οἰκέτης κ.τ.λ.*

22. **ἐπὶ δακτύλων** 'on his fingers', not *ψήφοις* 'with counters', cp. Aristoph. *Vesp.* 655 *καὶ πρῶτον λογίσαι φαύλως μὴ ψήφοις ἀλλ' ἀπὸ χειρός.* Hence the word *πεμπάζειν* or *πεμπάζεσθαι* 'to count', *Odyss.* 4, 412.

συμβαλλόμενος 'reckoning', cp. 2, 31 *τοσοῦτοι γὰρ συμβαλλομένῳ μῆνες εὑρίσκονται.*

23. **ἀπομόσας** 'denying on oath', p. 36, l. 1.

24. **πρῆγμα οὐδὲν ἐποιήσαντο** 'they took no especial notice of it', 'they did not regard it as important', cp. 7, 150.

26. **τὸ εἰρημένον μετέμελε** 'repented him of what he had said'. For *μεταμέλει* with nom. of thing repented of, cp. 9, 1, *τοῖσι τὰ πρὸ τοῦ πεπρηγμένα μετέμελε οὐδέν.* A more common construction is with participle, e.g. 1, 130 *μετεμέλησέ σφι ταῦτα ποιήσασι.*

27. **ἐς τὰ μάλιστα**, p. 51, l. 18.

35 2. **διὰ πάντων δὴ** 'above every single king', cp. 1, 25 *θέης ἄξιον διὰ πάντων τῶν ἐν Δελφοῖσι ἀναθημάτων.*

3. **ἀρὴν ἐποιήσαντο...γενέσθαι** 'offered a prayer for the birth of a son'. Demaratos therefore means 'prayed for by the people'. Rawlinson compares Louis le Désiré.

CHAPTER LXIV.

6. **ἔδεε** 'it was fated', cp. p. 79, l. 11; 5, 33 *καὶ οὐ γὰρ ἔδεε τούτῳ τῷ στόλῳ τὴν Νάξον ἀπολέσθαι.* 9, 109 *τῇ δὲ κακῶς γὰρ ἔδεε πανοικίῃ γενέσθαι.*

7. **ἀνάπυστα** 'generally known', cp. 9, 109.

8. **δι' ἃ** [a conjecture for the MS. reading *διὰ τὰ*, see Append. C. II., note 2] 'wherefore', i.e. because it was fated that such should be the case. But the reading is not satisfactory.

Κλεομένεϊ διεβλήθη 'incurred the dislike of Kleomenes'. Cp. 5, 35 *Μεγαβάτῃ διαβεβλημένος* and 9, 116 *Ξέρξεα διεβάλετο* 'he aroused the suspicions or anger of Xerxes'.

9. **πρότερον**, see 5, 75. It was in an invasion of Attica connected with an attempt to restore the Peisistratids.

10. **καὶ δὴ καί**, see p. 11, l. 13; p. 26, l. 6. **τότε**, see c. 50. **ἐπ' Αἰγινητέων τοὺς μηδίσαντας** 'to fetch away the medizers among the Aeginetans'; cp. p. 9, l. 18.

CHAPTER LXV.

14. **οἰκίης τῆς αὐτῆς**, sc. of the Eurypontids, or junior house. Elsewhere this Agis is called Agesilaus (8, 131). **ἐπ' ᾧ τε** 'on condition that', see on p. 5, l. 7.

17. **ἁρμοσαμένου** 'having betrothed to himself', 5, 32 *τοῦ τὴν θυγατέρα ἡρμόσατο Παυσανίης.*

20. **φθάσας...ἁρπάσας** 'having anticipated him in carrying off'. The ancient custom at Sparta, whereby a bridegroom carried off his bride by force, generally from the chorus of maidens, was still in use. See Plut. *Lycurg.* 15 *ἐγάμουν δι' ἁρπαγῆς.* The practice was probably in some way mitigated by convention; and the offence of Demaratos was that he availed himself in full of the old custom. See Müller's *Dorians*, vol. 2, p. 293, Engl. Tr.

23. **κατόμνυται Δημαρήτου** 'took oath against Demaratos'. In p. 38, ll. 10, 13 it is used without an object, ='to swear earnestly',

cp. *καταδοκέω*. [Stein reads *Δημαρήτῳ*, but against some of the best MSS., comparing the construction of *καταγελᾶν* in 3, 37.]

24. **ἰκνεομένως** 'rightfully', see p. 30, l. 28; p. 47, l. 8.

25. **μετὰ δὲ τὴν κατωμοσίην** 'after making the formal accusation on oath'. **ἀνασώζων** 'recalling to memory'.

1. **ἀπώμοσε** 'denied on oath', p. 34, l. 23. **36**

2. **ἐπιβατεύων** 'taking his stand upon'. In 3, 63 *ἐπιβατεύειν οὐνόματος*, 'to usurp a name'.

3. **ἀπέφαινε** 'tried to prove'.

4. **οὔτε...οὔτε**, with the participle *οὐ* not *μὴ* is the regular construction.

6. **πάρεδροι** 'sitting in council with him'.

CHAPTER LXVI.

11. **ἀνοίστου γενομένου** 'the matter having been referred (*ἀναφέρω*) by the deliberate design of Kleomenes'.

12. **προσποιέεται** 'he won over to his interest'. The Dorian inhabitants of Delphi were always closely connected with Sparta, on whose influence they depended for the maintenance of their claim to manage the temple of Delphi, as against the officers of the Phokian League, who represented the Aeolic and Achaean inhabitants of Phokis. Thus a Spartan king would be sure to have many men in Delphi on whom he could put pressure.

14. **τὴν πρόμαντιν** 'the Pythia', the girl who uttered the oracles which the *προφήτης* reduced to writing, generally in metre, and delivered to the applicants; see 7, 111; 8, 135 where the two are mentioned.

15. **ἀναπείθει** 'bribed', see p. 71, l. 21 and on 9, 33; 5, 63 *ἀνέπειθον χρήμασι*.

20. **ἐπαύσθη** 'was deposed'. Other instances of the Pythia being bribed are mentioned in 5, 63, 89, 90. Thucydides 5, 16.

CHAPTER LXVII.

21. **δή**, p. 28, l. 4.

23. **ἐκ τοιοῦδε ὀνείδεος** 'in consequence of the following insult'.

2. **ἦρχε αἱρεθεὶς** 'was holding an elective office', apparently one of **37** those connected with the management of the boys, for an account of which see Müller's *Dorians*, vol. 2, p. 310. Some have thought that he was an Ephor, which I think Baehr has given good reasons for

disbelieving. **γυμνοπαιδίαι** a festival held of great importance in Sparta, consisting of choruses of boys and men dancing and singing the warlike songs of Thaletas, Alkman and Dionysodoros. According to Pausanias it took place in the Agora near the statues of Apollo, Artemis and Leto (3, 11, 9); but it seems at one time to have been held at Thyrea, whence the garlands which served as prizes at it were called *Θυρεατικοί* [Athenaeus 678 B, C]. It took place at Midsummer.

5. **ἐπὶ γέλωτί τε καὶ λάσθῃ** 'by way of ridicule and insolence', cp. *ἐπὶ γέλωτι* 9, 82. The word *λάσθη* is rare, only occurring again in an epigram of the Samian poet Aischrion [Athen. 135 C], and in Aelian (Suid. s. v.) who joins it with *ὕβρις* and *γέλως* also. Hesychios has *λᾶσθαι· παίζειν· ὀλιγωρεῖν. λασθαίνειν· κακολογεῖν.* It seems to be connected with *λάσιος*, cp. *κομᾶν* 'to be insolent'.

6. **ὁκοῖόν τι εἴη** 'what it felt like to be a magistrate after being a king', cp. 1, 129 *εἴρετό μιν...ὅτι εἴη ἡ ἐκείνου δουλοσύνη ἀντὶ τῆς βασιληΐης.*

8. **οὐ**, sc. **πεπειρῆσθαι.** "The infinitive in indirect discourse regularly has *οὐ*, to retain the negative of the direct discourse". G. p. 308.

10. **μυρίης** 'very great', 'immense', cp. 2, 136 *μυρίη ὄψις.* For *ἄρξαι* 'to begin' with gen. cp. 1, 2 *ἄρξαι ἀδικημάτων.*

κακότητος 'misfortune', 8, 109.

11. **κατακαλυψάμενος** 'having wrapped his mantle closely round his head'; as a sign of vexation or shame, cp. *Od.* 8, 92 *ἂψ Ὀδυσεὺς κατὰ κρᾶτα καλυψάμενος γοάασκεν. id.* 10, 53 *καλυψάμενος δ' ἐνὶ νηῒ κείμην.* Cp. *περικαλυπτέα, ἐγκαλυψάμενος*, Arist. *Nub.* 727, 735. Thus Phaedra in her shame says *μαῖα, πάλιν μου κρύψον κεφαλάν· αἰδούμεθα γὰρ τὰ λελεγμένα μοι.* Eurip. *Hipp.* 243.

12. **ἔθυε** 'he set about sacrificing'.

CHAPTER LXVIII.

14. **ἐσθεὶς ἐς τὰς χεῖρας...σπλάγχνων** 'putting some of the entrails into her hands', as a solemn adjuration to her to speak the truth. This was a custom at a sacrifice, in order solemnly to connect all attending it with the obligations it symbolised, or the future predicted from the entrails of the victim. See Polyb. 6, 11, where the entrails are brought to king Philip (*κατὰ τὸν ἐθισμὸν*) and handed by him to others. Stein also quotes Lykurgos *v. Leocrit.* 20 *ἀξιοῦντε τοὺς μάρτυρας...λαβόντας τὰ ἱερὰ κατὰ τὸν νόμον ἐξομόσασθαι.*

16. **σε.** Obs. the place of the pronoun in such solemn appeals.

Cp. Eur. *Hipp.* 605 ναὶ πρός σε τῆς σῆς δεξιᾶς. **καταπτόμενος** 'appealing to', cp. 8. 65 Δημαρήτου τε καὶ τῶν ἄλλων μαρτύρων καταπτόμενος.

17. **ἑρκείου Διὸς** 'Zeus the god of our household', lit. the God of the ἕρκος or front court, where there would be a statue of him.

τοῦδε 'about this'; objective genitive.

19. **ἐν τοῖσι νείκεσι** 'in the course of his suit against me', cp. Hom. *Od.* 12, 440 κρίνων νείκεα πολλὰ δικαζομένων αἰζηῶν.

20. **ἐλθεῖν παρὰ Ἀρίστωνα** 'came to Ariston's house', 'married Ariston'.

23. **μετέρχομαι** 'I beseech', cp. p. 38, l. 2. The word is used in four connexions by Herodotos, (1) threats, p. 49, l. 31, (2) prayers, as here, (3) worship, 4, 7, (4) punishment, p. 50, l. 5: all deducible from the original meaning 'to pursue' or 'come after'.

τῶν θεῶν 'in the name of the Gods', usually πρὸς τ. θ.

26. **πολλὸς**, *creber*, 'prevalent'.

27. **τεκεῖν ἄν**, the infinitive with ἂν in indirect discourse, for the aorist indicative with ἂν in direct, p. 26, l. 21.

CHAPTER LXIX.

2. **μετέρχεαι**, p. 37, l. 23. 38

3. **ἐς σὲ**, cp. p. 50, l. 14. Another instance of Herodotos' use of ἐς nearly allied to that of πρός, cp. p. 1, l. 13; p. 10, l. 10.

4. **ἠγάγετο ἐς ἑωυτοῦ**, *me duxit*, 'took me to his house as his wife'.

5. **εἰδόμενον Ἀρίστωνι** 'in the shape of Ariston', cp. 7, 56 Ζεὺς ἀνδρὶ εἰδόμενος Πέρσῃ.

10. **κατωμνύμην** 'protested with an oath', cp. p. 35, l. 23. **οὐ καλῶς**, *male*, the negative is closely connected with the adverb, not with the infinitive, cp. p. 5, l. 20.

13. **ἔμαθε** 'he understood'.

14, 16. **τοῦτο μὲν...τοῦτο δὲ** 'in the first place'...'in the second place', p. 23, l. 14.

15. **ἡρωΐου** 'chapel of a hero', or deified man. See note on 9, 25, and on p. 20, l. 3. **θύρῃσι...αὐλείῃσι** 'the front door', leading from the street into the first αὐλή or court of the house.

17. **εἶναι**, sc. τὸν συνευνηθέντα.

18. **ἔχεις** 'you are in possession of', 'you know', cp. p. 28, l. 11. **ὅσον τι καὶ βούλεαι** 'as much as you can possibly wish to know'.

25. **οὐδέκω**, p. 34, l. 22.

26. **ἀπέρριψε τὸ ἔπος** 'threw out this insulting remark', cp. 8, 92 *ταῦτα...ὁ Πολύκριτος ἀπέρριψε ἐς Θεμιστοκλέα.*

39 1. **λόγους** 'stories', 'account'.

CHAPTER LXX.

7. **ἐπόδια** 'journey money'. As the currency in Lakonia was iron and not useable abroad, Demaratos doubtless provided himself with other money—silver or gold, which as king he would be able to get from the exchequer, but his doing so probably could not be concealed and aroused suspicion.

8. **τῷ λόγῳ φὰς** 'giving out as his ostensible object'; for no member of the royal family might leave Sparta for permanent residence abroad (*ἐπὶ μετοικισμῷ*) under penalty of death. Plut. *Agis* 11.

11. **ἔφθη διαβὰς** 'crossed before they could catch him'. He crossed to Zakynthos, as being inhabited by Achaeans, who were not likely to assist the Spartan Government.

12. **ἐπιδιαβάντες** 'having crossed over to Zakynthos in pursuit'. **αὐτοῦ ἅπτοντο** 'were laying hands on Demaratos himself', 3, 137 *εὑρόντές μιν ἀγοράζοντα ἅπτοντο αὐτοῦ.*

13. **ἀπαιρέονται**, the historic present for an aorist. They actually took away his slaves, but they only *tried* to arrest him (imperfect).

14. **μετὰ δὲ** adverbial, p. 6, l. 8; **γὰρ** introducing the reason before stating the action, see p. 33, l. 1.

16. **ὁ δὲ ὑπεδέξατο...ἔδωκε.** The king gave him Pergamum, Halisarta and Teuthrania, where his descendants were settled in the time of Xenophon (*Anab.* 2, 1, 3), and one was married to a daughter of Aristotle. (St.) Compare the king's liberal conduct to the son of Miltiades, c. 41. The gift of a *πόλις* was practically a gift of an income arising from a certain share in the taxation and other dues.

18. **καὶ τοιαύτῃ...ἀπολαμπρυνθεὶς** 'and in this position of disgrace, after having, among other distinguished services to Sparta in field and council, won to its honour an Olympic prize with his four-horse chariot'.

19. **Λακεδαιμονίοισι ἀπολαμπρυνθεὶς** 'having become renowned in the eyes of the Lakedaemonians', cp. 1, 41.

20. **ἐν δὲ δὴ** 'and among them, as we know', cp. p. 26, l. 17. **σφι...προσέβαλε**, sc. *Ὀλυμπιάδα*, 'brought home to them an Olympic

victory'. Readers of Pindar know well how the success of the Olympic victor was held to reflect its chief glory on his country.

22. **πάντων δὴ**, emphatic. 'Of all the kings that ever reigned in Sparta'.

CHAPTER LXXI.

1. **διεδέξατο** 'received in succession to him', cp. 8, 141 διαδεξάμενοι, sc. λόγον. 40

2. **μετεξέτεροι** = ἔνιοι 'some', a word peculiar to Ionic, cp. 1, 95; 8, 7, 87, etc.

8. **ἔρσεν**, App. II. 1 (*a*).

9. **τὴν Ἀρχίδημος...γαμέει.** Archidamos therefore married his aunt, though younger than himself. The marriage of uncle and niece was common in Greece. See 5, 39; 7, 239; Lysias *Orat.* 32 § 4.

δόντος, i.e. having given her willingly. Archidamos did not carry her off, as in the case mentioned in c. 65.

CHAPTER LXXII.

13. **ἐς Θεσσαλίην.** The expedition into Thessaly was for the purpose of punishing the medizing party in that country. Being generally the rich men, such as the Aleuadae (9, 58), they would be well able to offer the Spartan king a bribe. Pausanias 3, 7, 9.

παρεὸν 'when it was in his power'.

This absolute use of the neuter participle is commonest with impersonal verbs, but is not confined to them; see p. 25, l. 13. We have also for instance παρασχόν Thucyd. 1, 120, παρέχον Herod. 5, 49; προσταχθὲν Lysias *Orat.* 30 § 2; αἰσχρὸν ὂν Xen. *Cyrop.* 2, 2, 20; ἀπόρρητον Soph. *Antig.* 44. To be distinguished from this usage is that of participles both nominative and accusative with ὡς: for instance, Demosthenes 14, 15 ἀπεβλέψατε πρὸς ἀλλήλους ὡς αὐτὸς μὲν ἕκαστος οὐ ποιήσων τὸ δόξαν, τὸν δὲ πολέμιον πράξοντα. The Grammarians have been divided in their explanation. Matthiae § 564 and Donaldson § 445 call it the 'nominative absolute', Madvig § 182, Goodwin § 278, Clyde § 64 the 'accusative absolute'. There is no certain means of settling the question, since it occurs only in the neuter. It does not occur in writers earlier than Herodotos: and Clyde seems justified in looking upon it as a step towards indeclinability, from the synthetic to the analytical stage in language.

14. **ἐδωροδόκησε ἀργύριον πολλόν** 'he received a large sum of money as a bribe'. *δωροδοκέω* (*δῶρον δέχομαι*) is used absolutely at p. 45, l. 16, and *ἀργύριον πολλὸν* must be regarded as a cognate accusative. Susceptibility to bribes was a common weakness in Greece; and the Spartan kings seem to have been especially open to the temptation, owing no doubt to the severe restrictions which their laws placed on the possession of money. See 3, 148.

15. **ἐπ' αὐτοφώρῳ** 'in the very act', a legal term, like the old English law term 'with the manoir', cp. p. 80, l. 26. **αὐτοῦ** 'on the spot', i.e. not after returning to Sparta.

16. **χειρὶς** 'a sleeve', hanging loose to the outer garment, and into which the hands could be thrust for warmth when desired. As it hung down the king could sit on it if he chose. See Xen. *Hellen.* 2, 1, 8.

17. **ὑπὸ...ὑπαχθεὶς**, that is by the Ephors. **ἔφυγε** 'was banished'.

CHAPTER LXXIII.

20. **δὴ**, dismissing a subject, cp. p. 36, l. 21.

21. **ὡς τῷ Κλεομένεϊ...πρῆγμα** 'when Kleomenes' intrigue against Demaratos had succeeded'. **ὁδοῦν**='to guide', 'to manage', cp. 4, 139 *τὰ ἀπ' ὑμέων ἡμῖν χρηστῶς ὁδοῦται*.

23. **ἐπὶ τοὺς Αἰγινήτας** 'to fetch the Aeginetans', p. 9, l. 18.

24. **προπηλακισμὸν** 'insulting language'. The origin of the word is doubtful. The received derivation was from *πηλός* 'mud', giving the general sense of 'pollution'. Dr Rutherford [*New Phryn.* p. 127] derives it from *πηλίκος* 'how old?', as though *προπηλακίζειν* was to ask a man how old he was before you knew him, i.e. to take a liberty. There are however objections to this derivation, besides its somewhat far-fetched character. It generally refers to language rather than action. **οὕτω δὴ** 'in these circumstances', cp. p. 18, l. 29. **οὔτε** followed by **τε** p. 9, l. 15.

41 1. **ἀμφοτέρων τῶν βασιλέων**. The Aeginetans do not venture to resist a demand made by Sparta against which no technical objection can be made; thus confirming the high position of that town in Hellenic politics implied by the original application of Athens.

2. **ἐκεῖνοι**, sc. Kleomenes and Demaratos.

4. **ἦγον** 'took them away', as prisoners, p. 47, l. 20. **καὶ δὴ καί**, cp. p. 11, l. 13; p. 26, l. 6; p. 35, l. 10.

7. **παραθήκην παρατίθενται** 'deposited them as a charge with the Athenians', cp. p. 48, l. 11. [Stein reads *καταтίθενται* with some MSS.] Observe **ἐς** which Herodotos uses after any verb which implies movement.

CHAPTER LXXIV.

10. **ἐς** 'in reference to', p. 1, l. 13. **δεῖμα...Σπαρτιητέων** 'a fear of the Spartans', objective genitive. **ὑπεξέσχε** 'secretly removed to Thessaly', cp. 8, 132. See on p. 39, l. 8.

12. **νεώτερα ἔπρησσε πρήγματα** 'tried to raise a party against Sparta' Equivalent to *ἐνεωτέριζε* 'adopted revolutionary measures'.

13. **ἐπὶ** 'against', p. 26, l. 7.

14. **ἦ μὲν** like *ἦ μήν*, the formula introducing an oath. Cp. 3, 74.

15. **καὶ δὴ καὶ**, l. 4.

17. **ἐξορκοῦν τὸ Στυγὸς ὕδωρ** 'to swear them by the water of Styx', the most binding of oaths; *Il.* 15, 37:

ἴστω νῦν τόδε Γαῖα καὶ Οὐρανὸς εὐρὺς ὕπερθεν,
καὶ τὸ κατειβόμενον Στυγὸς ὕδωρ, ὅστε μέγιστος
ὅρκος δεινότατός τε πέλει μακάρεσσι θεοῖσιν.

Cp. *Il.* 2, 755 and Vergil, *Aen.* VI. 324:

Cocyti stagna alta vides Stygiamque paludem
Di cujus jurare timent et fallere numen.

19. **τοιόνδε τι**, see Geographical Index s.v. Styx.

20. **ἄγκος**, a natural basin or hollow. **αἱμασιῆς κύκλος** 'a circle of a wall', 'a circular wall', thus making it into a well or fountain, like the various Holywells in every country.

CHAPTER LXXV.

2. **κατῆγον** 'restored him' (p. 3, l. 8) to his royal position, which he had forfeited by leaving Lakonia. **ἐπὶ τοῖσι αὐτοῖσι...τοῖσι** 'on conditions exactly the same as those on which he had ruled before'. 42

4. **μανιὰς νοῦσος** 'a fit of madness seized him'. Soph. *Aj.* 59 *φοιτῶντα...μανιάσι νόσοις.*

5. **ὑπομαργότερον** 'somewhat crazy'. The word applied to Kambyses, 3, 29. **ὅκως...ἐντύχοι** 'as often as he met', cp. p. 16, l. 9.

7. **ποιεῦντα...παραφρονήσαντα**, obs. the two tenses, 'as he was in the habit of acting thus and was now gone clean mad'.

8. **οἱ προσήκοντες** 'his relations', into whose guardianship he would pass as being insane, by a similar law apparently to that which existed in Athens and at Rome (*deducere ad gentiles et agnatos*), see Cicero, *de Senect.* § 22. **ἐν ξύλῳ** 'in wooden stocks', which apparently left the hands free; see the case of Hegesistratos in 9, 37, where only one leg seems to have been fastened.

12. **ἦν γὰρ τῶν τις εἱλωτέων** 'for he was a helot', and therefore would have no redress against anything the king might do afterwards.

14. **ἤρχετο...ἑωυτὸν λωβώμενος** 'began mutilating himself', cp. 3, 155.

19. **τὴν Πυθίην ἀνέγνωσε.** See c. 66. This sense of *ἀναγινώσκειν* 'to persuade' is peculiar to Ionic Greek, cp. p. 46, l. 18.

21. **ἐς Ἐλευσῖνα.** See c. 64. Madness was looked upon as in a special sense a visitation of Providence and a punishment of impiety.

ἔκειρε τὸ τέμενος 'began cutting down the trees in the sacred enclosure', cp. 8, 65; 9, 15. **τῶν θεῶν** of Demeter and Persephone. See Historical Index s.v. *κόρη*.

22. **ἐξ ἱροῦ...Ἄργου.** According to Pausanias [2, 20, 8] the Argives who took refuge in the temple of Argos were tempted out by a promise of their lives; and when the first who came out were killed, the rest shut themselves up in the temple and were burnt in it: while the town of Argos itself was saved by the heroic defence of the women under the leadership of the poetess Telesilla, see c. 79.

24. **καταγινέων** 'inducing them to come down', from the temple and enclosure. Stein points out that the word is appropriate, as the temple and grove were on a hill, *Ἄργου λόφος*.

CHAPTER LXXVI.

27. **Ἄργος αἱρήσειν** 'that he would take Argos', which was fulfilled by his taking the temple of Argos, not the town, p. 45, l. 5. For such equivocal utterances of Oracles we may compare the Oracle which about B.C. 416 told the Athenians that they would 'take the Syracusans', fulfilled by the capture of the lists of Syracusan men-at-arms; and the advice of the Oracle to the Athenians *ἄγειν* the priestess of Klazomenae, whose name was *Ἡσυχία* 'Peace' [Plutarch *Nicias*, cc. 14 and 13].

43 3. **ἐς χάσμα ἀφανὲς** 'into a cavern of invisible depth'; a deep limestone chasm into which the waters of this lake do actually disappear.

6. **δ' ὦν** 'however that may be', resuming the thread of his narrative after the parenthesis.

7. **αὐτῷ**, sc. to the river Erasinos. For sacrifices to rivers, see 7, 113; 8, 138. **καὶ οὐ γάρ**, cp. p. 33, l. 1. **ἐκαλλιέρεε** 'favourable omens were not obtained'. Impersonal, or with *τὰ ἱρὰ* understood, cp. 9, 19 *καλλιερησάντων τῶν ἱρῶν*, p. 46, l. 1.

8. **οὐ προδιδόντος** 'for not betraying'.

9. **οὐδ' ὧς** 'not even so', cp. p. 26, l. 19. **χαιρήσειν** 'get off scot free', 'escape punishment'.

11. **ταῦρον**, the regular offering to the Sea God, according to Homer *Odyss.* 3, 6 *ἐπὶ θινὶ θαλάσσης ἱερὰ ῥέζον, Ταύρους παμμέλανας ἐνοσίχθονι κυανοχαίτῃ.*

CHAPTER LXXVII.

16, 17. **τῷ οὔνομα κέεται** 'which has the name', *κεῖμαι* being used as the passive of *τίθημι*. **μεταίχμιον** 'space between the armies', p. 65, l. 21. From this the word passed to mean any space, as Aesch. *Choeph.* 55 *μεταίχμιον σκότου* 'the debateable space between light and darkness'. **ἵζοντο**, see on p. 3, l. 17.

18. **ἐνθαῦτα δή**, see p. 7, l. 22.

20, 21. **καὶ γὰρ δή...χρηστήριον** 'For in fact it was to this latter case that the oracle they had (*σφι*) referred', i.e. to their being taken by treachery. **καὶ γάρ**, cp. p. 13, l. 7. **εἶχε...ἐς**, p. 1, l. 13. **τὸ ἐπίκοινα ἔχρησε**, see p. 10, l. 9.

23. But when the woman has conquered the male and driven him headlong
Sheer from the land, and glory has earned for herself among Argives,
Many an Argive cheek shall be torn and bleeding for sorrow.
Then in the days to come, may be, shall they say to each other:
Slain by the spear fell the snake, thrice coiled, dreadful, and perished.

ὅταν ἡ θήλεα. Herodotos seems to interpret this mysterious oracle as prophesying the victory of Sparta over Argos (feminine over masculine); and to refer the *ὄφις* to the national emblem of *Ἄργος*, with which he connects the old interpretation of *Ἀργειφόντης* 'the slayer of Argus', or of the 'snake'. Stein points out that Sophocles (*Ant.* 125) calls an Argive warrior *ἀντίπαλος δράκων*, and that in Euripides (*Phoen.* 1137) the Argive king has the figure of a snake on his shield. Pausanias (2, 20, 6—10) interprets it of the victory of the Argive women under the poetess Telesilla, who according to him defended the town of Argos successfully against Kleomenes, when he attacked it after burning the sacred Grove. Herodotos says nothing of this incident, and some have supposed it to be a tale growing out of the words of the oracle itself. But we must consider (1) that though Herodotos does not say anything of this heroic defence of their town by the Argive women, he yet describes Kleomenes as retiring from Argos without any assignable motive, (2) and secondly that if Kleomenes were beaten by women, both he and his men would prefer any explanation of their retreat to be given rather than one so derogatory to their valour. There appears to me therefore some reason for suspecting that what Pausanias and Plutarch (*de Virt. Mulier.*) found in their authorities had some foundation in fact, though the resistance of the women may have been supplemented by a bribe also.

25. **ἀμφιδρυφέας** 'torn on both cheeks', that is, in mourning for their brethren and husbands slain in the war. *Il.* 2, 700 *τοῦ δὲ καὶ ἀμφιδρυφὴς ἄλοχος Φυλάκῃ ἐλέλειπτο.*

28. **συνελθόντα** 'coinciding', i.e. the oracle and the Spartan invasion.

44 1. **καὶ δὴ** 'and so', 'and accordingly', p. 16, l. 24.

τῷ κήρυκι...χρᾶσθαι 'to use the enemy's herald' is to follow his signals and orders, as though he were their own, cp. 2, 13 *χρᾶσθαι γνώμησί τινος* 'to adopt anyone's opinions', and similar phrases.

3. **ὅκως...προσημαίνοι**, see on p. 16, l. 9; p. 42, l. 5.

CHAPTER LXXVIII.

6. **ὁκοῖόν τι** 'whatsoever', the *τι* adds to the indefiniteness of the pronoun, cp. p. 48, l. 17.

8. **χωρέειν ἐς=ἐπὶ** 'to charge', cp. 9, 62 *οἱ Τεγεῆται ἐχώρεον ἐς τοὺς βαρβάρους,...ἐχώρεον καὶ οὗτοι ἐπὶ τοὺς Πέρσας...*

9. **ἐκ=ὑπό**, cp. p. 11, l. 20; p. 22, l. 13.

11. **ἐκ τοῦ κηρύγματος**, to be taken closely with *ποιευμένοισι*, 'who were getting their breakfast in accordance with the proclamation of the herald'. **ἐπεκέατο**, App. D. II. c.

13. **περιιζόμενοι**, see p. 43, l. 17. **ἐφύλασσον** 'they kept a watch on', p. 28, l. 9, equivalent to 'they beleaguered', though *φυλάσσειν* is more often used of watching for protection.

CHAPTER LXXIX.

18. **ἄποινα δὲ...Πελοποννησίοισι δύο μνέαι** 'now the ransom customary among the Peloponnesians was two minae (about £8) a man'. *ἄποινα* is a poetical word for the prose *λύτρον* (see Homer *Il.* 1, 13 etc.). Herodotos seems to infer that the amount of ransom customary was different in different parts of Greece. But in 5, 77 we hear of Boeotians and Euboeans ransomed for two minae a man also. The regular sum was one mina. In Aristotle *Eth.* 5, 10 *τὸ μνᾶς λυτροῦσθαι* is an instance of 'convention', cp. Diodor. Sic. XIV. 102. For the Roman prisoners after Cannae Hannibal demanded 3 minae a head [Polyb. 6, 58]. The custom was not it appears for the state as such to pay the ransom of its own citizens, but for the prisoners to appeal to the various towns with which they were connected in blood to contribute. See Polybios 9, 42, cp. Her. 91, 99.

20. **κατὰ πεντήκοντα δὴ ὦν** 'about fifty in all, then'. This method of enticing men out one by one, and killing them in detail, was practised again by the 30 tyrants at Eleusis in B.C. 404, Xenoph. *Hellen.* 2, 4, 8. **δὴ ὦν = δ' οὖν**, resuming the thread of a story.

24. **ὅ τι ἔπρησσον** 'how they were faring', 'what was happening to them', cp. 8, 21 *οἱ μὲν δὴ ταῦτα ἔπρησσον, παρῆν δὲ ὁ ἐκ Τρηχῖνος κατάσκοπος.*

πρίν γε δή, *donec tandem*, 'until at length'.

25. **κατεῖδε** 'looked down upon and saw'. Cp. Arist. *Equit.* 169

ἀλλ' ἀνάβηθι κἀπὶ τοὐλεὸν τοδὶ
καὶ κάτιδε τὰς νήσους ἁπάσας ἐν κύκλῳ.

οὐκ ὦν δὴ 'accordingly, of course, they did not go on coming out when called', p. 3, l. 4.

CHAPTER LXXX.

26. **ἐνθαῦτα δὴ** 'thereupon', or, 'in these circumstances', p. 7, l. 22; p. 9, l. 15.

28. **περινέειν ὕλῃ τὸ ἄλσος** 'to pile up (dry) wood round the grove'.

1. **καιομένου δὲ ἤδη ἐπείρετο** 'and it was already catching fire when he asked'. **45**

5. **συμβάλλομαι** 'I guess', cp. p. 61, l. 11. **ἐξήκειν** 'has reached its fulfilment', *evenisse*. Cp. 1, 120 *πλεῖστος γνώμην εἰμὶ...ἐξήκειν τὸν ὄνειρον.* Cp. the somewhat similar quasi-fulfilment of Hippias' prophetic dream. c. 107 and p. 42, l. 27.

CHAPTER LXXXI.

9. **ἀριστέας** 'the nobles', used generally in the plural. It apparently refers here to the *ὁμοῖοι* or peers, i.e. the genuine Spartans; the perioeki, and those helots who were not in immediate attendance on the Spartans, having been sent home. **τὸ Ἡραῖον,** the famous temple of Herè stood about 1½ miles from Mycenae and 5 miles N. E. of Argos.

10. **ὁ ἱρεύς.** The temple of Herè was under the care of a priestess; but there would doubtless be a priest also to do the hard work of an actual sacrifice.

CHAPTER LXXXII.

15, 16. **ὑπῆγον...ὑπὸ τοὺς ἐφόρους** 'brought him before the Ephors'. Cp. the case of Leotychides p. 40, l. 17. The Ephors, five in number, were elected annually, at first probably for the humbler duties of superintending the markets; but they had by this time obtained a paramount power in the state, as forming a tribunal before which every officer, from the kings downwards, could be tried.

16. **δωροδοκήσαντα,** cp. p. 40, l. 14.

17. **παρεὸν** 'though he might', see on p. 40, l. 13.

19. **δ' ὦν** 'be that as it may'.

20. **ἐξεληλυθέναι=ἐξήκειν** in l. 5.

21. **πειρᾶν,** see p. 47, l. 4.

22. **πρὶν...χρήσηται καὶ μάθῃ** 'until he should have sacrificed and learnt'. *πρὶν* with subj. regularly has *ἄν*. The latter however is often omitted in poetry. Cp. Soph. *Phil.* 917 *μὴ στέναζε πρὶν μάθῃς,* and occasionally in prose, cp. 1, 136 *πρὶν δὲ ἢ πενταέτης γένηται, οὐκ ἀπικνέεται ἐς ὄψιν τῷ πατρί.* So also *πρότερον ἢ...γένηται* 7, 54. Goodwin, *M. and T.* § 66—7, Madvig, § 127.

23. **εἴτε παραδιδοῖ** (sc. *πειρᾶν*)...**εἴτε ἔστηκε,** 'whether God allows him to make the attempt, or forbids him' (stood in his way). For **διδοῖ** see App. D. IV. *a*.

46 1. **καλλιερευμένῳ** 'but as he was sacrificing for good omens'. The sacrifices were said *καλλιερέειν*, p. 43, l. 7, the man *καλλιερέεσθαι*, cp. 7, 113 *ἐκαλλιερέοντο σφάζοντες ἵππους λευκούς.* Arist. *Plut.* 1181 *ὃ δ' ἂν ἐκαλλιερεῖτό τις.* For the dative, cp. p. 14, l. 11.

3. **αἱρέει** 'that he would not take'. The present is used for the future dramatically to give greater vividness, cp. p. 48, l. 16.

5. **αἱρέειν ἂν** 'he would have taken', see on p. 26, l. 21. **κατ' ἄκρης** 'entirely', p. 10, l. 2.

6. **πάν οἱ πεποιῆσθαι** 'everything had been accomplished by him'. The dative of the agent is almost confined to perfects and pluperfects passive, G. § 188, 3.

8. **διέφυγε πολλὸν** 'he escaped being convicted by his prosecutors by a large majority of votes'. πολλὸν is adverbial, cp. p. 58, l. 17; 5, 1 πολλὸν ἐκράτησαν. The sense here given is derived from considering that both διώκοντας and διέφυγε are used in their legal sense.

CHAPTER LXXXIII.

10. **ἐχηρώθη** 'was widowed of its men'. Solon *fr.* 37 πολλῶν ἂν ἀνδρῶν ἥδ' ἐχηρώθη πόλις.

11. **ὥστε οἱ δοῦλοι...πρήγματα** 'so that the slaves obtained possession of the government'. Plutarch *de Virt. Mulier.* denies this, and says that the places of the slain citizens were filled up by the best of the perioeki, who were however despised by their wives. So also Aristotle, *Pol.* 5, 3 καὶ ἐν Ἄργει τῶν ἐν τῇ ἑβδόμῃ ἀπολομένων ὑπὸ Κλεομένους τοῦ Λάκωνος ἠναγκάσθησαν παραδέξασθαι τῶν περιοίκων τινάς.

12. **ἐπήβησαν** 'grew up', became ἔφηβοι.

13. **ἀνακτώμενοι** 'recovering possession of', cp. 1, 61 ἀνακτᾶσθαι ὀπίσω τὴν τυραννίδα.

15. **Τίρυνθα.** The hostility of Argos and Tiryns continued, and caused the latter perhaps to take the Hellenic side against the Persians in B.C. 480 when the Argives medized (9, 28, 31).

16. **ἄρθμια** 'friendly relations' [ἀρ-, cp. ἀραρίσκω, ἄρθρον], elsewhere applied to persons ἄρθμιοι 'on friendly terms', 7, 101; 9, 9.

17. **ἀνὴρ μάντις** 'a professional soothsayer', probably of one of the numerous mantic families in the Peloponnese, in whom this art was hereditary. See on 9, 33.

18. **ἀνέγνωσε**, p. 42, l. 19.

CHAPTER LXXXIV.

22. **μέν νυν**, p. 24, l. 13; p. 29, l. 11.

24. **ἐκ δαιμονίου**, p. 11, l. 20.

25. **ἀκρητοπότην** 'a drinker of strong (unmixed) wine'. Herodotos does not mention elsewhere this habit of the Skythians; but Plato *de Leg.* 1, 20 says *Σκύθαι ἀκράτῳ παντάπασι χρώμενοι.* For the temperate Greek to drink wine unmixed with water was wholly alien to his habits; though the inhabitant of a colder climate might do so with impunity.

47 1. **μεμονέναι** 'were very desirous'. This perfect (of which no primary tense was in use) is elsewhere confined to poetry.

2. **συντίθεσθαι** 'to agree', cp. p. 67, l. 11.

3. **παρὰ Φᾶσιν ποταμὸν**, that is, after crossing the Caucasus to keep close to the shore of the Black Sea, a route which is said to be impracticable for an army.

4. **πειρᾶν**, as in p. 45, l. 21. Elsewhere Herodotos always uses the middle *πειρᾶσθαι.* **σφέας...κελεύειν** 'and (to say) that they bade the Spartans start from Ephesos etc.'. The two parallel clauses are *ὡς χρεὸν εἴη* and *σφέας κελεύειν*, and Herodotos varies the construction as he does frequently the moods and tenses of two parallel clauses. See on p. 2, l. 9. Stein quotes a somewhat similar variation from 3, 53 *ὁ Περίανδρος κήρυκα πέμπει βουλόμενος αὐτὸς μὲν ἐς Κέρκυραν ἥκειν, ἐκεῖνον δὲ ἐκέλευε ἐς Κόρινθον ἀπικόμενον διάδοχον.*

6. **ἐς τὠυτὸ ἀπαντᾶν** 'to come together', 'to meet'.

7. **ἐπὶ ταῦτα** 'with this object'.

8. **τοῦ ἱκνεομένου** 'than was suitable', cp. p. 30, l. 28; p. 35, l. 24.

10. **ἐκ τε τοῦ** 'and ever since'. So *πρὸ τοῦ* p. 82, l. 10.

11. **ζωρότερον** 'somewhat stronger wine than usual', i.e. to drink merum or unmixed wine: the mark of *ἀγροικία* in Theophrastus *Char.* 4, cp. *Il.* 9, 203 *ζωρότερόν τε κέραιε.*

12. **ἐπισκύθισον** 'pour out Skythian-wise', i.e. unmixed wine. Athenaeus 427 C relates this story of the cause of Kleomenes' madness, and quotes some lines of Anakreon

Ἄγε δεῦτε, μηκέθ' οὕτω
πατάγῳ τε κἀλαλητῷ
Σκυθικὴν πόσιν παρ' οἴνῳ
μελετῶμεν.

τὰ περὶ Κλεομένεα 'the story of Kleomenes'.

CHAPTER LXXXV.

17. **καταβωσομένους** 'to denounce', followed by the genitive, as

κατηγορέω and other compounds of *κατά* with the meaning of 'against'. Madvig, § 59.

18. **τῶν ὁμήρων**, see c. 73. **δικαστήριον** 'a court of enquiry', which in such a case was composed, according to Pausanias (3, 5, 2), of the 28 members of the Gerusia, the 5 Ephors, and the other king.

19. **ἔγνωσαν** 'decided', cp. p. 63, l. 3. **περιυβρίσθαι** 'had been treated with outrageous violence'. *ὕβρις* was a legal term for assault, and the word here implies that the proceeding was contrary to law or custom.

20. **ἄγεσθαι**, p. 41, l. 4. For **κατακρίνειν** followed by infinitive, cp. 9, 93 *κατέκριναν...τῆς ὄψιος στερηθῆναι*. Sometimes it takes the dative of the punishment, as *κατακρίνειν θανάτῳ*. In 2, 133 the person is in the dative, *κατακεκριμένων οἱ τούτων*. Cp. 7, 146 *τοῖσι κατακέκριτο θάνατος*.

2. **ὅκως...μὴ** 'take care lest, if you carry this out, they may hereafter inflict upon your country some utterly ruinous mischief'. Before *ὅκως* must be understood *ὁρᾶτε* or *εὐλαβεῖσθε* or some such word. The future indicative is more common than the subjunctive in this phrase. Madvig, § 124, Rem. 1. Cp. 5, 79 *ἀλλὰ μᾶλλον μὴ οὐ τοῦτο ᾖ τὸ χρηστήριον*.

5. **ἔσχοντο τῆς ἀγωγῆς** 'abstained from the forcible removal'. This peculiar sense of *ἔχεσθαι* (which with the genitive more often means 'to hold on to') occurs again in 7, 169 *ἔσχοντο τῆς τιμωρίης* and ib. 237 *ἔχεσθαι κακολογίης*.

6. **ἐπισπόμενον** 'who accompanied them voluntarily', opposed to *ἀγόμενον*.

48

CHAPTER LXXXVI.

9. **τὴν παραθήκην**, p. 41, l. 7.

10. **προφάσιας εἷλκον** 'made excuses to delay the business'. Abicht quotes Aristoph. *Lysist.* 726 *πάσας τε προφάσεις ὥστ' ἀπελθεῖν οἴκαδε ἕλκουσιν*.

11. **οὐ δικαιοῦν**, the negative is closely connected with the verb, 'they did not think right', p. 5, l. 20.

12. **οὐ φαμένων δὲ** 'so, as the Athenians refused'.

16. **ποιέετε** 'ye will be doing what religion demands'. The vivid present for the future, cp. p. 46, l. 3. **ὅσια** refers to the religious sanctity of a deposit: cp. 9, 79 in regard to the treatment of the dead *ὅσια μὲν ποιέειν ὅσια δὲ καὶ λέγειν*.

17. **ὁκοῖόν τι**, cp. p. 44, l. 6.

22. **πάντα περιήκειν τὰ πρῶτα** 'had attained the first position in all other respects'. So 7, 16 *τά σε...περιήκοντα ἀνθρώπων κακῶν ὁμιλίαι σφάλλουσι.*

23. **καὶ δὴ καὶ**, p. 11, l. 13. **ἀκούειν ἄριστα** 'had the highest reputation', cp. 9, 79 *φὰς ἄμεινόν με ἀκούσεσθαι.*

25. **ἐν χρόνῳ ἱκνεομένῳ** 'in due time', 'at the time destined by providence': as though the whole affair was part of a fate which he could not avoid, like that which dogged the house of Atreus, and others, p. 30, l. 28; p. 47, l. 8.

26. **λέγομεν** 'we (Spartans) say'.

27. **προϊσχόμενον τοιάδε** 'offering the following explanation' or 'proposal', p. 5, l. 17.

49 2. **ἀπολαῦσαι** 'to enjoy the benefit of'. For the position of *γὰρ* see on p. 33, l. 1; p. 39, l. 15. **ἀνά**, see p. 25, l. 22.

3. **ἐν δὲ** 'and among the rest', p. 6, l. 10.

4. **ἐμεωυτῷ λόγους ἐδίδουν** 'I began to consider in my own mind', cp. p. 81, l. 22; 8, 9 *λόγον σφίσι αὐτοῖσι ἐδίδοσαν* 'consulted with each other'.

5. **ἐπικίνδυνος...Ἰωνίῃ** 'always has been and is subject to dangers', from the interference of Alyattes and Kroesos and then of Kyros, and the constant inclination of the Ionians subsequently to revolt.

6. **καὶ διότι...τοὺς αὐτοὺς** 'and that one can never see (in Ionia) the same people continuing in the possession of any property', i.e. that property is always changing hands irregularly.

9. **ἐξαργυρώσαντα** 'having converted it into money'.

11—13. **δέξαι...σῶζε...ἀποδοῦναι**, notice the tenses; the receiving and paying were single acts (aorist), the preserving the tallies was a continuous one (present).

12. **σύμβολα** 'tallies', of which the Milesian kept counterparts (l. 18). They were given sometimes, not for the reclaiming of a deposit, but simply to demand the mutual kindnesses of two **ξένοι**, cp. Eurip. *Med.* 613 *ξένοις τε πέμπειν σύμβολ' οἳ δράσουσί σ' εὖ.*

15. **ἐπὶ τῷ εἰρημένῳ λόγῳ** 'on the terms mentioned', p. 42, l. 2.

19. **διωθέετο** 'he tried to put them off'.

ἀντυποκρινόμενος τοιάδε 'returning them the following answer'. The compound in this sense does not occur again and it does not appear to differ materially from *ὑποκρινόμενος*. The *ἀντὶ* may mean 'in place of simply restoring the money'; or, 'as against their demand'.

20—22. Obs. the negatives **οὔτε...οὔτε...** followed by **τε**. Cp. p. 9,

l. 15; p. 15, l. 23. **περιφέρει** 'it does not occur to me': a variation of the phrase by which memory is said *περιφέρειν τινά* to some object: cp. Plato, *Laches*, 180 E *περιφέρει δέ τίς με καὶ μνήμη ἄρτι τῶνδε λεγόντων*. The nominative case to *περιφέρει* may be *μνήμη* implied in *μέμνημαι*, or simply *ἃ λέγετε*, cp. the Latin phrase *redire in memoriam alicujus*, Cicero *de Sen.* § 21.

23. **ἀποδοῦναι**, sc. *βούλομαι*. **ἀρχὴν** 'at all', p. 17, l. 17.

24. **νόμοισι** 'customs', i.e. as to persons making fraudulent demands. **ἐς ὑμέας** 'against you'.

25. **ταῦτα ἀναβάλλομαι κυρώσειν** 'I postpone the settlement to the 4th month, when I will confirm it'. For the future infinitive, which is a consolidation of two separate clauses, *ἀναβάλλομαι κυρῶσαι ἐς τέταρτον μῆνα*, and *κυρώσω τετάρτῳ μηνί*, cp. 5, 49 *ἀναβάλλομαι ἐς τρίτην ἡμέρην ἀποκρινέεσθαι*.

26. **συμφορὴν ποιεύμενοι** 'regarding it as very grievous'. Cp. 9, 76; p. 33, l. 11.

27. **ὡς ἀπεστερημένοι** 'fully convinced that they had been cheated out of the money'.

30. **εἰ...ληΐσηται** 'if he was to plunder'. For the deliberative subjunctive with *εἰ* in indirect discourse, see on p. 18, l. 28.

31. **μετέρχεται**, historic present 'addressed him', cp. p. 37, l. 23. It has here a sense of 'reproving'. The editors all quote Hesiod *Op.* 319 sqq.:

> *εἰ γάρ τις καὶ χειρὶ βίῃ μέγαν ὄλβον ἕληται*
> *ἢ ὅ γ' ἀπὸ γλώσσης ληίσσεται, οἷά τε πολλὰ*
> *γίγνεται, εὖ τ' ἂν δὴ κέρδος νόον ἐξαπατήσῃ*
> *ἀνθρώπων, αἰδῶ δ' ἀναιδείη κατοπάζῃ,*
> *ῥεῖα δέ μιν μαυροῦσι θεοί, μινύθουσι δὲ οἶκοι*
> *ἀνέρι τῷ, παῦρον δέ τ' ἐπὶ χρόνον ὄλβος ὀπηδεῖ.*

1—7. Better it were for the moment, oh Glaukos, to act as thou askest,— 50

Carry thy will by an oath, and take the money as booty.
Swear: for the keeper of oaths must die no less than the perjured.
Yet hath the Oath-god a son: unnamed is he: handless and footless:
Who on the track of the sinner still follows swift, till he catches
Race and house and wealth, and whelms them all in destruction.
But of the true to his oath a goodlier offspring remaineth.

4. **ἔπι = ἔπεισι.**

6. **γενεὴν καὶ οἶκον.** The former refers especially to persons, the latter to the estate.

7. This line is from Hesiod *Op.* 283.

10. **τὸ ποιῆσαι** 'the committal of the crime'. **ἴσον δύνασθαι,** *idem valere,* i.e. one was as bad as the other.

11. **δὴ** 'accordingly', p. 3, l. 4.

13. **ὁρμήθη λέγεσθαι** 'was originally mentioned', lit. 'proceeded to be told'. Cp. 4, 16 *τῆς πέρι ὅδε ὁ λόγος ὥρμηται λέγεσθαι.*

14. **ἐς ὑμέας,** cp. p. 38, l. 3.

15. **ἱστίη,** App. A. II. (6), 'hearth', and therefore central home and family. So in 1, 177 it means the family itself, *ὀγδώκοντα ἱστίαι ἔτυχον ἐκδημέουσαι.*

16. **πρόρριζος** 'utterly', 'from the very root', 'root and branch'. Cp. 3, 40 *κακῶς ἐτελεύτησε πρόρριζος.*

17. **μηδὲ διανοέεσθαι...ἄλλο γε ἢ** 'not to have any other thought even about a pledge except to restore it'.

19. **οὐδὲ οὕτω** 'not even after this address'. **ἐσήκουον** 'gave ear to him', 'obeyed', with dat. cp. 1, 214 *ὥς οἱ ὁ Κῦρος οὐκ ἐσήκουε.*

CHAPTER LXXXVII.

21. **τῶν,** relative attracted to the case of the antecedent *ἀδικημάτων,* G. § 153.

22. **Θηβαίοισι χαριζόμενοι** 'by way of gratifying the Thebans'. Athens and Thebes were always inclined to be at enmity; but the Thebans were now still smarting under a severe defeat sustained some years before at the hands of the Athenians on the Euripus, whither they had advanced to aid the Chalkidians [5, 77].

24. **καὶ ἦν γὰρ δή.** See on p. 6, l. 11; p. 33, l. 1.

25. **πεντετηρίς** 'a quinquennial festival'. This seems to refer to some ship-races which took place off Sunium during the greater Panathenaea, which occurred every fifth year, i.e. at intervals of four years. This is mentioned in connexion with the festival in Lysias 21, § 5 *νενίκηκα δὲ τριήρει μὲν ἁμιλλώμενος ἐπὶ Σουνίῳ, ἀναλώσας πεντηκαίδεκα μνᾶς· χωρὶς δὲ ἀρχιθεωρίας καὶ ἀρρηφορίας καὶ ἄλλα τοιαῦτα.*

The reading of most MSS. is **πεντήρης**. But this presents a great difficulty. The Greeks are not known to have possessed quinqueremes until B.C. 325; and it is unlikely that this vessel should have been one. On the other hand we have no information about these ship-races except the allusion in Lysias. This is definite enough; but does not seem conclusive as to their existence so much earlier, and before the Athenians had become possessed of a fleet, or made themselves conspicuous as a naval power.

26. **τὴν θεωρίδα**, the ship carrying the state deputation (*θεωροί*) to view the naval games at Sunium.

CHAPTER LXXXVIII.

2. **οὐκέτι ἀνεβάλλοντο...μὴ οὐ...Αἰγινήτῃσι** 'no longer hesitated to put in practice every kind of device against the Aeginetans'. A verb containing a negative idea as 'forbidding' or 'hindering' is followed by *μὴ* with the infinitive; if this verb is negatived the infinitive has *μὴ οὐ*, G. § 283, 6—7. For *ἐπὶ* 'against', see p. 26, l. 7; p. 41, l. 14. 51

3. **καὶ ἦν γάρ**, see p. 50, l. 24.

We possess no other information as to this movement of Nikodromos in Aegina. From c. 91 it seems to have been a popular rising against the Dorian oligarchs; joined perhaps with an attempt to assert the position of Ionian or Achaean settlers against the Dorians who formed the bulk of the inhabitants. The Athenians never found Aegina a peaceful possession till they removed the Dorian land-owners and put in Ionians in their place [Thucyd. 2, 27; 7, 37], who again were displaced by Lysander in B.C. 404 [Xenoph. *Hellen.* 2, 2, 5]. Probably the question of race was mixed up with that of political constitution.

7. **ἀναρτημένους** 'prepared', 'resolved', cp. 7, 8 § 3 *τούτων εἵνεκεν ἀνάρτημαι ἐπ' αὐτοὺς στρατεύεσθαι.*

CHAPTER LXXXIX.

11. **κατὰ**=*καθ' ἃ* 'according as', cp. 5, 11, 22. But Stein reads *κατὰ τὰ* in both places.

12. **τὴν παλαιὴν.** This has been variously explained to mean the citadel of the town of Aegina, and another ancient town called Oea (5, 83), about 3 miles farther inland than Aegina.

13. **ἐς δέον** 'up to time', 'at the right moment'. So *ἐς τὸ δέον* 'when wanted' 2, 173; but cp. 1, 186 *ἐς τὸ δέον γεγονέναι* 'to serve its purpose'. It is a general expression and may refer to other things than time, see Aristoph. *Nubes* 859 *ὥσπερ Περικλέης εἰς τὸ δέον ἀπώλεσα,* cp. Plutarch, *Pericl.* c. 21.

14. **ἀξιόμαχοι** 'good enough to fight a battle with the Aeginetans'. The people of Aegina had long been one of the first naval powers; the Athenians were but beginning.

15. **ἐν ᾧ...χρῆσαι** 'while they were engaged in begging the Korinthians to lend them ships': i.e. triremes, which had been first built at Korinth, Thucyd. 1, 13, 2. Cp. 1, 41, where they are called *νῆες μακραί*. See note on p. 25, l. 26.

16. **τὰ πρήγματα** 'their undertaking'.

17. **γάρ**, cp. p. 6, l. 11.

18. **ἐς τὰ μάλιστα**, p. 34, l. 27. Herodotos says the Korinthians and Athenians were *at that time* friends, because he is writing during the Peloponnesian war, when they were at enmity.

19. **πενταδράχμους**, sc. *νέας*, selling them at a nominal sum of five drachmae apiece.

22. **ἑβδομήκοντα τὰς ἁπάσας** 'seventy in all'. The fifty ships of the Athenians' own were probably for the most part of smaller bulk than triremes. They were not however entirely without triremes, see p. 20, l. 13. The Korinthian orator in Thucyd. 1, 41 says only that they were *σπανίζοντες νεῶν μακρῶν* 'short of ships of war'; and we know that they sent a fleet of 20 ships, presumably triremes, to Asia at the beginning of the Ionian revolt in B.C. 501 (5, 99).

23. **ἡμέρῃ μιῇ** 'by one day', dative of time before or after, or of quantity, p. 4, l. 18; p. 14, l. 14. **τῆς συγκειμένης**, gen. after the comparison involved in *ὑστέρησαν*.

CHAPTER XC.

52 3. **Σούνιον**, the town of Sunium.

4. **ἔφερόν τε καὶ ἦγον**, p. 22, l. 7.

6. **ταῦτα**, i.e. the plundering of Aeginetan coasts by these exiles. **ὕστερον**, after the battle of Marathon.

CHAPTER XCI.

7. **ἐπαναστάντος** 'having revolted against them'.

9. **ἐξῆγον** 'were leading them out of the town'.

10. **τὸ ἐκθύσασθαι...ἐπιμηχανεόμενοι** 'which they were never able to wipe out by sacrifices, though they tried various means of doing so'.

11. ἔφθησαν ἐκπεσόντες 'but they were expatriated before'. Referring to their removal by the Athenians in B.C. 431 mentioned in the note to p. 51, l. 3; when the Spartans assigned them lands in the district of Thyrea (Thucyd. 1, 82).

15. πρόθυρα, see on p. 18, l. 18. The temples of Demeter were usually outside the town in some lonely spot; and it was therefore probably not far from the place selected for such an abominable execution. For **θεσμοφόρου**, see note on p. 9, l. 13.

16. ἐπισπαστήρων εἴχετο 'held on to the handles of the door', which was closed, so that he could not get into the temple. But any connexion with the sacred building would be held to retain him under the goddess's protection. See the account of the Kylonian conspirators coming down from the Akropolis holding on to a cord attached to the image of the Goddess. [Plutarch, *Solon* 12.]

17. ἀπέλκοντες 'by pulling at him'.

18. ἦγον 'took him off to execution', p. 41, l. 4.

19. ἐπισπάστροισι (*ἐπίσπαστρον*): the MSS. give this form, which is recognised by Pollux 10, 22.

CHAPTER XCII.

23. ἐπεκαλέοντο, cp. p. 12, l. 18.

24. τοὺς αὐτοὺς καὶ πρότερον 'the same people as they had called to their aid before', see 5, 86. For *καὶ* following *ὁ αὐτὸς*, cp. p. 82, l. 2; 8, 45 *τὠυτὸ πλήρωμα παρείχοντο καὶ ἐπ' Ἀρτεμισίῳ*. **Ἀργείους** 'namely Argives', the definite article omitted, cp. p. 12, l. 2. **καὶ δὴ** 'now', introducing a continuation of the story.

26. ἔσχον τε...καὶ συναπέβησαν 'not only put in on the shore of Argolis, but also landed with the Lakedaemonians', c. 76, 77.

53

4. ζημίη...ἐκτῖσαι in apposition: 'a penalty was imposed on them, namely that they should pay'. Madvig, § 165 b.

5. μέν νυν, p. 24, l. 13. **συγγνόντες ἀδικῆσαι** 'acknowledging that they had done a wrong'.

6. ὡμολόγησαν...εἶναι 'they agreed that they should be free on the payment of a hundred talents'.

7. οὔτε...τε, p. 9, l. 15; p. 15, l. 23. 'They did not acknowledge it (sc. ἀδικῆσαι) and they were contumacious in their attitude'.

8. δὴ ὦν resuming from l. 24. 'It was, I say then, on this account'. **ἀπὸ τοῦ δημοσίου** 'on the authority of the state', p. 32, l. 6.

11. **πεντάεθλον ἐπασκήσας** 'who had practised the pentathlum'. He won in the pentathlum at the Nemean games [Paus. 1, 29, 5]. The pentathlum was a contest in five things,—leaping, running, throwing the discus, throwing the javelin, and wrestling. As to the requirements for winning the prize, see on 9, 33; but whether a man had to win all five or only three out of the five, he would train for all of them; and the victory, which seems to have been comparatively rare, made him a marked man, cp. 9, 33, 75.

13. **μουνομαχίην ἐπασκέων** 'turning his attention to single combats', cp. 2, 166 *τέχνην ἐπασκῆσαι*, 3, 82 *ἀρετὴν ἐπασκέων*. It seems necessary to take it in a slightly different sense from the technical use in l. 11.

CHAPTER XCIII.

17. **ἀτάκτοισι** 'when they were in no settled order'. The Athenian sailors had not yet learnt the naval manoeuvring, for which they became famous afterwards.

19. **αὐτοῖσι ἀνδράσι** 'with their crews', cp. p. 16, l. 25.

CHAPTER XCIV.

20. **Ἀθηναίοισι πόλεμος συνῆπτο πρὸς Αἰγινήτας** 'so then Athens and Aegina were at war'. *συνάπτειν=committere*, cp. 7, 158 *ὅτε μοι πρὸς Καρχηδονίους νεῖκος συνῆπτο.* For **δὴ** summing up a series of statements, cp. p. 52, l. 6.

21. **ὁ δὲ Πέρσης** 'but the king of Persia', Darius. So the Sultan used to be called 'the Great Turk'. **τὸ ἑωυτοῦ ἐποίεε** 'was carrying out his design'.

22. **ἀναμιμνήσκοντος τοῦ θεράποντος.** When Darius heard of the burning of Sardis, he ordered a slave every day at dinner to remind him of the Athenians, 5, 105. **μεμνῆσθαι** 'namely, that he should remember both by his slave always reminding him (*μιν*) and because the Peisistratids always etc.' **ὥστε**=*ἅτε*, p. 27, l. 22, and goes with **ἀναμιμνήσκοντος**. Cp. 8, 118 *ὥστε ἐπὶ τοῦ καταστρώματος ἐπεόντων συχνῶν Περσέων.* 9, 37 *ταῦτα ποιήσας ὥστε φυλασσόμενος ὑπὸ φυλάκων.*

54 1. **ἅμα δὲ βουλόμενος** 'and at the same time because Darius

himself wished, by fastening on this pretext, to subdue those in Greece who had refused him earth and water'.

2. **ἐχόμενος**, cp. p. 8, l. 2.

3. **τῆς Ἑλλάδος**, a topographical genitive, cp. l. 13.

4. **φλαύρως πρήξαντα** 'as having been unsuccessful', i.e. owing to the loss of the fleet on Athos, see c. 44—5.

6. **ἐπὶ** 'against'. Eretria and Athens were especially singled out for attack as having assisted the Ionians in their assault upon Sardis [5, 99].

7. **ἐόντα Μῆδον** 'though he was a Mede by birth'. The Medes were seldom employed in positions of importance, which were usually filled by Persians. Grote, *H. of Greece*, vol. 4, p. 256.

CHAPTER XCV.

13. **τῆς Κιλικίης**, cp. l. 3.

17. **ὁ ἐπιταχθεὶς ἑκάστοισι** 'which had been levied upon the several states'. See c. 48.

25. **Ἰκάριον**, sc. *πέλαγος* 'the Icarian sea', the southern part of the Aegean, from the Island of Icaria. **καὶ διὰ νήσων** 'and by the Island course', i.e. going from Island to Island; as opposed to the coasting voyage round the N. of the Aegean. 8, 108; 9, 3. **ὡς ἐμοὶ δοκέειν** 'as I think', or 'in my opinion'. Cp. 8, 66 *ὡς ἐμοὶ μὲν δοκέειν οὐκ ἐλάσσονες ἐόντες ἀριθμὸν ἐσέβαλον*. Sometimes without *ὡς*, cp. 8, 22 *Θεμιστοκλέης ταῦτα ἔγραψε, δοκέειν ἐμοί, ἐπ' ἀμφότερα νοέων*.

1. **πρὸς δὲ** 'and besides', p. 72, l. 28. 55

ἡ Νάξος...ἁλοῦσα 'the fact that they had formerly failed to take Naxos'. That is, the Persians having failed to subdue Naxos (5, 34) thought it dangerous to leave it in their rear, as a centre of revolt, while their army and navy were engaged in Greece. It was also a good starting place for Euboea [5, 31].

CHAPTER XCVI.

5. **ἐπεῖχον** 'were minded', cp. 1, 153 *ἐπ' οὓς ἐπεῖχε στρατηλατέειν αὐτός*.

6. **τῶν πρότερον** 'what had happened before'.

9. **ἐπὶ** 'to attack', p. 54, l. 6.

CHAPTER XCVII.

15. **πέρην** 'on the opposite side of the channel', cp. p. 15, l. 4. Rhenaea was the chief place of residence of the Delians, and was used as a burial place and for other secular purposes, Delos itself being properly reserved for sacred celebrations and other things connected with the worship of Apollo, though there was a constant tendency to secularize it, leading to periodical purifications.

16. **ἠγόρευε**, see p. 6, l. 10.

17. **οὐκ ἐπιτήδεα...κατ' ἐμεῦ** 'condemning me of wrong intentions towards you'. *κατὰ* is a pleonasm with *καταγνόντες*, which would have the same sense with the genitive without it; but it serves to emphasize the sentiment.

19. **ἐπὶ τοσοῦτό γε φρονέω** 'I have so much reverence in me at least'. So *τὸ φρονεῖν* is contrasted with impiety in Eurip. *Bacch.* 389.

20. **ἐν τῇ χώρῃ**, *in qua terra.* **δύο θεοὶ**, Apollo and Artemis, see 4, 35. The antecedent *χώρῃ* is attracted into the clause and case of the relative, G. § 154.

22. **ἄπιτε**, leave Rhenaea and return to Delos.

24. **μετὰ δὲ**, adverbial, p. 2, l. 18. **λιβανωτοῦ...καταvήσας** 'having made a heap of incense on the altar weighing 300 talents'. The talent = 57 lbs. *avoirdupois.*

25. **ἐθυμίησε** 'he burnt it'. Such an enormous amount of incense could hardly have been burnt at once. The *βωμὸς* was doubtless the large altar outside the temple.

CHAPTER XCVIII.

56 1. **μὲν δή**, p. 52, l. 6.

3. **Αἰολέας**. Herodotos has not mentioned them before in this connexion. They had not been involved in the Ionian revolt, and had long been tributary to Persia (2, 90); he appears to mention them here only to draw attention to the fact that the fleet sailing against Greece was partly manned by Greeks. **μετὰ δὲ τοῦτον ἐξαναχθέντα** 'and after his departure'.

5. **καὶ πρῶτα καὶ ὕστατα μέχρι ἐμεῦ** 'for the first and last time up to my time'. Thucydides says that there was an earthquake at Delos 'shortly before the Peloponnesian war and *never before*'.

These two statements of course cannot both be true, and we may suppose (1) That Thucydides meant to contradict Herodotos, or (2) That while Herodotos had got his information from the Delians before going to live in Italy (i.e. before B.C. 443) and had not heard of the second earthquake (just as at p. 22, l. 6, he writes about the φόρος paid by the Ionian cities to Persia as though he were not fully acquainted with minute details of events since the Persian wars), Thucydides on the other hand had also got his information from the Delians, who either forgot that they had given Herodotos this information, or wished to maintain the reputation of their island as supposed to be free from earthquakes. Stein supposes that the truth may lie between the two, and that the earthquake was neither just after the departure of Datis nor just before the Peloponnesian war, but somewhere between them. I cannot see that this is very helpful as an explanation.

8. **Δαρείου—Ἀρταξέρξεω.**

Darius...........	reigned	B.C. 522—485.
Xerxes...........	,,	B.C. 485—465.
Artaxerxes......	,,	B.C. 465—425.

The period, therefore, indicated by Herodotos includes the Ionian revolt, the Persian wars, the Helot revolt in Laconia (464—455), constant wars between Sparta, Athens, Corinth, Boeotia, and Euboea (445) and Samos (440); and finally the outbreak of the Peloponnesian war (432).

12. **ἀπ' αὐτῶν τῶν κορυφαίων** 'from the leading states themselves', p. 12, l. 31, i.e. Athens and Sparta.

περὶ τῆς ἀρχῆς 'for the supremacy'. We have seen (c. 49) that before the Persian wars Sparta's supremacy was generally acknowledged. But the rapid growth of Athens as a naval power, her brilliant achievements in the Persian wars, and her commanding position as the head of the Confederacy of Delos, entirely altered the state of affairs, and Athens became the most powerful state in Greece. Still the old prestige clung to Sparta: and when she had at last suppressed her revolting Helots, it soon became evident that the question of the supremacy would have to be decided by arms. After various acts of hostility a five-years truce was made in B.C. 452. But the events of the Sacred War B.C. 448, in which Sparta supported the Delphians and Athens the Phokians, made it plain that they would soon be at war again, as they were in B.C. 445; and the 30 years peace then made only lasted till B.C. 432, the quarrel of Korkyra and Korinth serving only as a pretext for the war which was really to decide the question of the supremacy of Athens or Sparta.

14. **ἀκίνητον.** So Pindar (fr. 58) calls Delos *πόντου θύγατερ, χθονὸς εὐρείας ἀκίνητον τέρας.* The legend was that it had once been a floating island, but, on the birth of Apollo there, had been fixed, *Quam pius Arcitenens oras et litora circum Errantem Mycono e celsa Gyaroque revinxit Immotamque coli dedit et contemnere ventos*, Vergil, *Aen.* 3, 74: and this legend seems to point to a tradition of volcanic disturbance, which nevertheless may have been rare in historic times.

17—19. **δύναται κατὰ Ἑλλάδα γλῶσσαν.** According to modern Orientalists Darius means 'the holder' or 'possessor'; Xerxes 'king' (Shah) or 'mighty man'; Artaxerxes 'Great King'.

Ἑλλάς, as an adjective, cp. 5, 93 Ἑλλὰς πόλις, 9, 16 Ἑλλάδα γλῶσσαν.

18. **ἐρξίης** (ἔργω) 'worker' 'accomplisher'. [Herodotos seems to form the word after the analogy of some proper names such as Βλεψίας, Κινησίας, Σωσίας. St.]

CHAPTER XCIX.

23. **προσίσχον πρὸς τὰς νήσους** 'kept putting in at the Islands', i.e. touched at one after the other. **προσίσχον**, sc. τὰς νέας. There is a variety in this phrase: p. 57, l. 1; p. 69, l. 4 προσίσχειν ἐς. 4, 156 προσίσχειν τῇ γῇ. 9, 99 προσχόντες τὰς νέας. So also the simple ἔχειν, Thucyd. 7, 1, 3 σχόντες Ῥηγίῳ. But 4, 3, 1 ἐς τὴν Πύλον σχόντας.

57 1. **οὐ...οὔτε...οὔτε.** 'A composite negation, which follows another with the same predicate, does not cancel the former but extends it'. Madvig, § 209 b. For **γὰρ** see p. 6, l. 11.

2, 3. **οὔτε ἔφασαν στρατεύεσθαι** 'refused to serve'. "Verbs expressing *to hope, to expect, to promise*, and the like, after which the future infinitive stands regularly in indirect discourse, sometimes take the present". Goodwin, *Moods and Tenses*, p. 14.

4. **ἐνθαῦτα** 'in these circumstances', p. 9, l. 15.

5. **ἔκειρον**, p. 42, l. 21.

6. **παρέστησαν ἐς...γνώμην** 'they submitted to the will of the Persians', 'they submitted to take the Persian side', cp. p. 83, l. 11.

CHAPTER C.

8. **Ἀθηναίων...βοηθοὺς γενέσθαι** 'that they would be their helpers'. The accusative and infinitive is properly used after words of asking

etc., unaffected by the case of the person to which it refers. Madvig, § 164.

10. **τοὺς τετρακισχιλίους κληρουχέοντας** 'the four thousand cleruchs who had assigned to them the land of the Chalkidian knights'. This refers to the 4000, among whom the Athenians had divided the lands of the Chalkidian rich men, or Hippobotae (men who kept horses), in B.C. 506, after their successful invasion, undertaken to avenge raids of the Chalkidians on Attica [5, 74, 77]. Cleruchs differed from colonists (*ἄποικοι*) in that they did not cease to be citizens of the mother city, when receiving their allotments of land in a conquered country, and indeed did not necessarily reside on them. The object was to make Euboea as much as possible a part of Attica.

13. **ἄρα** 'as it turned out'. **ὑγιὲς** 'honest', opposed to *σαθρὸν*, p. 64, l. 12.

14. **διφασίας ἰδέας** 'two different plans'. *ἰδέα=genus*, p. 69, l. 14.

16. **τὰ ἄκρα τῆς Εὐβοίης.** The mountainous district which formed the centre of Euboea.

17. **οἴσεσθαι** 'that they would get for themselves', p. 59, l. 3.

18. **τούτων ἑκάτερα** 'both of these designs', i.e. that of retiring into the interior, and of making terms with the Persians.

19. **τὰ πρῶτα** 'the leading man among the Eretrians', 9, 78 *Λάμπων...Αἰγινητέων τὰ πρῶτα.* Cp. 3, 157 *πάντα εἶναι* 'to be all in all'.

20. **τοῖσι ἥκουσι τῶν Ἀθηναίων** 'those of the Athenian cleruchs who came from the Chalkidian territory to help in the defence of Eretria'.

22. **ἵνα μὴ προσαπόλωνται** 'that they might not perish as well as themselves'. The *dramatic* subjunctive in a clause depending on a verb in an historic tense. Aeschines' own words would have been *ἵνα μὴ προσαπόλησθε.* See on p. 5, l. 5.

CHAPTER CI.

26. **κατέσχον τὰς νέας** 'put in their ships', does not differ materially from *προσέσχον* in l. 1. Like *κατάγειν* it means to bring *down* to shore, as opposed to *ἀνάγειν* to put to sea, as though the sea were higher than the land, cp. 8, 41 *οἱ ἄλλοι κατέσχον ἐς Σαλαμῖνα.*

τῆς Ἐρετρικῆς χώρης 'in the territory of Eretria', a topographical genitive, p. 54, ll. 3 and 13.

58 1. **ἐξεβάλλοντο** 'set about disembarking'.

3. **οἱ δὲ Ἐρετριέες...ἔμελε πέρι** 'but it was not the design of the Eretrians to sally out and give battle; rather, they were intent upon defending their walls if they could by any means'. For **ἐποιεῦντο βουλὴν** = *ἐβουλεύοντο*, see on p. 32, l. 22; p. 60, l. 11. For **εἴ κως... ἔμελε**, cp. p. 27, l. 26.

5. **τούτου ἔμελε πέρι**, cp. 8, 65 *περὶ στρατιῆς τῆσδε θεοῖς μελήσει*. **ἐνίκα** *placuit* 'it was decided', cp. 8, 9 *πολλῶν λεχθέντων ἐνίκα...νύκτα μέσην παρέντες πορεύεσθαι*. With a nominative case, p. 63, l. 16.

7. **ἐπὶ ἓξ ἡμέρας** 'extending over 6 days', cp. 9, 8 *τοῦτο δὲ καὶ ἐπὶ δέκα ἡμέρας ἐποίεον*.

11. **τοῦτο μὲν...τοῦτο δὲ**, p. 14, l. 10; p. 38, l. 14.

12. **τῶν ἐν Σάρδισι κατακαυθέντων**, i.e. when the Athenians took Sardis, see 5, 102 *καὶ Σάρδιες μὲν ἐνεπρήσθησαν, ἐν δὲ αὐτῇσι καὶ ἱρὸν ἐπιχωρίης θεοῦ Κυβήβης, τὸ σκηπτόμενοι οἱ Πέρσαι ὕστερον ἀντενεπίμπρασαν τὰ ἐν Ἕλλησι ἱρά*. B.C. 500. See 8, 33—5, 130. For **κατακαυθέντων**, see on p. 17, l. 12.

CHAPTER CII.

17. **κατέργοντές τε πολλὸν** 'making much haste', 'pressing on eagerly': in 5, 63 it is transitive, and some explain it here as governing *τοὺς Ἀθηναίους* 'pressing the Athenians closely'. For **πολλὸν**, cp. p. 46, l. 8. **ταὐτὰ τὰ καί**, see on p. 52, l. 24.

19. **καὶ...γάρ**, p. 6, l. 11; p. 51, l. 17, and Index.

19, 20. **ἐπιτηδεώτατον...ἐνιππεῦσαι** 'the most suitable district in Attica for cavalry to ride in'. Cp. 9, 2 *χῶρος ἐπιτηδεότερος ἐνστρατοπεδεύεσθαι*. Attica as a rule is mountainous and not fitted for cavalry, 9, 13 *οὔτε ἱππασίμη ἡ χώρη ἦν ἡ Ἀττική*, the only good plain being that of Athens itself. The plain of Marathon, though not very large, was naturally selected for this purpose; and Hippias knew of it from experience, having landed there with his father Peisistratos when he came to recover his rule, more than 30 years before [1, 61—2]. **τῆς Ἀττικῆς**, topographical genitive, cp. p. 57, l. 26.

CHAPTER CIII.

24. **ὁ δέκατος.** The Strategi were elected annually, one from each of the ten tribes. The order in the list of each Strategus

would depend on that of his tribe, which was probably decided by lot.

25. **κατέλαβε** 'it befell', cp. p. 20, l. 6.

2. **Ὀλυμπιάδα ἀνελέσθαι**, p. 18, l. 30. **τεθρίππῳ** 'with a four-horse chariot', cp. p. 18, l. 14. 59

3. **ἀνελόμενον...ἐξενείκασθαι**, sc. *συνέβη*, 'and in carrying off this victory it was his fortune to win exactly the same as his half-brother'. For the change to the accusative with infinitive, cp. p. 11, l. 21.

5. **μετὰ δὲ**, *adverbial* 'and afterwards', p. 2, l. 18, and Index. **τῇ ὑστέρῃ** 'at the next Olympiad'. **τῇσι αὐτῇσι ἵπποισι** 'with the same mares'. Mares were preferred to horses for racing purposes.

6. **ἀνακηρυχθῆναι** 'to be proclaimed victor', with the name of his father and country, as we said on p. 18, l. 14.

9. **ἀποθανεῖν** 'to be put to death'. The word is used as passive of *κτείνω*. For Kimon's murder cp. p. 20, l. 15.

10. **οὐκέτι περιεόντος...Πεισιστράτου**, that is, after B.C. 427, the year in which Peisistratos died.

11. **κατὰ τὸ πρυτανήϊον**, see on p. 20, l. 7. The Prytaneium at Athens, originally on the Akropolis, and afterwards near the Agora, was the office of the **Πρυτανεῖς** or bouleutae of the presiding tribe. It was also the common house or home of the city, where a meal was provided for such persons as were assigned that honour, and where a fire was always kept burning on the *ἑστία* of the city.

12. **ὑπείσαντες** 'having placed in ambush'. The obsolete verb *ἕω* 'to set' is only used in the aorist active, and in the future (*εἴσομαι*) and aorist (*ἑσσάμην*) middle. The perfect passive is *ἧμαι* 'I sit'. The idea of secrecy is given by *ὑπό*. Cp. 3, 126 *κτείνει μιν ὄπισθε κομιζόμενον ἄνδρας οἱ ὑπείσας*.

13. **πέρην...ὁδοῦ** 'across the road leading through the place called the Hollow', apparently on the S.W. of the town, approached from what was afterwards the Gate of Melite. **Κοίλη** was also the name of an Attic deme, and this may be meant. The author of the life of Thucydides says that the monument of Kimon (**Κιμώνια** *μνήματα*) was *ἐν* **Κοίλῃ**. Tombs were erected mostly in the Keramikos through which the Sacred Way went.

14. **τετάφαται**. This perf. passive of *θάπτω* seems to occur nowhere else. The MSS. vary between *τετάφαται* and *τεθάφαται*. App. D. II. *a*.

17. **δὴ**, resuming the thread of the story, cp. p. 53, l. 8.

CHAPTER CIV.

23. δὴ ὦν = δ' οὖν, cp. p. 44, l. 20.

24. ἐστρατήγεε Ἀθηναίων 'was holding the office of general at Athens', i.e. one of the ten generals for the year.

26. ἀναγαγεῖν παρὰ βασιλέα, cp. p. 15, l. 22.

28. ἐς τὴν ἑωυτοῦ, sc. γῆν, cp. p. 9, l. 8.

29. ὑποδεξάμενοι 'having received him with hostile feelings', 'lay in wait to attack him'. The word is used by Xenophon (*Cyr.* 2, 4, 20) of the hunters waiting outside the covert which the others are beating.

30. ἐδίωξαν τυραννίδος 'prosecuted for tyranny'. The prosecution for τυραννίς was probably in virtue of some decree (ψήφισμα) of the Ecclesia passed after the expulsion of the Peisistratidae. We have no example of such a procedure, but the declaration against a τυραννίς was part of the oath of the dicasts [Demosth. 746]. The very fact however of Miltiades being subject to Athenian law shows that he had not been a τύραννος in the ordinary sense. He ruled the Chersonese as a representative of the Athenians (who always claimed it afterwards), and not ceasing to be an Athenian citizen could be prosecuted for misuse of his office, as any other magistrate could be on his audit. See on p. 20, l. 12.

60 1. αἱρεθεὶς 'elected'. The Strategi were always elected, unlike the Archons and members of the Boulè who were appointed by lot.

CHAPTER CV.

3. ἐόντες ἔτι ἐν τῷ ἄστεϊ, i.e. before starting for Marathon.

5. ἄλλως...τε καὶ τοῦτο μελετῶντα 'who, besides that he was a professional runner, made a special practice of this sort of feat', i.e. of doing long distances quickly. See on 9, 12, and cp. Livy 31, 24 *hemerodromos vocant Graeci ingens die uno cursu emetientes spatium.*

8. Πὰν was a god of the mountains and woods, and was worshipped especially in Arkadia. περιπίπτει 'fell in with', cp. 8, 94.

10. ἀπαγγεῖλαι δι' ὅ τι 'to take a message to the Athenians (asking) them why'. δι' ὅ τι = διὰ τί.

11. ἐπιμέλειαν ποιεῦνται = ἐπιμελοῦνται, cp. p. 32, l. 22; p. 58, l. 3.

12. **καὶ πολλαχῇ...τὰ δ' ἔτι καὶ** 'who had already been serviceable to them on many occasions, and was likely to be so on others besides'.

13. **καταστάντων...πρηγμάτων** 'when their affairs were safely established'; i.e. after the Persian wars.

15. **ὑπὸ τῇ ἀκροπόλι**, in a grotto on the N.W. side of the Akropolis.

16. **ἀπὸ** 'in consequence of', p. 76, l. 6.

17. **λαμπάδι** 'with a torch race'. The torch races, or *λαμπαδηφορίαι*, were principally in honour of Hephaestos (8, 98) or of the Thracian Artemis or Bendis. Sometimes it was a race between individuals, each carrying a torch, the winner being the first man to arrive at the goal with his torch alight; at others it consisted of a trial of speed between two or more lines of men stationed at definite intervals, and handing on a torch. Each man carried it along the distance between him and the next, and handed it over. That line won in which the torch was first carried still lighted to the end. Hence the Latin lines referring metaphorically to the generations of men: Persius 6, 61 *Qui prior es cur me in decursu lampada poscis?* 'Why do you, who are a post farther on, ask me for my torch while I am still running my lap?' Lucret. 2, 77 *et quasi cursores vitai lampada tradunt.* The race was from an altar in the Academy to the city gate (Pausan. 1, 30, 2), and was started by throwing down a torch, or by the sound of a trumpet [Schol. on Aristoph. *Ran.* 129]. Sometimes it was run on horseback. Plato, *Rep.* 328 A.

CHAPTER CVI.

19. **ὅτε περ** 'which was the occasion on which he said that Pan appeared to him'.

20. **δευτεραῖος.** The distance was about 150 miles; which was a considerable run for less than two days, as it must have been, seeing that he arrived on the 2nd day early enough to visit the magistrates at once.

24. **πρὸς** 'at the hands of', cp. p. 5, l. 7.

26. **πόλι λογίμῳ** 'weaker by the loss of a considerable city'. Dative of quantity or amount by which.

4, 5. **εἰνάτῃ...μὴ οὐ πλήρεος** 'it was impossible without a breach of their law to go out on an expedition on the 9th until the moon was

61

full'. It is not possible to decide how far this was really dictated by fidelity to an ancient custom; and how far it was an excuse for selfish delay. Plutarch, *de Malign. Her.* 26, denies the whole story, and asserts that the battle was fought on the 6th of Boedromion (21 Sept.), and if the full moon was on the 12th of September [17 Metageitnion] the Spartans would have had ample time to reach Marathon. Böckh however has shown that Plutarch has in all probability confused the day of the commemorative feast with the actual day of the battle. The most probable account is that the month was the Spartan Karneios, the Attic Metageitnion (August—September), in which the Dorians celebrated the Karneia from the 7th to the 15th (7, 206). The 9th of Karneios would be the 5th of September, and the Spartans refused to start till the full moon (the 12th). They started the next day, 13th, and arrived at Marathon on the fourth day's march (16th), the day after the battle, the 15th of September. For **ἱσταμένου** see on p. 30, l. 11.

5. **μὴ οὐ...ἐόντος**, cp. p. 5, l. 7. "*μὴ οὐ* is sometimes put (but in Attic prose rarely) with participles, or other accessory definitions denoting an exception from the negative or quasi-negative statements of the principal sentence". Madvig § 211 c.

CHAPTER CVII.

9. **τῆς παροιχομένης νυκτὸς**, i.e. in the night before the disembarkation.

12. **κατελθών**, see p. 3, l. 5. Cp. Arist. *Ran.* 1165 *φεύγων δ' ἀνὴρ ἥκει τε καὶ κατέρχεται.* **ἀνασωσάμενος** 'having regained', cf. 3, 65 *μὴ ἀνασωσαμένοισι τὴν ἀρχὴν μηδ' ἐπιχειρήσασι ἀνασώζειν.*

13. **ἐν τῇ ἑωυτοῦ**, sc. *γῇ*, cp. p. 59, l. 28.

14. **συνεβάλετο**, cp. p. 45, l. 5.

15, 17. **τοῦτο μὲν...τοῦτο δὲ** 'in the first place'...'in the second place', p. 14, l. 10; p. 23, l. 15.

19. **διέτασσε** 'set about posting them'.

20. **οἷα**, with participle, 'as was natural', p. 33, l. 9.

21. **ἐσείοντο** 'were loose', 'were liable to be shaken'.

22. **ὑπὸ βίης** 'violently'. In such phrases *ὑπὸ* indicates the attendant circumstances, cp. Eur. *Hipp.* 1299 *ὑπ' εὐκλείας θανεῖν.*

23. **ἐποιέετο πολλὴν σπουδὴν** 'was very eager', see on p. 10, l. 22.

27. **μετῆν** 'was fated to be mine'. For such disappointing fulfilments, see on p. 45, l. 5.

CHAPTER CVIII.

2. **ἐξεληλυθέναι** 'had had its fulfilment', p. 45, l. 20. 62

ἐν τεμένεϊ Ἡρακλέος. The precinct of Herakles is believed to have been on the plain below the hill on which the temple stood. The people of Marathon were especially devoted to the worship of Herakles, and claimed to have introduced it into Greece [Paus. 1, 32, 4]. The sons of Herakles were said to have resided there for some time when exiled from the Peloponnese [Apoll. 2, 8, 2].

4. **ἐδεδώκεσαν σφέας αὐτοὺς** 'had committed themselves to the protection of Athens'. According to Thucydides (3, 86) this took place 29 years before, i.e. B.C. 519. Grote (vol. 4, p. 94) maintains that it must have been after the expulsion of Hippias B.C. 510; principally on the ground (not a very good one) that before that date Kleomenes would have had no motive for wishing to embarrass Athens, he being a close friend of Hippias.

9, 10. **παρατυχοῦσι Κλεομένεϊ...Λακεδαιμονίοισι** 'to Kleomenes and the Lakedaemonians who happened to be in their neighbourhood' We only hear of two invasions of Attica by Kleomenes, the first in B.C. 510 to expel the Peisistratids in accordance with the corrupt oracle [5, 63—4]: the second, which is Grote's date for this occurrence, in B.C. 508—7 undertaken to restore Hippias [5, 74—6]. The date B.C. 519, if it is correct, must refer to some earlier invasion of which we have no account, and to which we can assign no known motive. The Plataean envoys probably visited Kleomenes in the neighbourhood of Eleusis, where the road descended from Kithaeron leading from their town.

12. **ἐπικουρίη ψυχρὴ** 'but a cold (i.e. ineffective) defence'. Cp. 9, 4 *ψυχρὴ νίκη*.

13. **φθαίητε...ἤ** 'you would be enslaved before any one of us heard of it'. Cp. Hom. *Il.* 23, 444 *φθήσονται τούτοισι πόδες καὶ γοῦνα καμόντα ἢ ὑμῖν*, cp. p. 52, l. 11 *ἔφθησαν ἐκπεσόντες...πρότερον...ἤ*.

15. **πλησιοχώροισι.** The Attic frontier along Kithaeron coincided with that of Plataea on the Northern slope.

17—19. **οὐ...οὕτω...ὡς** 'not so much from goodwill to the Plataeans as from a wish to embarrass the Athenians by bringing them into hostility with the Boeotians'. **συνεστεῶτας**, p. 15, l. 12.

21. **τοῖσι δυώδεκα θεοῖσι** 'engaged in a sacrifice to the twelve gods', accompanied as usual by a feast. An altar of the twelve gods

was erected by Peisistratos, grandson of the tyrant (Thucyd. 6, 54), and was subsequently enlarged. It was in the Agora, and served as the starting-point for the measurement of distances from Athens. The twelve gods are the chief Olympian Deities.

22. **ἱκέται** 'as suppliants', that is to the whole people. Grote argues that had this been before B.C. 510 it would have been to Hippias that they would have addressed themselves. There does not however seem anything in the words of Herodotos inconsistent with the prayer having been addressed to Hippias, who, if he were tyrant at the time, would in all probability have been presiding at the sacrifice. This rejection by the Lakedaemonians, and this advice to the Plataeans to put themselves under the Athenian protection, was appealed to in their justification by the Plataean orator when pleading before the Spartan tribunal in B.C. 427 [Thucyd. 3, 55].

23. **Θηβαῖοι.** The Thebans attacked the Plataeans on the ground of their quitting the Boeotian league by joining Athens; thereby assuming the position they always, and often successfully, claimed of head of the league.

26. **ἐπιτρεψάντων ἀμφοτέρων** 'on both sides having submitted the matter to their arbitration'. The Korinthians had before acted as mediators, and shown a friendly spirit to Athens [5, 7, 92—3].

63 2. **ἐᾶν** 'that the Thebans should not use compulsion to those of the Boeotians who did not wish to belong to the Boeotian league'. **μὴ βουλομένους**, indefinite negative, *qui nollent.* **τελέειν ἐς**, see p. 29, l. 3. This decision of the Korinthians is noticeable as showing the view in Greece, at this period, of these combinations or leagues. There was as yet no distinct idea of a central government, whose first duty is to secure the loyalty of its component parts. They are voluntary combinations for special purposes, primarily religious; each member retaining full independence. Athens after the Persian wars endeavoured to act in a different spirit towards the subject allies: but the time was not come even then for forming a political league; and her claim to empire was in the end successfully resisted.

3. **ταῦτα γνόντες** 'having given this decision', p. 47, l. 19.

4. **ἐπεθήκαντο**, cp. p. 32, l. 13.

8. **πρὸς Πλαταιέας καὶ Ὑσιάς** 'in the direction of Plataea and Hysiae'. The latter was a village a few miles to the East of Plataea, also on the N. slope of Kithaeron, and through it the road from Thebes to Athens passed. It was therefore important to have it well within a friendly district.

9. **δή**, summing up and dismissing a story, cp. p. 40, l. 20.

11. **τότε** 'on the occasion we are now speaking of'.

CHAPTER CIX.

13. **οὐκ ἐώντων** *vetantium*. **ὀλίγους συμβαλεῖν** 'too few to engage'. Cp. 7, 207 *ἐόντων αὐτῶν ὀλίγων στρατὸν τὸν Μήδων ἀλέξασθαι*, Thucyd. 1, 50, 6 *μὴ αἱ σφέτεραι δέκα νῆες ὀλίγαι ἀμύνειν ὦσι*, 2, 61, 2 *ταπεινὴ ὑμῶν ἡ διάνοια ἐγκαρτερεῖν ἃ ἔγνωτε* 'too broken to endure'.

16. **ἐνίκα** 'was about to prevail', see on p. 58, l. 5. The votes appear to have been equal among the ten, and therefore the result would naturally be inaction.

17. **γάρ**, cp. p. 6, l. 11; p. 51, l. 17. **ψηφιδοφόρος** [*ψηφίς, -ῖδος*] 'possessed of a right of voting'.

ὁ τῷ κυάμῳ λαχὼν πολεμαρχέειν 'he who had been assigned by lot to the office of Archon Polemarchos', that is the 3rd of the 9 Archons; the first of all being *the* Archon who gave his name to the year (Eponymus), and the second the Archon Basileus. The other six were called Thesmothetae. He speaks of him as being appointed by lot (*τῷ κυάμῳ λαχών*) in contradistinction to the Strategi who were elected, p. 60, l. 1.

18. **τὸ παλαιόν** 'formerly'. When Herodotos wrote, the change, inevitable in the case of a people constantly engaged in military expeditions, whereby the command in the field was wholly in the hands of the elective officers, had already taken place. The Polemarch became, like the other archons, a civil magistrate; his particular function being to act as magistrate in the preliminary trials in suits in which aliens or foreigners were engaged.

22. **ἐν σοὶ νῦν ἐστὶ** 'on you depends', cp. 8, 60 *ἐν σοὶ νῦν ἐστὶ σῶσαι τὴν Ἑλλάδα*. See 8, 118.

23. **μνημόσυνα λιπέσθαι** 'to leave behind a memorial *of yourself*'.

24. Harmodios and Aristogeiton were of Aphidna also, like Kallimachos.

25. **δή**, emphatic, 'for at this moment above all others'.

64

2. **ὑποκύψωσι** 'submit', p. 13, l. 20.

3. **περιγένηται** 'conquer', cp. l. 14. It also means 'survive' 5, 46 *μοῦνος περιεγένετο τούτου τοῦ πάθεος*.

4. **οἵη τε** 'it is capable of becoming', see on p. 5, l. 7.

6. **καὶ κῶς ἐς σέ τοι...ἀνήκει** 'and how it has fallen precisely

upon you'. **σέ τοι** 'you and no one else'. Soph. *Aj.* 360 *σέ τοι, σέ τοι μόνον δέδορκα.*

7. **τὸ κῦρος ἔχειν,** 'to have the decision', 'the power of determining'.

9. **τῶν δὲ οὔ,** sc. *κελευόντων.*

10. **ἔλπομαι** 'I expect', 9, 113 *ὁ δὲ ἐλπόμενός τί οἱ κακὸν εἶναι.*

12. **τι καὶ σαθρὸν** 'some dishonourable sentiment', opposed to *ὑγιές* p. 57, l. 13. This is the ordinary meaning of the word, and it does not seem necessary to explain it as a metaphor from a leak in a ship as Stein does. [Connected perhaps with *σήθω* to sift.]

13. **μετεξετέροισι** 'to some of the Athenians', p. 40, l. 2. There was always a medizing party even at Athens. In fact the Athenians had been the first to set the example of appealing to Persia [5, 73]; some treason was afterwards (c. 115) shown to exist at this time; and eleven years later, some of the oligarchical Athenians met at Plataea and designed to submit to Persia [Plut. *Aristid.* 13].

14. **θεῶν τὰ ἴσα νεμόντων,** cp. p. 6, l. 20. **οἷοί τε...περιγενέσθαι,** see ll. 3 and 4.

15. **ἐς σὲ νῦν τείνει** 'depends on you',—much the same as *ἀνήκει* in l. 6. Cp. Eurip. *Phoen.* 438 *ἐς σὲ τείνει τῶνδε διάλυσις κακῶν* (St.). **καὶ ἐκ σέο ἤρτηται** 'and hangs upon your decision', cp. 9, 68 *πάντα τὰ πρήγματα τῶν βαρβάρων ἤρτητο ἐκ Περσέων.*

16. **προσθῇ** 'if you give your adherence to', cp. 1, 109 *οὔ οἱ ἔγωγε προσθήσομαι τῇ γνώμῃ.*

18. **τὴν τῶν ἀποσπευδόντων,** sc. *γνώμην.*

19. **ἕλῃ** 'adopt'.

CHAPTER CX.

21. **προσκτᾶται** 'gained over', cp. 8, 136 *τοὺς γὰρ Ἀθηναίους οὕτω ἐδόκεε μάλιστα προσκτήσεσθαι.*

23. **ἐκεκύρωτο** 'it had been decided'. The force of the pluperfect is explained by the next clause.—'Though it had been decided to fight, yet the four generals, to make it still more certain, surrendered their days to Miltiades'. Why then did he not fight at once? Mr Grote thinks it could have been no mere punctilio about acting on his own day. And perhaps that alone would not have decided it: although he may have considered the extra danger of prosecution at home if he acted unsuccessfully on a day other than his one of legal command; just as afterwards Aratos was placed in some danger by taking over the

seal of the Achaean league some days before he was legally in office and failing in his movements [see Polyb. IV. 14]. But still there was doubtless something else. What Miltiades wanted, I think, was not necessarily to make the attack at once; but the power of making it whenever he chose. He was apparently aware from spies, or other sources, of a movement about to take place in the Persian camp, which would give him the opportunity of charging them when in the hurry and disorder of an embarkation. The traitorous signal of the flashing shield (c. 115) afterwards displayed shows that some intrigues were already going on, of which he was likely to have got intelligence; and he would be therefore waiting for the right moment to strike. This view of his action is supported by the article in Suidas s.v. χωρὶς ἱππεῖς, who says that the Ionians affirmed that, when Datis had gone on board with the cavalry, they signalled to Miltiades χωρὶς οἱ ἱππεῖς 'the cavalry are gone', and that then he charged.

24. **ἔφερε** 'inclined', 5, 118; 8, 100.

25. **πρυτανηΐη** 'chief command'. Herodotos uses the word in a general sense of being first, without reference to any technical title; just as he used πρυτανήϊον p. 20, l. 7 for a public hall or court-house.

26. **συμβολὴν ἐποιέετο** = συνέβαλλε, see on p. 10, l. 22.

27. **πρίν γε δὴ** 'absolutely until', p. 44, l. 24.

CHAPTER CXI.

65

1. **ἐνθαῦτα δὴ** 'thereupon', p. 7, l. 22.

4. **νόμος τότε εἶχε...δεξιόν** 'it was then the law at Athens that the Polemarch should command the right wing',—that is, as the post of honour formerly occupied by the king, to the warlike part of whose functions the Polemarch succeeded. Herodotos says τότε, as at the time of his writing these functions of the Polemarch had fallen into disuse, see on p. 63, l. 18.

6. **ἐξεδέκοντο...αἱ φυλαὶ** 'the other tribes were stationed successively according to their order'. The tribe of Kallimachos, *the Aantis*, would be with him; the others under their several Strategi in the order in which they had been numbered by lot, for the various purposes for which the tribal magistrates and bouleutae had to serve in turn.

9. **θυσίας ἀναγόντων** 'when offering sacrifices', cp. 2, 60 μεγάλας

ἀνάγοντες θυσίας. The word perhaps originally referred to the leading up the victims to the altars placed on high spots, such as the Akropolis.

10. **ἐν τῇσι πεντετηρίσι** 'at the quinquennial festival', i.e. at the Great Athenaea, see p. 50, l. 25.

12. **γίνεσθαι** depends upon *κατεύχεται*, 'prays that blessings may befall the Plataeans as well as the Athenians'. **λέγων** is pleonastic. He prays saying: 'May blessings etc.'. Cp. ἔλεξε φάμενος p. 45, l. 19. The festival, like other public business at Athens (as at Rome also), was opened by a solemn prayer. See a serio-comic version of such a prayer in Aristoph. *Thesmoph.* 295. **τὰ ἀγαθὰ** 'such things as were good for them'. St. compares Xen. *Memor.* 1, 3, 2 *εὔχετο πρὸς τοὺς θεοὺς ἁπλῶς τἀγαθὰ δοῦναι.*

13. **ἐγίνετο τοιόνδε τι** 'a result was obtained something like this'. The numbers of the Persian army have been stated variously by different authors from 110,000 to 600,000. But, whatever they were, they were doubtless vastly superior to those of the Athenians: and that the line of the latter should have been of equal length (**ἐξισούμενον**) with theirs is to be accounted for by the narrowness of the ground available for the Persians; and also probably by the fact that a large number were already reembarked, and especially the cavalry, which would have been on the two wings, thus greatly extending the line.

15. **ἐπὶ τάξιας ὀλίγας** 'very few deep', cp. 9, 31 *ἐπὶ τάξεις πλεῦνας.*

17. **ἔρρωτο** [*ῥώννυμι*] 'had been strengthened'. This word is generally used (1) of physical health 'to be well', Thucyd. 7, 15, 2; (2) of mental feelings 'to be encouraged', Thucyd. 2, 8, cp. 8, 15 *ἐπέρρωσαν* 'encouraged them'.

CHAPTER CXII.

18. **τὰ σφάγια ἐγίνετο καλὰ** 'the sacrifices became favourable', i.e. for attack. For this waiting to charge until the sacrifices showed good omens cp. the conduct of the Lakedaemonians at Plataea 9, 61—2. There however the Tegeans appear to have begun the charge before the sacrifices became favourable. It may perhaps be suspected that a good commander took care that these favourable omens should come at the time he thought best for moving. In Xen. *Hell.* 4, 2, 18 we are told that in a battle, while the Boeotians were opposite the Spartans,

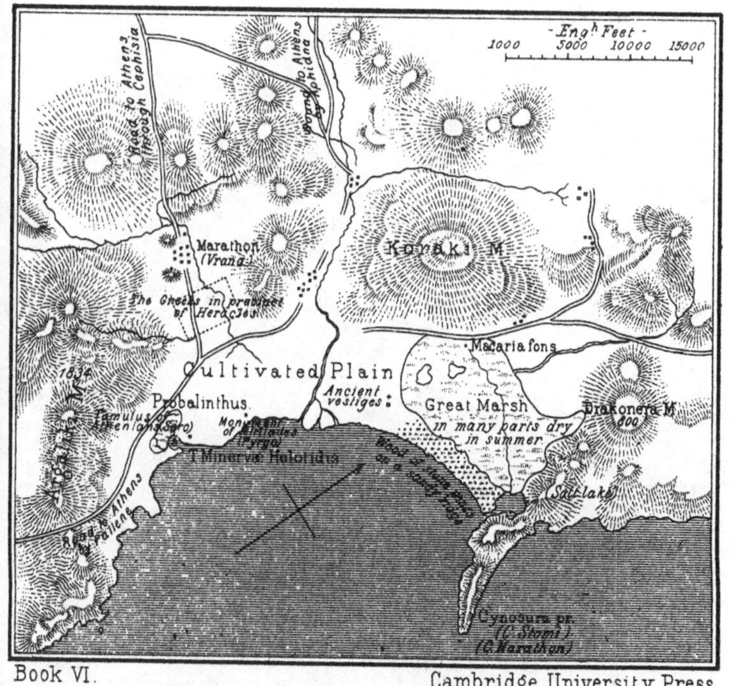

Book VI.

Cambridge University Press.

SITE OF THE BATTLE OF MARATHON

the omens continued unfavourable for a charge; but directly, by a change of order, they became opposite the Achaeans, the omens at once became favourable: which at any rate was a convenient coincidence.

19. **ὡς ἀπείθησαν** (ἀπίημι) 'when they were let go', as though they had been longing to charge, and at last got the order, cp. 7, 122 ὁ ναυτικὸς στρατὸς ὡς ἀπείθη ὑπὸ Ξέρξεω.

20. **δρόμῳ ἵεντο** 'they started at the double to charge the barbarians'. This of itself would tend to show that the Persians were not in battle array, but in all probability engaged in the embarkation; for no general would lead his army at the double for a mile before reaching an enemy drawn up to receive them. Arriving in a panting and partly exhausted state they would be in a poor condition for attacking an enemy standing fresh and in good order.

στάδιοι ὀκτώ, nearly a mile, p. 19, l. 7. **μεταίχμιον**, p. 43, l. 16.

22. **παρεσκευάζοντο** 'began to get ready'. So that they were not in line, and were taken by surprise by the movement.

23. **καὶ πάγχυ ὀλεθρίην**, sc. μανίην, 'they attributed madness to the Athenians, and a madness that would undoubtedly lead to their utter destruction'.

25. **οὔτε ἵππου...τοξευμάτων** 'though they had neither cavalry nor archers' to cover their charge. The Persian method was to begin by pouring in volleys of arrows before charging, see 9, 61.

26. **μέν νυν**, p. 24, l. 13. **κατείκαζον** 'they surmised', i.e. that they were mad, cp. καταδοκέω, p. 9, l. 17.

66 2. **πρῶτοι...δρόμῳ.** What Herodotos regards as an innovation—charging at the double—seems, as I have said, only a measure adopted under particular circumstances: from the desire, that is, of reaching the enemy while in disorder, and before they had time to form.

4, 5. **πρῶτοι...ἐσθημένους.** It has naturally been often pointed out that Herodotos has already himself described several instances of battles between Hellenes and Persians: see 1, 169; 5, 2, 102, 110, 113, 120; 6, 28. At the same time we may notice that, in all these cases, it was Asiatic Hellenes who were engaged, and that they were in every case signally defeated. Herodotos' words are not an exaggeration therefore as far as the Hellenes of Europe were concerned, who had never come face to face with Persians, and had heard of their constant victories over the Ionians.

CHAPTER CXIII.

9. **τὸ μὲν μέσον** 'in the centre'.

11. **ἐτετάχατο**, App. D. II. *a*.

12. **ῥήξαντες**, sc. *τὸ μέσον*, see l. 15.

15. **συναγαγόντες τὰ κέρεα.** The right and left wing of the Athenians closed in, not apparently so as to get between the Persian centre and the sea, but so as to charge them on either flank as they were returning from the pursuit of the Athenian centre.

18. **πῦρ τε αἴτεον** 'they called for fire and began laying hands on the ships', which were being filled and pushed off by the fugitive Persians as well as those who had gone on board before. See note on p. 67, l. 1. This particular scene in the battle was represented in the picture in the Stoa Poikile. There some of the Persians were shown, some pushing each other into the marsh, others being killed by the Athenians as they crowded into the ships. The two most conspicuous figures were those of Kallimachos and Miltiades [Paus. 1, 14].

CHAPTER CXIV.

20. **τοῦτο μὲν...τοῦτο δὲ**, p. 61, l. 15.

22. **ἀπὸ δ' ἔθανε.** Herodotos frequently separates preposition and verb, cp. 8, 33 *κατὰ δ' ἔκαυσαν.* 8, 89 *ἀπὸ μὲν ἔθανεν ὁ στρατηγὸς...ἀπὸ δὲ ἄλλοι πολλοί.* 9, 5 *κατὰ μὲν ἔλευσαν αὐτοῦ τὴν γυναῖκα, κατὰ δὲ τὰ τέκνα.*

67 1. **ἐπιλαβόμενος τῶν ἀφλάστων νεὸς** 'laying hands on the curved sterns (*aplustria*) of the ships'. The scene is described almost in the words of Homer, *Il.* 15, 717:

> *Ἕκτωρ δὲ πρύμνηθεν ἐπεὶ λάβεν, οὐχὶ μεθίει,*
> *ἄφλαστον μετὰ χερσὶν ἔχων, Τρωσὶν δὲ κέλευεν*
> *οἴσετε πῦρ, ἅμα δ' αὐτοὶ ἀολλέες ὄρνυτ' ἀϋτήν.*

Kynegeiros, the hero of this anecdote, was a brother of the poet Aeschylos. For later additions to the story, see Biographical Index.

2. **τοῦτο δὲ** 'and in the third place', an extension of the ordinary *τοῦτο μὲν—τοῦτο δὲ*, cp. p. 71, l. 3.

CHAPTER CXV.

6. **ἐξανακρουσάμενοι** 'having shoved themselves off from land', cp. *ἐξαναχθεὶς* 8, 84. The ships were on shore with their sterns as usual

toward land, cp. Eurip. *Iph. T.* 1349 sqq. ἀνακρούεσθαι is not therefore used in the sense of 'to back water', as in 8, 84 ἐπὶ πρύμνην ἀνεκρούοντο. It refers to the pushing off by long poles (κοντοί).

ἐκ τῆς νήσου, sc. Styra, see cc. 101, 107.

8. **περιέπλωον Σούνιον** 'they began rounding the headland of Sunium'.

9. **αἰτίη ἔσχε** 'an accusation arose at Athens'. ἔσχε=κατέσχε 'prevailed', or perhaps we may take it ἔσχε, sc. αὐτοὺς as in 5, 70 [Stein reads αἰτίην ἔσχε and translates 'it roused suspicion'].

11. **συνθεμένους** 'having agreed with the Persians on a signal', p. 72, l. 2.

12. **ἀναδέξαι ἀσπίδα** 'displayed a shield'. A bright shield serving as a flashing signal, cp. 7, 128 ἀνέδεξε σημήϊον. Vergil, *Aen.* 10, 260 *Iamque in conspectu Teucros habet et sua castra, Stans celsa in puppi, clipeum cum deinde sinistra Extulit ardentem.* So also Lysander at Aegospotami orders his men ἆραι ἀσπίδα κατὰ μέσον τὸν πλοῦν [Xenoph. *Hell.* 2, 1, 27]. Diodor. Sic. 20, 51 Δημήτριος...ἦρε τὸ συγκείμενον πρὸς μάχην σύσσημον, ἀσπίδα κεχρυσωμένην. See on signalling generally note to 9, 3.

ἐοῦσι ἤδη ἐν τῇσι νηυσί 'when they were already on board their ships'. They must however have had some previous information to have induced them to embark; and were only waiting for the signal to start.

CHAPTER CXVI.

14. **ὡς ποδῶν εἶχον** 'as quick as they could march', cp. 9, 59 ὡς ποδῶν ἕκαστος εἶχον. Aesch. *Suppl.* 837 σοῦσθε, σοῦσθε ἐπὶ βᾶριν ὅπως ποδῶν. Herodotos does not say that this return was on the same day as the battle; and it seems unlikely that it should have been so. Plutarch, *Arist.* 5, says that nine of the tribes, on seeing that the Persians were rounding Sunium, hurried back to Athens and arrived home on the same day (αὐθήμερον), but he seems to mean that they did the march (about 26 miles) in one day, not on the same day as the battle; for elsewhere (*de Glor. Ath.*) he says that Miltiades arrived home on the day *after* the battle. But whatever the authorities may say, a march of 26 miles after a long day's battle (μαχομένων... χρόνος ἐγίνετο πολλός, p. 66, l. 8) seems impossible.

18. **τῇσι νηυσί**, instrumental dative, p. 6, l. 27. **ὑπεραιωρηθέντες** 'having laid to out at sea opposite Phalerum'.

19. **τοῦτο γὰρ...τότε** 'for this was then the harbour of the Athenians', that is, before the formation of the great harbour of the Peiraeus and its fortification by Themistokles.

20. **ὑπὲρ τούτου**, repeating the last clause after a parenthesis, cp. p. 74, l. 4. **ἀνακωχεύσαντες**, sc. *νέας*, 'after riding at anchor', cp. 7, 100, 168.

CHAPTER CXVII.

68 1. **κατὰ ἑξακισχιλίους...ἄνδρας** 'about 6400 men'. These numbers were much exaggerated in later times. The epigram under the pictures in the Poikile said 200,000, and some even asserted that 300,000 were killed. Plutarch *de Mal. Herod.* 26 speaks of an *ἀνήριθμον πλῆθος*. But though at one point of the battle, when the Persians were driven into the marsh, the slaughter was probably large; it is evident that the bulk of the men got away, as only seven of the ships were taken.

6. **τῶν ὀμμάτων στερηθῆναι.** The sight of a supernatural being, god, nymph, or hero, or even of an animal to which any superstitious feeling attached, was believed to be attended with danger to mind or body. Thus madness was caused by the nymphs, dumbness by the sight of a wolf [Theocr. 14, 152]; and when Aristomenes passed the tree on which the Dioskouri were sitting, at the battle of the Boar's Pillar, he immediately lost his shield and nearly his life [Pausan. 4, 32, 4].

9. **λέγειν**, imperf. infinitive, 'was accustomed to relate', cp. p. 70, l. 17.

10. **ἤκουσα** 'I have been told'. The usual way of indicating that the writer is stating a common report, which he has not cared or been able to verify. See Demosth. *de Coron.* § 8. Lysias 19, § 5. **λέγειν** 'that he was accustomed to narrate'.

ἄνδρα...ὁπλίτην. The Athenians believed this to be the hero Theseus [Plutarch, *Themist.* 35], and accordingly he was represented in the picture in the Stoa Poikile as fighting among other gods and heroes [Pausan. 1, 15, 4]. For the appearance of national heroes on the field at times of national danger, see note on 8, 38. Such appearances are related as happening at Delphi, Salamis, Leuktra and other battles.

11. **τὸ γένειον**, the great length of beard is a mark of antiquity.

13. **τὸν παραστάτην** 'the man next him', p. 61, l. 25.

14. **μὲν δή**, p. 63, l. 9.

CHAPTER CXVIII.

16. **πορευόμενος** 'in the course of his voyage', a word generally used of land journeys, *κομιζόμενος*, or some such word, of those by sea.

20. **ἄγαλμα Ἀπόλλωνος κεχρυσωμένον** 'a figure of Apollo gilded', and which the thief probably imagined to be gold. Such statuettes of the gods were generally offerings of the faithful. So Plutarch says that in his time a small figure of Pallas [Παλλάδιον] on the Akropolis given by Nikias was still standing *with the gilding rubbed off* [*Nic.* 3].

21. **ὁκόθεν** i.e. from what temple. That the temple of Delium should have been robbed is the only indication which we have that the ships of the Persian fleet committed acts of hostility along the coast previous to the battle of Marathon.

23. **ἀπίκατο** 'they had arrived' i.e. from Rheneia where they had taken refuge. See c. 97.

29. **δι' ἐτέων εἴκοσι** 'after an interval of 20 years'. The Thebans would demand the restoration of the image as representing the Boeotian League.

1. **ἐκ θεοπροπίου** 'in consequence of an oracle', p. 36, l. 16. 69

CHAPTER CXIX.

4. **προσίσχον ἐς**, see p. 56, l. 23.

5. **ἀνήγαγον**, p. 15, l. 22 note.

7. **ἐνεῖχε...χόλον** 'was entertaining violent resentment', cp. 8, 27 *ἅτε σφι ἐνέχοντες αἰεὶ χόλον.*

ἀρξάντων...Ἐρετριέων, i.e. by furnishing ships to the Ionians when they attacked Sardis [5, 99].

10. **κακὸν οὐδέν**. For this conduct of the King, cp. cc. 30 and 40. The descendants of these men were occupying the same territory, it is said, in the third century A.D., and still retained their native language. The transference of large bodies of men from a conquered country was not an unusual measure in the East. Cp. the case of the Paeonians [5, 13], and the captivity of the Jews, about 80 years before this. See p. 2, l. 13.

11. **ἐν σταθμῷ ἑωυτοῦ** 'in a town of his own'. The σταθμὸς was a 'station', 'resting place' along a road. Thence, as a town grew up round such stations, the word is used for it and its immediate territory. Some of these σταθμοί belonged especially to the King, βασιλήϊοι σταθμοί [2, 152; 5, 52]. Stein quotes an epigram of Plato in the Anthology [7, 259]:

> Εὐβοίης γένος ἐσμὲν Ἐρετρικόν, ἄγχι δὲ Σούσων
> κείμεθα· φεῦ, γαίης ὅσσον ἀφ' ἡμετέρης.

14, 15. **τοῦ φρέατος.** Bitumen pits are common in the district of *Kir-Ab*, where Rawlinson places this Ardericca. **τριφασίας ἰδέας** 'three sorts of produce', p. 57, l. 14.

ἔλαιον. Some such pit of petroleum seems to be referred to in 2 Macc. 1, 19 where the Priests who were of the Captivity are said to have taken 'the fire of the altar privily, and hid it in a hollow place of a pit without water, where they kept it sure, so that the place was unknown to all men. Now after many years, when it pleased God, Neemias, being sent from the king of Persia, did send of the posterity of those priests, that had hid it, to the fire: but when they told us they found no fire, but thick water; then commanded he them to draw it up, and bring it; and when the sacrifices were laid on, Neemias commanded the priests to sprinkle the wood and the things laid thereon with the water. When this was done, and the time came that the sun shone, which afore was hid in a cloud, there was a great fire kindled, so that every man marvelled'.

16. **κηλωνηΐῳ.** A κηλώνειον or κήλων is a switch-pump used for draining purposes, consisting of a beam working on a pivot.

17. **γαυλοῦ**, see on p. 9, l. 24. The γαυλὸς is 'a bucket', instead of which 'a wine-skin cut in half' is used [ἥμισυ ἀσκοῦ].

ὑποτύψας 'having dipped into the well with this', 2, 136 κοντῷ ὑποτύπτοντες ἐς λίμνην.

18. **ἀντλέει** 'the workman draws it up'. This use of a singular verb to describe a process in which many are engaged is found again in 1, 195; 5, 16. **ἐς δεξαμενήν** 'into a reservoir', cf. 3, 9 δεξαμενὰς ὀρύξασθαι, ἵνα δεκόμεναι τὸ ὕδωρ σώζωσι.

19. **ἐς ἄλλο** 'into another vessel with three inlets'. The different specific gravity of the three substances would suffice to carry them off through the different channels. Rawlinson translates **τράπεται τριφασίας ὁδούς** 'here takes three different shapes'. But it seems to me that Herodotos means to describe the reservoir (δεξαμενή) as having three outlets which send the three substances out by different channels into the second vessel (ἐς ἄλλο) where they arrive divided.

23. **ὀδμὴν βαρέαν** *gravem odorem* 'an unpleasant smell'. The name *ῥαδινάκη* for this petroleum does not I believe occur elsewhere, nor is it known what Persian word it represents.

24. **μέχρι ἐμέο**, and many centuries afterwards, see note on l. 10. Herodotos seems to have visited the place.

26. **μὲν δή**, p. 68, l. 14.

CHAPTER CXX.

2. **μετὰ τὴν πανσέληνον**. That is, they started on the 13th of September, arrived at Athens on the 15th, the very day of the battle, and proceeded to Marathon presumably on the next day, to see the dead Medes before they were buried. From Sparta to Athens was about 1500 stades (about 187 miles). 70

3. **καταλαβεῖν**, the exact meaning of the words seems rather doubtful. Stein explains sc. *'Αθήνας* 'to reach Athens'. I am more inclined to accept Abicht's explanation, sc. *τὰ πρήγματα* 'to be in time to take part in the action'.

6. **μετὰ δέ**, p. 2, l. 18.

7. **τὸ ἔργον**, p. 15, l. 14

CHAPTER CXXI.

9. **τὸν λόγον** 'the assertion'. See c. 115.

10. **ἄν κοτε ἀναδέξαι** 'that they ever displayed'. The infinitive aorist with *ἄν* here stands for the modest or dubitative aorist optative after a verb *declarandi* or *sentiendi*, Madv. § 173. Cp. p. 71, l. 11 where *ἄν* is omitted, and l. 27. The *κοτε* adds to the indefiniteness of the clause without a distinctly temporal meaning, much as we use the word 'ever', cp. p. 26, l. 21.

12. **μᾶλλον ἢ ὁμοίως** 'equally, if not more so'; lit. 'more or equally', cp. p. 71, l. 9 *ὁμοίως ἢ οὐδὲν ἔσσον*.

14. **τε...καὶ** 'as'...'so', *ut...ita*.

15. **ὅκως...ἐκπέσοι** 'whenever he was driven out', for Peisistratos was twice banished. See p. 16, l. 9 for *ὅκως* with optative.

17. **ὑπὸ τοῦ δημοσίου** 'by the public slave' employed as auctioneer, this among other things being the function of the slaves of the state. See Boeckh's *Economy of Athens*, p. 207. Besides the danger of purchasing the property of a Tyrant who might return, there was always a certain discredit attaching to the buying of confiscated

property. See Polybios 39, 15, who dissuaded his countrymen from doing this after the Roman conquest in B.C. 146. And compare Cicero's bitter reproaches to Antony in the Second Philippic for having been the *sector* of Pompey's property.

ὠνέεσθαι, the imperfect infinitive, 'used to buy', cp. p. 68, l. 9.

18. **ἐμηχανᾶτο** 'was ever contriving'.

CHAPTER CXXII.

[This chapter is rejected by many editors, and regarded as an interpolation of some sophist desirous of paying court to some of the family of Kallias. The best MSS. omit it, and its phraseology is somewhat awkward and artificial, though, if it is by an imitator, it is by one who has studiously used Herodotean phrases. Stein points out also that Plutarch in his criticism on this portion of the book [*de Malign. Her.* 27] does not seem to have read it, though this does not appear to me certain.]

19. **πολλαχοῦ** 'on many accounts'.

20, 22. **τοῦτο μὲν...τοῦτο δὲ...τοῦτο δὲ**, see p. 66, l. 20. The three clauses refer (1) to his part in the liberation of Athens, (2) to his Olympic and Pythian triumphs, (3) to his conduct to his daughters.

21. **τὰ προλελεγμένα** for *κατὰ τὰ π.* 'in view of what I have said'; for omission of *κατὰ*, cp. p. 29, l. 19. **ὡς ἀνὴρ...πατρίδα** explain **τὰ προλελεγμένα.**

'In the first place on the ground of what I have mentioned (that he was a man who took a leading part in liberating his country); and in the second place by what he did in Olympia (having won the horse race, and been second in the four-horse chariot race, after having already won a race at the Pythian games) his extraordinary liberality became notorious to all Greece; and in the third place by his conduct to his three daughters,—on all these grounds it is right that every one should remember him'.

71 4. **οἷός τις ἐγένετο**, closely connected with **κατὰ τὰς θυγατέρας.**

5. **γάμου ὡραῖαι** 'of marriageable age', cp. 1, 196 *παρθένοι γάμων ὡραῖαι.*

δωρεὴν 'dowry'.

6. **ἐκείνῃσί τε ἐχαρίσατο.** This pronoun is awkward, if not wrong, after *σφι* in the line before. For **ἐχαρίσατο** without an accusative 'did them a favour', cp. p. 76, l. 9.

ἐκ γὰρ πάντων...ἐκλέξασθαι. The giving girls in marriage in Greece was the duty of their fathers or nearest male relative; but the taste of the girls does not seem generally to have been consulted.

CHAPTER CXXIII.

9. **ὁμοίως ἢ οὐδὲν ἔσσον,** cp. p. 70, l. 12.

11, 12. **οὐ προσίεμαι** 'I do not accept', p. 6, l. 4. **τούτους γε** 'that they, of all men in the world'. **ἀναδέξαι,** see p. 70, l. 10. **οἵτινες** *quippe qui* 'seeing that they were the men who'.

12. **ἔφευγον...ἐξέλιπον.** For the change of the subject of two verbs in coordinate clauses, see p. 15, l. 23. **ἔφευγον τοὺς τυράννους** 'were in exile at the instance of the Peisistratidae'. Cp. 5. 62 *Ἀλκμαιωνίδαι γένος ἐόντες Ἀθηναῖοι καὶ φεύγοντες Πεισιστρατίδας.*

16. **ὡς ἐγὼ κρίνω.** The judgment of Herodotos was that of Thucydides also, as we may gather from 6, 59. But in the popular view Harmodios and Aristogeiton were blameless heroes of supreme desert; and Herodotos would hardly have ventured to read this sentence at Athens.

17. **τοὺς ὑπολοίπους Πεισιστρατιδέων** 'those of the Peisistratids that were left'. The plural is used to include the whole family or dynasty; but the person really meant is Hippias, his brother Thessalos not taking part in politics. So in the next line *τοὺς λοιποὺς* is used, though Hippias was the sole *τύραννος.*

20. **εἰ δὴ οὗτοί γε** 'if it was really they, as they say'.

21. **οἱ ἀναπείσαντες** 'who bribed the Pythia', cp. p. 36, l. 15; 5, 62 *ὡς ὦν δὴ οἱ Ἀθηναῖοι λέγουσι οὗτοι οἱ ἄνδρες ἐν Δελφοῖσι κατήμενοι ἀνέπειθον τὴν Πυθίην χρήμασι.*

22. **μοι δεδήλωται.** For the dative of agent see p. 17, l. 3.

CHAPTER CXXIV.

23. **ἀλλὰ γάρ,** 'however', dismisses the previous point. Cp. Lysias 7, § 9. **ἴσως** introduces a supposed objection of an opponent, 'perhaps it will be said'; which is sometimes introduced by *ἀλλά*, or *ἀλλὰ νὴ Δία* *at enim.*

25. **μὲν ὦν** 'nay, rather', or, 'on the contrary',—introducing the answer to the previous objection.

26. **ἐτετιμέατο**, App. D. II. *a*.

27. **οὐδὲ λόγος αἱρέει** 'it is not even consistent with rational probability'. Cp. 3, 45; 4, 127 *ἢν μὴ ἡμέας λόγος αἱρῇ*. 7, 41 *ὅκως μιν λόγος αἱρέει*. **ἀναδεχθῆναι ἄν**, see on p. 70, l. 10.

ἔκ γε ἂν τούτων 'at least by these of all men in the world'. Observe the position of *ἂν* which belongs to *ἀναδεχθῆναι* but is placed between *ἐκ* and its noun for emphasis. For **ἐκ** = *ὑπὸ* see on p. 7, l. 21.

28. **ἐπὶ τοιούτῳ λόγῳ** 'on such grounds as this',—of their aristocratic dislike of the common people or democracy. He means that whatever may be the proved facts, there is no antecedent probability from a view of their position among the people on the side of believing the Alkmaeonidae guilty. **ἀνεδέχθη μὲν γάρ** 'for shown indeed a shield certainly was'. The *γὰρ* refers to *ἔκ γε ἂν τούτων*.

CHAPTER CXXV.

72 4. **καὶ τὰ ἀνέκαθεν** 'in earlier generations also', cp. p. 29, l. 5.

7. **τοῦτο μὲν** 'in the first place', the corresponding clause is introduced by *μετὰ δὲ* p. 73, l. 4.

8. **παρὰ Κροίσου ἀπικνεομένοισι**, when Kroesos sent to test all the Greek Oracles in B.C. 556. See 1, 53. Alkmaeon was the leader in the sacred war for the deliverance of Kirrha B.C. 595—586, Plut. *Solon* 11. It appears therefore that Herodotos has made an error in connecting Alkmaeon with Kroesos [560—546 B.C.], as he must have been rather the contemporary of his father Alyattes.

13. **τὸν ἂν δύνηται** 'as much as he could carry out at one time'.

τῷ ἑωυτοῦ σώματι 'about his own person'. The following story may be illustrated by the practice at Athens of men who were going into a house to search for stolen goods being forced to leave their *ἱμάτιον* behind. See Aristoph. *Nub.* 497—9.

15. **τοιάδε ἐπιτηδεύσας** 'having elaborately contrived the following plan'. So in the passive, 1, 98 *τὸ δὲ (χωρίον) καὶ μᾶλλόν τι ἐπετηδεύθη*.

16. **προσέφερε** 'he brought it to bear', 'he applied it'. **κόλπον** 'a fold' across the breast, to serve as a pocket.

17. **κοθόρνους** 'boots' with tops or buskins fastened high up the leg, as opposed to the ordinary sandals or shoes (*ὑποδήματα*): cp. 1, 155, where they are spoken of as a luxury; though they seem also to have been ordinarily worn by huntsmen. In Athens the term seems mostly reserved for the theatrical buskin.

20. **πρῶτα μὲν...μετὰ δὲ**: cp. l. 7 *τοῦτο μὲν* answered by *μετὰ δὲ* in p. 73, l. 4. For *μετὰ adverbial* see Index.

παρέσαξε παρὰ τὰς κνήμας 'be stuffed in on each side of his legs'.

23. **τοῦ ψήγματος**, *partitive genitive*, 'some of the gold dust'.

26. **τοῦ**, *quippe cujus*, 'for his mouth was stuffed up'. **πάντα** all parts of him. **ἐξώγκωτο** 'thoroughly puffed out',—used in a metaphorical sense in p. 73, l. 17.

27. **γέλως ἐσῆλθε** 'laughter took possession of him', 'he burst into a laugh'. Cp. 7, 46 *ἐσῆλθέ με λογισάμενον κατοικτεῖραι ὡς βραχὺς εἴη ὁ πᾶς ἀνθρώπινος βίος* 'I was struck by a feeling of pity when I calculated etc.'

28. **καὶ πρὸς** *adverbial* 'and besides', cp. p. 55, l. 1.

73

2. **τεθριπποτροφήσας**, see on p. 18, l. 14; cp. c. 122.

3. **Ὀλυμπιάδα ἀναιρέεται**, p. 39, l. 21. This again seems to be a mistake, as the first of the family to win an Olympic victory was Alkmaeon's son Megakles (Pind. *Pyth.* 7, 14), to whom perhaps the whole story may refer.

CHAPTER CXXVI.

4. **μετὰ δὲ** 'and at a later time', answering to *τοῦτο μὲν*, p. 72, l. 7. **δευτέρῃ** 'next', cp. p. 16, l. 7.

5. **μιν...ἐξῆειρε** sc. *τὴν οἰκίην* 'contributed to raise it'.

10, 11. **Ὀλυμπίων...κήρυγμα.** The Olympic and other games, being attended by men from all parts of Greece, offered convenient opportunities for the publication of any notice meant to apply to the Greeks at large, in or out of Greece proper. Another way of publishing such notices was to send them to Delphi and Delos, as places resorted to continually by visitors to the shrine or at their yearly festivals.

13. **ἥκειν** 'that he should come', the infinitive follows *κήρυγμα ἐποιήσατο* as equivalent to a verb of ordering. **ἐς ἑξηκοστὴν** 'by the 60th day'.

16. **ἐνθαῦτα** 'thereupon', p. 16, l. 18.

17. **ἐξωγκωμένοι** 'puffed up', 'proud of themselves or their country'.

19. **ἐπ' αὐτῷ τούτῳ εἶχε** 'devoted himself exclusively to this business', i.e. to entertaining his guests.

CHAPTER CXXVII.

23. **εἷς** refers to the superlative ἐπὶ πλεῖστον. It often strengthens a superlative with which it agrees. Here it may be translated 'who arrived at the greatest pitch of luxury that any single man ever did'. Cp. the use in Latin of *unus* with superlatives, e.g. *unus nequissimus*, *una pulcherrima* etc.

74 2. **Τιτόρμου**, sc. ἀδελφός.

3. **ὑπερφύντος τε...καὶ φυγόντος** 'who while surpassing all Greeks in physical strength, shunned mankind and fled to the farthest boundaries of the Aetolian territory'. Aetolia was always a country little known in the rest of Greece; its inhabitants were wild and believed to be semi-barbaric.

4. **τούτου**, for the repetition of the genitive, cp. p. 67, l. 20.

7. **τοῦ τὰ μέτρα ποιήσαντος** 'who introduced a system of weights and measures in the Peloponnese'. See Historical Index, s.v. *Pheidon*.

8. **ὑβρίσαντος μέγιστα δή**, notice the aorist, 'who committed the most outrageous acts of violence'. The Eleans had from the first had the control of the Olympic games, and from that circumstance had enjoyed by common consent immunity from violence, especially during the festival. To oust them from the management of the games, and to act as president himself, was an outrage on Greek feeling on Pheidon's part something analogous to what the violent removal of the Pope from his spiritual functions, and their assumption by the Emperor, would have been in the middle ages.

10. **τούτου δή**. For δή continuing a story after a parenthetical interruption, see p. 59, l. 17, and Index.

12. **'Αζὴν** 'an Azanian', see Geograph. Index, s.v. *Azania.*

14. **ἀπὸ τούτου** 'from that time forward'.

15. **ξεινοδοκέοντος πάντας ἀνθρώπους** 'entertained all comers', sc. lest he might be unawares again entertaining gods.

20. **ἀνθεύσης** 'at a high pitch of prosperity', i.e. as a mercantile and naval town.

CHAPTER CXXVIII.

26. **πρῶτα μὲν...μετὰ δέ**, cp. p. 72, l. 20.

29. **τῆς ὀργῆς** 'their disposition', or 'temper', in the old sense of the word, which does not refer merely or perhaps at all to anger, but to

the whole character of the mind. Stein quotes Theognis 963 μή ποτ' ἐπαινήσῃς πρὶν ἂν εἰδῇς ἄνδρα σαφηνέως Ὀργὴν καὶ ῥυθμὸν καὶ τρόπον ὅστις ἂν ᾖ.

3. **ἐν τῇ συνιστίῃ** 'by entertaining them at his own house', 75 opposed to ἰὼν ἐς συνουσίην 'visiting them or meeting them at outside gatherings'.

4. **τοῦτον**, sc. χρόνον 'during this time'. **πάντα**, sc. ταῦτα, all the things mentioned above.

5. **καὶ δή κου**, see p. 6, l. 9.

8. **ἐκρίνετο** 'was getting the preference' in his eyes. **τὸ ἀνέκαθεν** 'originally', used especially of pedigrees, see p. 29, l. 5; p. 72, l. 4.

9. **προσήκων** 'connected in blood', p. 31, l. 4.

CHAPTER CXXIX.

11. **τῆς κατακλίσιος τοῦ γάμου** 'for the marriage feast'. Hom. *Odyss.* 4, 4 τὸν δ' εὗρον δαινύντα γάμον (St.), cp. 1, 126 τοὺς Πέρσας κατακλίνας ἐς λειμῶνα εὐώχεε (Ab.).

ἐκφάσιος (ἐκ-φαίνω) 'for Kleisthenes to make known'.

14. **ὡς δὲ ἀπὸ δείπνου ἐγένοντο** 'and when they had finished dinner'. Indicating the time between the eating and the wine, cp. 9, 16 ὡς δὲ ἀπὸ δείπνου ἦσαν, διαπινόντων...So 5, 18 ὡς δὲ ἀπὸ δείπνου ἐγένοντο, διαπίνοντες κ.τ.λ.

15. **ἀμφί**, see on p. 34, l. 5.

16. **καὶ τῷ λεγομένῳ ἐς τὸ μέσον** 'on a subject proposed for general discussion'. The suitors recite or sing against each other, and hold also a kind of debate on some subject proposed. This is an interesting indication of an early habit of Hellenic society; as it suggests a reason for the later writers often putting their philosophical discourses in the form of a Symposium, as for instance Xenophon and Plato. See Plutarch *Cleom.* 12. Theognis is also quoted (493)

> ὑμεῖς δ' εὖ μυθεῖσθε παρὰ κρητῆρι μένοντες
> ἐς τὸ μέσον φωνεῦντες.

17. **κατέχων πολλὸν** 'by way of far outdoing'. Cp. Xen. *Hell.* 4, 6, 10 μάλα κατεῖχον βάλλοντες καὶ ἀκοντίζοντες οἱ Ἀκαρνᾶνες. Thucyd. 4, 92, 5 ἧσσον ἑτοίμως κατέχειν.

18. **ἐμμέλειαν** 'a dance tune'. According to a passage of Aristoxenos quoted by Stein from Bekker's *Anecd.* p. 101 a particular kind of tragic

dance was called an ἐμμέλεια, which would also be the tune on the αὐλὸς to which it was danced.

21. **ὑπώπτευε** 'began to have suspicions', i.e. as to his worthiness. That a Greek of good birth should dance was not only undignified, but rendered him open to charges of gross immorality. See Demosthenes' scathing description of the Makedonian court, as it had been reported to him, *Olynth.* 2, § 18—19, where, among other things, he says ἀνθρώπους οἵους μεθυσθέντας ὀρχεῖσθαι τοιαῦτα οἷα ἐγὼ νῦν ὀκνῶ πρὸς ὑμᾶς ὀνομάσαι. Music on the other hand was in many parts of Greece, especially in Arkadia, from very early times a necessary part of a gentleman's education, and he was expected to give evidence of his skill whenever called upon [Polyb. 4, 20].

μετὰ δὲ adverbial, see p. 2, l. 18.

24. **Λακωνικὰ σχημάτια** 'Lakonian steps', i.e. a war dance. The ancient Kretans and Spartans, according to Polybios 4, 20, first introduced the pipe and rhythmic movement in war. See also Historical Index. **πρῶτα μὲν...μετὰ δὲ...τρίτον δὲ**, for another way of marking three stages or clauses cp. p. 66, l. 20; p. 71, l. 3.

26. **τὰ μὲν πρῶτα καὶ τὰ δεύτερα**, that is the Lakonian and Attic steps.

27. **ἀποστυγέων** 'rejecting the idea with disgust'.

28. **ἂν γενέσθαι** for ὅτι ἂν γένοιτο, cp. p. 70, l. 10.

76 3. **ἀπωρχήσαο** 'you danced away', a transitive verb coined to express a special thing; cp. such a word as κατακυβεύειν τὰ ὄντα 'to gamble away one's property'.

CHAPTER CXXX.

6. **ἀπὸ τούτου μὲν τοῦτο ὀνομάζεται** 'from this circumstance these words have become proverbial', sc. οὐ φρόντις Ἱπποκλείδῃ as a proverb for indifference of a rather foolish sort.

7. **ἐς μέσον** 'publicly', cp. p. 75, l. 16.

9. **εἰ οἷόν τε εἴη, χαριζοίμην ἂν** 'if it could be, I would gratify you all': The supposition is an impossible one and might have been expressed by indicative in protasis and apodosis; but the optative puts the matter more modestly and less offensively, as though the conditions were still future. Goodwin, *Moods and Tenses*, p. 105. For χαρίζεσθαι cp. p. 71, l. 6.

11. **ἀλλ' οὐ γὰρ οἷά τέ ἐστι** 'but seeing that it is impossible'.

For γὰρ introducing the reason by anticipation see p. 6, l. 11 and Index.

12. **κατὰ νόον ποιέειν**=*χαρίζεσθαι*.

15. **τῆς ἀξιώσιος εἵνεκεν τῆς ἐξ ἐμεῦ γῆμαι** 'in acknowledgment of your having done me the honour of wishing to take a wife from my family'. Cp. 3, 84 *γαμέειν ἄλλοθεν ἢ ἐκ τῶν συνεπαναστάντων*. The infinitive *γῆμαι* depends on the verbal meaning of **ἀξιώσιος**.

17, 18. **ἐγγυῶ** 'I betroth her', *despondeo*. **φαμένου ἐγγυᾶσθαι Μεγακλέος** 'upon Megakles saying that he accepted the engagement', *se conditionem accipere*.

19. **ἐκεκύρωτο...Κλεισθένεϊ** 'was ratified by Kleisthenes'. The pluperfect expresses the immediate consequence in the past, cp. *ἔρρωτο* p. 65, l. 17. For the dative of the agent see p. 17, l. 3; p. 71, l. 22, and on p. 46, l. 6.

CHAPTER CXXXI.

20. **ἀμφὶ** 'concerning'; cp. 1, 140. For its meaning with dative see on p. 34, l. 5.

21. **ἐβώσθησαν ἀνὰ τὴν Ἑλλάδα** 'became loudly talked of' or 'notorious throughout Greece'. He refers to what he said before p. 73, l. 5 *ὥστε πολλῷ ὀνομαστοτέρην γενέσθαι κ.τ.λ.* For **ἀνὰ** see on p. 25, l. 22.

22. **συνοικησάντων** 'having cohabited as man and wife'. Cp. 1, 37 *κοίλῳ ἐκείνῃ δόξει ἀνδρὶ συνοικέειν*;

23. **ὁ τὰς φυλὰς...καταστήσας** 'who established the tribes and democracy among the Athenians'. The ten tribes were called after various heroes connected with Attica—Erectheis, Aegeis, Pandionis, Leontis, Akamantis, Aiantis, Oeneis, Kekropis, Hippothoontis, Antiochis. The ten tribes consisted of all Athenians living in one of the 100 demes (a number afterwards raised) of Attica, care being taken that the demes of the same tribe should not be coterminous. From these tribes were taken, either by lot or election, members of the Boulè, Archons, Strategi and other officers. From them the national levy was made under the taxiarchs of the several demes. This division superseded the Solonian four tribes which had been communities of phratries or gentes, from whom alone the officers etc. had been taken, under the further restriction of certain *τιμήματα* or assessments of property. Kleisthenes' measure therefore transferred power from a comparatively close oligarchy of *gentes* to all Athenians. But it at first

fell short ot entire democratic equality in the fact that for some time the old τιμήματα, or property qualifications, were retained in regard to the Archonship. The system of phratriae also existed for certain purposes concurrently with the deme.

25. **οὗτός τε δὴ** 'both this man, I say'. For **δὴ** resumptive, see p. 74 l. 10.

77 3. **λέοντα.** Applied to a citizen powerful, but dangerous, cp. 5, 56, 92 § 2. So of Alkibiades, Aristoph. *Ran.* 1432:

> μάλιστα μὲν λέοντα μὴ 'ν πόλει τρέφειν
> ἢν δ' ἐκτρέφῃ τις, τοῖς τρόποις ὑπηρετεῖν.

Aristophanes also seems to refer to this dream in the comic oracle (*Eq.* 1037):

> ἔστι γυνή, τέξει τε λέονθ' ἱεραῖς ἐν 'Αθήναις
> ὃς περὶ τοῦ δήμου πολλοῖς κώνωψι μαχεῖται,
> ὥστε περὶ σκύμνοισι βεβηκώς· τὸν σὺ φυλάξαι
> τεῖχος ποιήσας ξύλινον πύργους τε σιδηροῦς.

CHAPTER CXXXII.

5. **τρῶμα** 'defeat' i.e. of the Persians, cp. 4, 160 etc.

9. **οὐ φράσας...ἐπιστρατεύσεται.** Though Miltiades did not distinctly state what was the object of his expedition, it seems to have been thoroughly understood that it was to punish some of the islanders for help given to the Persians [Nepos, *Milt.* 7], and in that point of view was in a certain sense a continuation of the struggle against Persia. A general who took such a roving commission necessarily depended for the ultimate approval of his acts on his success; and it is useless to discuss whether the measures taken against him afterwards were just or unjust. They were the inevitable result not of his attack, fair or unfair, on Paros, but of the failure of that attack, cp. Pausan. 1, 32, 4 συμβασης ὕστερόν οἱ τῆς τελευτῆς Πάρου τε ἁμαρτόντι καὶ δι' αὐτὸ ἐς κρίσιν 'Αθηναίοις καταστάντι. This is shown by the fact that the blockade of Paros lasted 26 days, during which he might have been easily recalled. A very similar incident occurred later on in B.C. 390 in the case of the hero of the restored democracy Thrasybulos, who went on a similar roving commission round the shores of the Aegean. He himself was killed at Aspendos, but one of his colleagues Ergokles was impeached and condemned to death. Lysias, *Or.* 28.

11. **τοιαύτην δὴ** 'of such wealth, as he pretended' [p. 21, l. 23].

12. **οἴσονται** 'they would get for themselves', p. 57, l. 17. The future indicative is generally retained in indirect discourse; but sometimes is represented by the future optative. Goodwin, *Moods and Tenses*, §§ 26, 69.

13. **τὰς νέας.** Seventy ships, according to Nepos *Milt.* 7: the whole available war fleet probably of Athens.

CHAPTER CXXXIII.

17. **ὑπῆρξαν,** sc. *ἀδικίας* 'had given the first provocation', cp. p. 69, l. 7 *οἷα ἀρξάντων ἀδικίης*. 1, 5 *τὸν ὑπάρξαντα ἀδίκων ἔργων*. 9, 78 *φυλάσσεται μὴ ὑπάρχειν ἔργα ἀτάσθαλα ποιέων*.

18. **πρόσχημα λόγου,** p. 23, l. 11.

19. **ἔγκοτον εἶχε τοῖσι Παρίοισι** 'was entertaining a grudge against the Parians', cp. p. 69, l. 7 *ἐνεῖχέ σφι δεινὸν χόλον*.

21. **διαβαλόντα μιν.** Probably in the later years of the Ionic revolt Lysagoras had denounced him to Hydarnes, as having shown an animus against Darius in the matter of the bridge of the Danube, and being therefore a dangerous man to leave in the Chersonese. It seems doubtful whether the Hydarnes meant is the man who was one of the Seven Magi (3, 70) or his son, afterwards Satrap of Asia Minor or part of it (7, 83).

23. **ἐπολιόρκεε Παρίους** 'he began blockading Paros', i.e. the town in the Island.

24. **αἴτεε ἑκατὸν τάλαντα.** Similar demands upon several of the Islands were made by Themistokles in B.C. 480 after the battle of Salamis [8, 108—112]: but then Themistokles was acting with the combined Greek fleet.

25, 26. **πρὶν ἢ ἐξέλῃ**: for the construction see p. 45, l. 22 and cp. 9, 86 *πρότερον ἢ ἐξέλωσι*. For the meaning of **ἐξελεῖν** see p. 17, l. 15.

26. **ὅκως δώσουσι,** after verbs of striving or contriving, *ὅπως* takes the future indicative regularly after a clause in present time, and generally also after a clause in past time also; though in the latter the fut. optative may be used. Goodwin, *M. and T.* pp. 73, 4.

78 1. **οὐδὲ διενοεῦντο** 'did not so much as give it a thought'. If *ἀργυρίου* is to stand it must depend on *τι*. Abicht however brackets it, as unnecessary; and Stein reads *ἀργυρίου οὐδὲν διενοεῦντο* 'had no

thought of money', understanding ἀργυρίου ἐχόμενον οὐδὲν, as in 5, 49, sub fin.

3. ἐπίμαχον 'capable of being stormed', i.e. from being too low.

CHAPTER CXXXIV.

8. λέγουσι, sc. τὸν αὐτὸν λόγον. They agree in giving the same account. αὐτοὶ 'alone', 5, 67, 85; 7, 49.

11. ὑποζάκορον 'under-priestess of the infernal Goddesses', sc. Demeter and Persephone.

12. περὶ πολλοῦ ποιέεται, cp. p. 33, l. 24.

13. ὑπόθηται 'suggest', p. 28, l. 6.

16. ἕρκος, the fence or low wall round the *τέμενος* of the temple. θεσμοφόρου, see on p. 9, l. 13. The temple of Demeter is as usual outside the city. See on p. 52, l. 15.

18. ὅ τι δὴ *nescio quid* 'something or other, I know not what'. It is more indefinite in the next line by the addition of *κοτε*, p. 34, l. 8. κινήσοντά τι τῶν ἀκινήτων 'with the intention of meddling with some of the things which might not lawfully be moved',—treasures or arms under the protection of the temple.

20. τε...καὶ expressing simultaneousness, cp. 8, 83 *ἠώς τε δὴ διέφαινε καὶ προηγόρευε*. The word πρόκατε [*πρόκα, πρό-*, with indefinite suffix *τε*] emphasises this, 'no sooner had he got near the door than a shuddering suddenly seizing upon him he hurried back the same way as he came'.

21. καταθρώσκοντα τὴν αἱμασιὴν 'as he was jumping off the wall', so *καταβαίνειν τὸ οὖρος* 7, 218. τὸν μηρὸν σπασθῆναι 'sprained his thigh', lit. was wrenched as to his thigh.

CHAPTER CXXXV.

24. μέν νυν, cp. p. 11, l. 18. φλαύρως ἔχων 'in a miserable state of pain', from his hurt. Cp. 3, 129.

79 6. ὡς σφεας ἡσυχίη τῆς πολιορκίης ἔσχε 'when they found themselves relieved from the siege'. 'To be quiet' or 'to enjoy immunity from trouble' is *ἄγειν ἡσυχίην*, cp. 1, 169 *Μιλήσιοι Κύρῳ ὅρκιον ποιησάμενοι ἡσυχίην ἄγον*. By varying the point of view *ἡσυχίη* is said *ἔχειν τινα* instead of the person being said *ἔχειν* or *ἄγειν ἡσυχίην*. See the note on *περιφέρει* p. 49, l. 20; cp. 1, 45 *ἡσυχίη ἐγένετο τῶν*

ἀνθρώπων περὶ τὸ σῆμα: and the phrase αἰτίη ἔχει τινὰ 5, 71; λόγος ἔχει τινὰ ('reputation') 9, 87.

7. **εἰ καταχρήσονται** 'whether they should kill'. The indicative future here represents the deliberative subjunctive in oratio recta; which Herodotos sometimes retains after εἰ, see p. 18, l. 28; p. 49, l. 30.

8. **ἐξηγησαμένην** 'for having pointed out to the enemy the way to capture her native town'. ἐξηγέεσθαι = (1) to lead the way, 9, 11 συστρατευσόμεθα ἐπὶ τὴν ἂν γῆν ἐκεῖνοι ἐξηγέωνται, (2) 'to demonstrate', 'to explain', with accusative, cp. 3, 4 ἐξηγέεται τὴν ἔλασιν.

9. **τὰ ἐς ἔρσενα γόνον ἄρρητα** 'things forbidden to the male sex', lit. 'not to be told to'. For the worship of Demeter was in some respects peculiar to women, see c. 16. For ἐς see p. 38, l. 3; p. 50, l. 14. For ἔρσην see App. A. II. (1).

10. **οὐκ ἔα** *vetabat*, p. 63, l. 13.

11. **οὐ Τιμοῦν εἶναι,** see p. 37, l. 8. **ἀλλὰ δέειν γὰρ** 'but since it was fated that'. For γὰρ anticipatory see p. 6, l. 11 and Index. For δέειν of the decrees of fate see p. 35, l. 6.

12. **φανῆναι,** sc. Τιμοῦν, i.e. her phantom appeared to lure him to his doom.

CHAPTER CXXXVI.

16. **ἔσχον ἐν στόμασι** 'talked about Miltiades', sc. to his discredit. The Aorist gives the sense of the immediateness of the scandal which arose. In 3, 157 the phrase is used in a favourable sense πάντες Ζώπυρον εἶχον ἐν στόμασι αἰνέοντες.

17. **θανάτου ὑπαγαγὼν ὑπὸ τὸν δῆμον** 'brought him before the people on a capital charge'. Cp. for the middle in the same sense p. 40, l. 17; p. 45, l. 15.

18. **ἐδίωκε τῆς...εἵνεκεν** 'and prosecuted him on the ground of his having deceived the people'.

I cannot accept Mr Grote's view that the charge was brought before a body of Dicasts. I hold with Curtius [*Hist. of Greece*, vol. 2, p. 227] that it was before the Ecclesia, as was usual in such public prosecutions (εἰσαγγελίαι); the Boulè, represented by the Prytanies, acting as εἰσαγωγεῖς, and presiding at the trial. Hence Plato [*Gorgias* 516 D, cp. Aristot. *Rh.* 232] tells us that he would have been ordered to be thrown into the Barathrum (i.e. he would have been refused sepulture) had not the πρύτανις, i.e. the president (*ἐπιστάτης*) of the prytanies, interfered. The prytanies would have had nothing to do with a trial before dicasts. Cp. Demosth. 1204 *νόμων ὄντων ἐάν τις τὸν δῆμον ἐξαπατήσῃ εἰσαγγελίαν εἶναι κατ' αὐτοῦ.* See Introduction. The distinction between *γραφαὶ τιμητοί* and *ἀτίμητοι* was probably not fully established at this time. The prosecutor in this case assessed the penalty at death: whether the fine of fifty talents was the counter-assessment of his friends, or was arrived at by merely calculating the expenses of the expedition, as Nepos [*Milt.* 7] seems to assume, cannot be fully made out. The penalty of *ἀπάτη* was settled as death, and what seems to have happened is that the people simply voted against the capital charge first, and then voted on a second motion that he should make good the expenses incurred.

23. **πολλὰ** 'at great length'.

τὴν αἵρεσιν. Herodotos constructs *ἐπιμιμνήσκεσθαι* in three ways: (1) with genitive *τούτων ἐπεμνήσθην* 8, 55; (2) with *περί* and genitive, *τοῦ μεγάθεος πέρι ἐπιμνήσομαι* 2, 101; (3) with accusative *τῶν ἐπεμνήσθην πρότερον τὰ οὐνόματα* 8, 66. In this case he uses both constructions, perhaps from that tendency to vary the construction of two co-ordinate clauses noticed on p. 2, l. 9; p. 47, l. 4.

24. **Λῆμνον...Πελασγοὺς**, see cc. 137—8.

25. **προσγενομένου...κατὰ τὴν ἀπόλυσιν** 'having taken his side so far as relieving him of the death penalty went'.

27. **κατὰ τὴν ἀδικίην** 'on the ground of his being guilty of the crime'. It is not quite clear whether Herodotos means that the people found him guilty of *ἀπάτη* and by a second vote assessed the punishment at a fine instead of death; or whether they voted him guilty of a less crime (*ἀδικίη*) and fined him for it. I think on the whole that we must not press too closely the procedure of a later date, and suppose a first vote of guilty or not guilty, and a second on the assessment and counter-assessment. *'Απάτη τοῦ δήμου* was punishable by law with death (Demosth. l. c.), and a suit resulting in condemnation for that would be *ἀτίμητος*, that is, no assessment would be necessary, the penalty being already fixed. It seems more likely that the *δῆμος* simply by its second vote declared him guilty of a malfaisance and at the same time fixed the penalty. For the dative **ταλάντοισι** see p. 16, l. 25.

80 1—3. Nepos and others say that he was imprisoned, and unless he could find securities for his debt to the state that would have been the ordinary course of law. Grote (vol. 4, p. 294) disbelieves in the imprisonment, on rather insufficient grounds, mainly because Herodotos does not mention it. Thirlwall and Curtius on the other hand believe

it; and the latter thinks that the estates of Miltiades in the Chersonese being in the hands of the Persians he was unable to pay so large a sum, as also for some time was his son Kimon. The latter as his father's heir would be *ἄτιμος* (because indebted to the state) until he paid.

CHAPTER CXXXVII.

5. **ἔσχε** 'got possession of', p. 19, l. 2.

6. **εἴτε ὦν δὴ δικαίως εἴτε ἀδίκως** 'no matter whether justly or unjustly'. *ὦν δὴ* have something of the force of our 'ever' in such a phrase, 'however justly or unjustly it may have been done'.

7. **πλὴν**, see on p. 3, l. 18.

8. **Ἑκαταῖος μὲν** answered by *ὡς δὲ αὐτοὶ Ἀθηναῖοι* in l. 17. **ἐν τοῖσι λόγοισι** 'in his history'.

9. **ἰδεῖν**, the infinitive of the oblique oration is preserved in subordinate clauses of it. Herodotos is giving the words of Hekataeos. So also the other infinitive *ὡς ἰδεῖν*, l. 13. Cp. 5, 84; 8, 111, 118.

10. **τὴν χώρην...ὑπὸ τὸν Ὑμησσὸν**, a district on the N.W. of Athens.

11. **μισθὸν** 'as a remuneration', in apposition to *οἰκῆσαι*, cp. p. 53, l. 4.

τοῦ τείχεος τοῦ περὶ τὴν ἀκρόπολιν 'the wall round the Akropolis which then existed', see 5, 64. It was apparently only round three sides of the Akropolis, the Western side being open (9, 51), until it was completed by Kimon s. of Miltiades [Paus. 1, 28, 3].

12. **ἐληλαμένου** 'built', for the meanings of *ἐλαύνω* see p. 34, l. 7.

14. **εἶναι** in apposition to *ἐξεργασμένην εὖ*. **λαβεῖν...τῆς γῆς** 'envy and a desire for the land seized them', cp. 1, 138 *τὸν λαμβανόμενον ὑπὸ τούτων (νούσων)*.

16. **προϊσχομένους**, p. 48, l. 27.

23. **οἰκέτας** 'domestic slaves'. This must have been in a period before that of the writings of Homer, in which slaves are constantly mentioned. The Editors quote a fragment of Pherekrates from Athenaeus 6, p. 263:

> *οὐ γὰρ ἦν τοτ' οὔτε Μάνης οὔτε σηκὶς οὐδενὶ*
> *δοῦλος, ἀλλ' αὐτὰς ἔδει μοχθεῖν ἅπαντ' ἐν οἰκίᾳ.*

Timaeos [Athen. 6, pp. 264, 265; Polyb. 12, 6] asserts that it was against Greek tradition originally to be served by purchased slaves, and that the Lakedaemonians and Thessalians and Chians were the first to practise it. The original slaves were no doubt captives in war, and the practise of purchasing barbarians was adopted as a means of avoiding the enslavement of captured Hellenes, who were put to ransom instead.

ὅκως δὲ ἔλθοιεν 'and whenever they came', see p. 16, l. 9.

24. **ὕβριος** 'wantonness', **ὀλιγωρίης** 'insolence'. The first has a sense of lewdness. **βιᾶσθαι** 'treated them with violence'.

26. **ἐπιβουλεύοντας...ἐπ' αὐτοφώρῳ φανῆναι** 'were detected in the act of making a plot to carry them off'.

81 1. **ἀμείνονας** 'more liberal and merciful'.

παρεὸν 'whereas they might lawfully have put them to death', p. 40, l. 13.

5. **καὶ δὴ καί**, see p. 11, l. 13. Here the second *καί* belongs closely to *Λῆμνον*, the first answers to **τε**. **ἐκεῖνα...ταῦτα** 'the former'...'the latter'. Notice the different tenses **ἔλεξε...λέγουσι**. Hekataeos *made a statement* in his history; the Athenians *continue to state* their view of the case.

CHAPTER CXXXVIII.

9. **τὰς Ἀθηναίων ὁρτὰς** 'being well acquainted with the dates at which the Athenians celebrated their festivals'. For the selection of such seasons for raids of this kind see on p. 9, l. 13.

11. **ἐν Βραυρῶνι**, for an account of this festival see Historical and Geographical Index.

14. **παλλακὰς** 'concubines', as opposed to *κουριδίαι γυναῖκες*, l. 24 and 1, 135; 5, 18.

20. **καὶ δὴ καὶ** 'nay, more', see p. 11, l. 13.

21. **ἐδικαίευν** 'claimed', p. 9, l. 3. **πολλὸν** 'by a great deal', p. 46, l. 8.

22. **ἑωυτοῖσι**=*ἀλλήλοις*, p. 7, l. 5. **λόγους ἐδίδοσαν** 'consulted with each other', p. 49, l. 4.

23. **δεινόν τι ἐσέδυνε** 'it struck them as something very formidable', lit. 'a certain terror was penetrating them'. Stein quotes Soph. *O. T.* 1317 *οἷον εἰσέδυ μ' ἅμα κέντρων τε τῶνδ' οἴστρημα καὶ μνήμη κακῶν.* **εἰ δὴ** *si quidem* 'if, as they were actually doing', p. 71, l. 20.

26. **τί δή**, *quid tandem*, 'what in the world they would do, when

they came to man's estate'. **δῆθεν** closely connected with *ἀνδρωθέντες*. There is no ironical meaning in it, as in p. 1, l. 7. It represents the thoughts of the Pelasgians and emphasises the contrast of *ἀνδρωθέντες* with *παῖδες*, they thought, 'If they did these things as children what would they do when they were *men*!' **ἐνθαῦτα** 'in these circumstances', p. 16, l. 18.

2. **ἀπὸ τούτου** 'from this circumstance', p. 76, l. 6. 82

3. **τοῦ προτέρου.** See Historical Index s. v. Lemnian deeds.

5. **ἀνά**, p. 25, l. 22.

CHAPTER CXXXIX.

9. **ὁμοίως...καὶ** 'as freely as', cp. *τοὺς αὐτοὺς καὶ* p. 52, l. 24.

10. **πρὸ τοῦ**, cp. p. 47, l. 10.

14. **δὴ** 'accordingly', p. 3, l. 4; p. 13, l. 25.

15. **ἐπηγγέλλοντο βουλόμενοι** 'proclaimed that they were willing'. Cp. 7, 27 *χρήματά τε ἐπηγγέλλετο βουλόμενος ἐς τὸν πόλεμον παρέχειν*.

16. **ἐν τῷ πρυτανηΐῳ.** See p. 59, l. 11.

21. **εἶπαν**, App. E (2).

24. **πρὸς νότον κέεται** followed by the genitive *τῆς Λήμνου* from its sense of separation, as though = *ἀπέχει*. **πολλὸν**, *adverbial*, 'far', 'by much', p. 81, l. 21.

CHAPTER CXL.

1. **ἔτεσι...πολλοῖσι**, dative of time how long after, p. 10, l. 3 and index. He refers to the incidents related in cc. 34—6. The expedition of Miltiades to Lemnos was probably contemporaneous with the outbreak of the Ionian revolt, about B.C. 501—500. 83

4. **ἐτησίων...κατεστηκότων** 'while the periodical (or *etesian*) winds were prevailing'—the N.W. winds blowing during July and August in the Aegean.

7. **τὸ χρηστήριον.** The oracle ordered them to pay the Athenians any satisfaction they should demand; what they were now reminded of was the condition which they had themselves laid down p. 82, ll. 20—22. For **ἀναμιμνήσκων** with a double accusative cp. Xenoph. *Hellen.* 3, 2, 7 *ἀναμνήσω ὑμᾶς τοὺς τῶν προγόνων κινδύνους* (Ab.). It seems to follow

the analogy of διδάσκειν. Cp. the construction of πείθεσθαι with genitive p. 7, l. 16.

τὸ οὐδαμὰ ἤλπισαν 'which they expected would never by any means be fulfilled upon them'. Cp. ἔλπομαι p. 64, l. 10.

11. παρέστησαν 'submitted', cp. p. 57, l. 6.

12. Ἀθηναῖοί τε καὶ Μιλτιάδης. Herodotos seems by these words again to represent Miltiades to be acting not as an independent tyrannus; but as an agent and representative of the Athenians. See on p. 59, l. 30.

HISTORICAL AND GEOGRAPHICAL INDEX.

ABDERA, cc. 46, 48.

A town on the S. coast of Thrace, according to Herodotos [1, 168] originally a colony from the Ionian city Klazomenae; and re-founded and occupied by the inhabitants of Teos in B.C. 541, who went thither to escape Harpagos, when he was left by Kyros to subjugate Ionia. It was some way to the east of the r. Nestos. In 7, 126 Herodotos seems to speak of the Nestos as flowing through the town, but a comparison with 7, 109 shows that he means through the territory. See also 8, 120. The town has completely disappeared. After the Persian invasion it seems to have lost its Hellenic character and become one of the cities of the Odrysae; but in B.C. 408 it was taken by the Athenians; and in B.C. 376 it suffered much from the Triballi, and never seems to have been of importance afterwards. Though the birthplace of several famous philosophers its inhabitants were proverbial for stupidity. Juv. 10, 50: Mart. 10, 25, 4.

ABYDOS, c. 26.

In Mysia, on the Asiatic coast of the Hellespont, opposite Lesbos, where the channel is about a mile broad. Xerxes placed the head of his bridge of boats there [7, 33]. It had taken part in the Ionian revolt, but was recaptured by Daurises in B.C. 498 [5, 117]. It afterwards was included in the confederacy of Delos; but in B.C. 411 revolted from Athens to Derkylidas and Pharnabazos [Thucyd. 8, 62]. It was a colony from Miletos.

AEAKES, cc. 13, 14, 22, 25.

Son of Syloson, who had been made tyrant of Samos by the Persians [3, 149]. Aeakes soon succeeded him, and was one of the Greek tyrants who voted against breaking the bridge over the Danube when Darios was on his Skythian expedition [4, 138]. After his restoration mentioned in c. 25 we know nothing more of him.

AEAKOS, c. 35.

Son of Zeus and Aegina. His reputation for justice as ruler of Aegina caused Aeakos to be made one of the judges in the lower world.

His descendants (the Aeakidae), Achilles, Peleus, Telamon, Ajax and Teuker were the national heroes of Thessaly, Aegina, and Salamis [8, 64, 84].

Aegileia, c. 107.

A small island between Euboea and Attica, at the entrance of the Myrtoan sea, belonging to Styra.

Aegilia, c. 101.

A place in the territory of Eretria in Euboea. It is nowhere else mentioned. There was an Attic deme of the same name.

Aegina, cc. 35, 50, 61, 88—90.

Aeginetans, cc. 49—50, 65, 73, 85, 87—93.

An island in the Saronic gulf, containing about 40 square miles. It is about 12 miles from the coasts of Attica, Megaris, and Epidauros. From the latter town it was said to have been peopled, and for some time remained under its control [5, 83; 8, 46]. The tyrant of Argos, Pheidon, about B.C. 748, took possession of it, and its merchants quickly acquired great wealth by their activity. So much so that the coinage introduced by Pheidon was either first produced there, or at any rate got the name of Aeginetan from the fact of its merchants making it most widely known. In Naukratis, the earliest Greek settlement in Egypt, the Aeginetans had a temple of their own [2, 178]; and by 500 B.C. its supremacy at sea was universally acknowledged. We find the Aeginetans in this book on bad terms with Athens; the origin of which feeling is stated by Herodotos [5, 82] to have been an attempt of the Athenians to carry off the olive-wood images of the national heroes, the Aeakidae. However, this was probably only one in a series of mutual provocations, in which the Aeginetans from their superiority at sea would act the more effectively by descents upon the Attic coast. In B.C. 505 the Aeginetans had further irritated the Athenians by helping the Boeotians against them [5, 81]. Their action in regard to the Persians detailed in this book was partly perhaps prompted by their jealousy of Athens. The war between Aegina and Athens was renewed after B.C. 489; and to meet it Athens for the first time built a fleet of considerable importance [7, 144]. For the services of the Aeginetans in the 2nd Persian invasion see 8, 60, 63, 79—84; 9, 75—85. During the administration of Perikles the island was subjected to Athens [Plut. *Perik.* 8], and in B.C. 431 its Dorian inhabitants were expelled and allowed by Sparta to occupy Cynuria, while Attic settlers were put in [Thucyd. 2, 27; 7, 57], who were in their turn expelled by Lysander in B.C. 405 [Xen. *Hell.* 2, 2, 5—9].

Aenyra, c. 47.

A mining district in Thasos opened by the Phoenikians.

Aeolians, cc. 8, 28, 98.

One of the great branches of the Hellenic race, according to the myth, descended from Aeolus, the second of the three sons of Hellen. It was originally the widest spread of any of the Hellenic immigrations. Settling first in Thessaly they partly removed to Boeotia when driven out by the Thessalians, and are also found in Aetolia, Lokris, Korinth, Elis and Messenia, and in Asia Minor and Lesbos. Their dialect was that used by the Lesbian school of Lyric poetry, of which the best known are Sappho and Alkaeus. [*Lesbous barbitos* Hor. *Od.* 1, 1, 34, *Aeolium carmen* 3, 30, 13.] They were apparently an eminently seafaring folk, and as such particularly devoted to the worship of Poseidon. To them also belongs the legend of the first Greek naval expedition of Jason in the Argo from Iolkos. Herodotos [7, 96] says that the Aeolians in Asia Minor were formerly called Pelasgi, which points to a legend of the antiquity of this settlement in Greece.

AESCHINES, c. 100.

Son of Nothon, an Eretrian of high position.

AETOLIA, AETOLIANS, c. 127.

The Aetolians, though Hellenic, lived much apart from the rest of Greece, and were little known to the Greeks generally, among whom the wildest reports of their fierceness and barbarity prevailed [Thucyd. 3, 94]. Herodotos only once again mentions them [8, 73]. They were best known and most hated by the Messenians and Eleans, because of their predatory habits: it having long been looked upon by them as a natural thing to devastate the coasts of Messenia. With the Eleans they had some hereditary connexion, symbolised by the myth of their having acquired their name from Aetolos, a king of Elis, who being obliged to fly his country for homicide settled in the valley of the Achelous. They became more important in Hellenic politics in the 3rd and 2nd centuries B.C., and were the first of the Greek states to form an alliance with Rome.

AGAEOS, c. 127.

A native of Elis, father of Onomastos (q. v.).

AGARISTE (1), cc. 126, 130.

Daughter of Kleisthenes, king of Sikyon, married to Megakles, son of Alkmaeon of Athens, by whom she became the mother of Kleisthenes the reformer.

(2) c. 131, grand-daughter of the last, m. to Xanthippos, by whom she became the mother of Perikles, having shortly before dreamed that she was delivered of a lion [Plut. *Perikl.* 3].

AGETOS, cc. 61—2.

A Spartan, who was cozened into surrendering his wife to king Ariston.

AGIS, c. 65.

One of the junior royal family at Sparta (Eurypontidae) descended from Prokles, the fifth in descent from Hercules [see note on Herakleidae Bk IX.]. Eighth in descent from Prokles was Theopompos [8, 131].

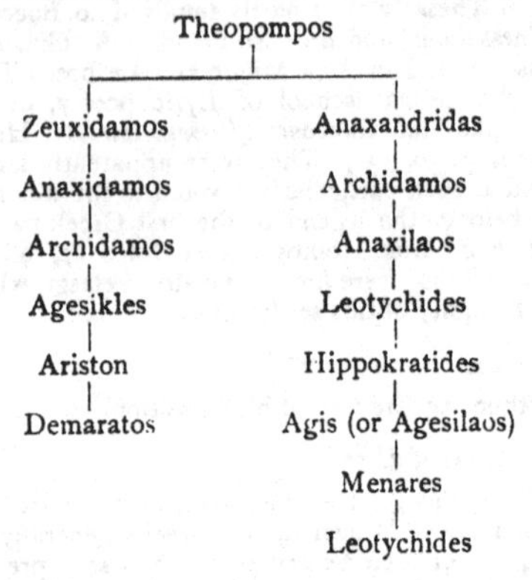

AJAX, c. 35.

Son of Telamon and grandson of Aeakos (q. v.). He was one of the national heroes of Salamis, and tradition said that his sons Eurysakes and Philaeas surrendered Salamis to Athens and became citizens of Attica; and a deme of the tribe Aegeis was named after the latter Philaidae [Plut. *Sol.* 10].

AKANTHOS, c. 44.

On the E. of the Isthmus connecting Acte with the mainland of Chalkidice, about 1½ miles above the canal of Xerxes [7, 115—116]. Mod. *Erisso.* It was a colony from Andros [Thucyd. 4, 84].

AKRISIOS, c. 53.

Father of Danae, son of Abas, the son of Danaos tyrant of Argos, and Hypermnestra.

ALEIAN PLAIN, the, c. 95.

A plain in Kilikia between the rivers Saros and Pyramos. Its mention in Homer (*Il.* 6, 201) shows it to have been a lonely, uninhabited tract of country. It is even more so now, for the course of the Pyramos has changed, leaving it sandy and arid.

ALKEIDES, c. 61.

Of Sparta, the father of Agetos (q. v.).

ALKIMACHOS, c. 101.

A man of high position in Eretria, father of Euphorbos. [Pausan. 7, 10, 2.]

ALKMAEON, cc. 125, 127, 130.

Alkmaeon traced his descent to an Alkmaeon son of Amphiaraus, one of the besiegers of Thebes. The Alkmaeon here mentioned may be regarded as the first historical founder of the family greatness, though his father Megakles appears in the list of eponymous archons at Athens for the year B.C. 612. For the credibility of the story told of him here, see note on the passage. According to Plutarch (*Sol.* 11) he was the general in the Kirrhaean war, undertaken at the instance of the Amphiktyonic League for the liberation of Kirrha, the port of Delphi, for the benefit of the visitors to the shrine [B.C. 595].

ALKMAEONIDAE, cc. 115, 121, 123—4, 131.

The whole family of the Alkmaeonidae had been in banishment since B.C. 598, owing to the curse brought upon them, and through them upon the city, by their violation of the sanctuary in the case of the conspirators of Kylon [τὸ Κυλώνειον ἄγος] B.C. 612. Kylon and his confederates had seized the Akropolis during the Olympic games; and were there besieged by the people, who eventually committed the business to Megakles the Archon, head of the family of the Alkmaeonidae. Kylon and his brother escaped, but the other conspirators being reduced by hunger took up their position as suppliants at the altar of Athena. When induced to come down on promise of their lives being spared, they fastened a cord to the image of the goddess to which they all held; but as they came down the cord broke, and the Archon immediately put to death those who were detached as well as some who had taken refuge at the altar of the Eumenides. Some years of scarcity and pestilence following, a judicial investigation was held and all living members of the family banished, and even the bones of those who had meanwhile died were removed. In B.C. 548 the temple of Delphi was burnt and the Alkmaeonidae undertook its restoration, and performed it with such liberality and splendour that the Pythian priestess constantly gave out oracles to the effect that they must be restored. The Pythia really gave voice to the aristocratic party, supported by Sparta, who wished to get rid of the Peisistratidae; and their deposition accordingly was followed by the restoration of the Alkmaeonidae [5, 62—3]. They were however never free from suspicion, owing to the old 'curse', and a favourite device of the enemies of Athens was to demand that the accursed race (τὸ ἄγος) should be put away. The Spartans did this in regard to Kleisthenes in B.C. 510 (Herod. 5, 70) and Perikles in B.C. 432 (Thucyd. 1, 126). [The authorities on the subject are Herod. 5, 71; Thucyd. 1, 126; Plutarch, *Solon* 12; Pausanias 1, 28, 40; 7, 25; Suidas, s. v. Κυλώνειον ἄγος; Schol. in Aristoph. Eq. 443.]

ALKON, c. 127.

A Molossian, one of the suitors for the hand of Agariste. The Molossians were not usually regarded as Hellenic, and it is somewhat remarkable that a suitor for the hand of a daughter of an Hellenic king should have been found among them. He probably claimed Hellenic or Pelasgic descent.

AMIANTOS, c. 127.

An Arkadian of Trapezūs, one of the suitors for Agariste.

AMPE, c. 20.

On the Persian Gulf, bordered by the Tigris. Pliny *N. H.* 6, 28, probably means this place by Ampelore, which he calls *colonia Milesiorum*.

AMPHIMNESTOS, c. 127.

Of Epidamnos, a suitor of Agariste.

AMPHITRYON, c. 54.

The husband of Alkmena and putative father of Herakles. He was said to be the s. of Alkaeos, king of Troezen, and Hippomene. He, accompanied by Herakles, delivered Thebes from a tribute to Erginos, king of the Minyans; but he perished in the battle and was buried at Thebes, where his tomb was still shown in the time of Pausanias (1, 41, 1).

AMYRIS, c. 127.

A native of Siris in Magna Graecia, half-way between Sybaris and Tarentum. He appears to have got the title of 'the Wise' from the following circumstance, related among others by Suidas [s. v. Ἄμυρις μαίνεται]. An answer from Delphi had declared that the Sybarites would be destroyed when they 'honoured men before gods'. Now Amyris chanced to see a slave being scourged, who had fled to the temple and failed to obtain exemption from punishment; and who then fled to the tomb of his tormentor's father and was forthwith exempted. Amyris thereupon understood that the saying of the oracle was fulfilled; he therefore sold all his property and migrated to the Peloponnese. The Sybarites said he was mad, but events justified him, for Sybaris was destroyed by the Krotonians in B.C. 510.

ANAXANDRIDAS, cc. 50, 108.

Father of Kleomenes, king of Sparta, of the line of the Agidae, B.C. 560—520. He was the son of Leon. His first wife had no children, and, rather than put her away as the Ephors wished, he married another by whom he had Kleomenes. Soon afterwards his first wife had three sons in quick succession Dorieus, Leonidas and Kleombrotos [5, 39—41]. See also 1, 67; 7, 204; 9, 10, 64.

ANAXILAOS or -LAS, c. 23.

Tyrant of Rhegium; he was son of Kretines, and married to Kydippe, a daughter of Terillus, king of Himera, and did much to bring the forces of Carthage into Sicily in 481—80 B.C. [7, 165]. He was by extraction a Messenian [Thuc. 6, 4, 5], his ancestor Alkidamidas having migrated to Rhegium on the fall of Ithome towards the end of the 2nd Messenian war [B.C. 724]. Anaxilaos, after constant quarrels with the people of Zancle, finally by the help of his kindred Messenians drove them out of their town and settled the Messenians in it, changing its name to Messene [Pausan. 4, 23, 5 sq.]. He had established the tyranny himself, Rhegium having been before ruled by an oligarchy [Arist. *Pol.* 5, 12]. He was noted also for his victory in the chariot race at Olympia, and for having introduced hares into Sicily. [Pollux 5, 75.]

ANDREOS, c. 126.

Of Sikyon, great-grandfather of the tyrant Kleisthenes.

APHIDNAE, c. 109.

A fortified and very ancient town a few miles beyond Dekeleia, on the road from Athens to Oropos [9, 73]. It is supposed to have been on the hill now called Kotroni, where there are some ancient remains. It was the native town of the poet Tyrtaeos, as well as of Harmodios and Aristogeiton.

APOLLO, cc. 57, 80, 118.

The Sun-god, and the god of prophecy. The Dorians were especially devoted to Apollo. In his first capacity the Persian invaders showed a special respect to his shrine at Delos, because of their own national Sun-worship [c. 97]; and in his second capacity the Spartan king offered up a public sacrifice to him every month. A large number of towns were called after his name, principally among the Makedonians, whose kings claimed descent from the Dorian Argos.

APOLLOPHANES, c. 26.

A native of Abydos, father of Bisaltes.

APSINTHIANS, cc. 34, 36—37.

A Thrakian tribe living east of the Hebros, bordering on the Thrakian Chersonese. The town of Aenos was also called Absinthos or Apsynthos.

ARCHIDAMOS II., c. 71.

Son of Zeuxidamos, and married to Lampito daughter of Leotychides. He was king of Sparta (of the junior house) B.C. 469—427. For his character see Thucyd. 1, 79, who says that he was a man of clear intellect and steady prudence [*ξυνετὸς δοκῶν εἶναι καὶ σώφρων*]. He was a friend of Perikles, and incurred grave suspicion at Sparta for his slackness in the invasions of Attica [Thucyd. 2, 10—72].

ARDERIKKA, c. 119.

A district or town on the Euphrates, in Kissia (*Khuzistan*).

ARGEIA, c. 52.

Wife of Aristodemos, fifth in descent from Herakles. She was daughter of Autesion, and sister of Theras, the founder of Thera [4, 147–8].

ARGIVES, the, cc. 75—9, 83, 92, 127.

ARGOLIS, c. 92.

ARGOS, cc. 75—6, 82—3.

Argos in Homeric times was the seat of the chief power in Greece, and the Argeioi (Lat. *Argivi*) are spoken of as almost synonymous with the whole nation of Achaioi. When the Achaeans were replaced by the Dorian invaders, Argos, which was believed to be the most ancient city in Greece, still retained its position for some time. But the rivalry of Sparta gradually proved too strong for it. There was a long-standing dispute between the two countries for the possession of Kynuria (or Thyreatis) a district between their frontiers. We hear of a battle as early as B.C. 669 [Paus. 2, 24, 7] for this territory, in which the Argives were victorious. But about B.C. 547 the Spartans finally obtained it by the victory of their 300 champions [1, 82]. The national antipathy between the two states however long continued, and, being farther embittered by the events narrated in this book, led to the Argives taking the anti-patriotic side in the Persian war [7, 150—2, 9, 12; Thucyd. 2, 67]. This also doubtless suggested to Alcibiades the idea during the Peloponnesian war of making a diversion in the Peloponnese by an Argive confederacy against Sparta [Thucyd. 1, 102; 5, 44, 47. Plutarch *Alc.* c. 14, *Nikias* c. 10].

ARGOS, c. 75.

Son of Zeus and Niobe, succeeded, according to the legend, his grandfather Phoroneus as king of the land called Argos after him. His tomb was still shown in the time of Pausanias [2, 16, 1; 22, 5].

ARIPHRON, cc. 131, 136.

Father of Xanthippos, and grandfather of Perikles, married to Agariste, granddaughter of Kleisthenes king of Sikyon, and daughter of Alkmaeon and Agariste.

ARISTAGORAS, cc. 1, 5, 9, 13, 18.

Son of Molpagoras, nephew and son-in-law of Histiaeos, tyrant of Miletos. Histiaeos was summoned to Susa by Darios, and kept there under pretence of the king's desire for his society, but really in a kind of honourable confinement [5, 23—4]. Meanwhile his son-in-law Aristagoras was left in charge of Miletos; and about B.C. 505 incurred the enmity of the Persian satrap Artaphernes, by first raising and then defeating his hopes of subduing Naxos and other islands [5, 30—5]. While in alarm at this he received a message from his father-in-law Histiaeos,—communicated by letters branded on a slave's head,—urging him to raise a revolt in Ionia. Histiaeos' motive was a hope that a disturbance among the Ionian cities would cause the king to send him down to the coast to settle it; but Aristagoras welcomed the proposal because it relieved him from an embarrassing position [5, 35]. Measures were immediately taken to stir a revolt in all the cities against their several tyrants, who were all dependent for their position on the support of the Persian Government [5, 36—7]. Aristagoras himself went to Sparta to beg assistance, and nearly succeeded in persuading king Kleomenes to promise it; but more prudent counsels prevailed, and the request was finally refused [5, 38—9, 49—51]. He was more successful at Athens, where there was a bitter feeling against Artaphernes, caused by his protection and support of the banished tyrant Hippias [5, 96]; and accordingly the Athenians voted to send 20 ships to the aid of the Ionians [5, 97]. Aristagoras next sent a messenger to Phrygia, where certain Paeonians had been brought from their native country and settled by Darios, promising to restore them to their home; who forthwith set out. The revolt was now in full swing. But Aristagoras remained at Miletos while the Ionians, having been joined by the 20 Athenian ships, with 5 from Eretria, advanced upon Sardis, which was taken and burnt, though the citadel still held out [5, 99—101]. The Persian officers acted with energy and promptitude, one after another the revolting towns were reduced, and before long Aristagoras made up his mind that he must seek some safety elsewhere than at Miletos. Committing the government of Miletos to one of the citizens, named Pythagoras, he sailed away to Myrkinos in Thrace, which had formerly been given to Histiaeos by Darios. There he fell in a battle with the Thracians [5, 124—126], according to Thucydides [4, 102] and Diodorus Siculus [12, 68] while endeavouring to found Amphipolis.

ARISTODEMOS, c. 52.

A descendant of Herakles, son of Aristomachos, and father of Eurysthenes and Prokles, who being twins started the double line of the Spartan sovereigns [4, 107].

ARISTOGEITON, cc. 109, 123.

One of the assassins of Hipparchos. He as well as Harmodios were members of an Athenian family called Gephyraei, who professed to have come to Attica from Eretria, but whom Herodotos [5, 57—8] asserts to have been Phoenikians who first settled at Tanagra in Boeotia, and afterwards migrated to Athens. The account of their assassination of Hipparchos, the younger son of Peisistratos, as given by Thucydides 6, 54—59, does not tend to exalt their characters or motives, or explain the enthusiastic reverence in which their memory was held at Athens. Their names became the symbol of liberty, and their descendants for many generations were exempted from all public burdens [Demosth. *c. Lept.* 462]. Their statues were set up on the ascent of the Akropolis where no other statues of mortals were allowed [Pausan. 1, 8, 5], and yearly sacrifice was made to them by the Polemarch, and they formed the theme of the most popular songs. It is difficult for us to understand the feelings entertained towards the slayers of a tyrannus in Greece. So far from being guilty of a crime, they were regarded as having performed an action of the purest and most unalloyed glory. No question occurred to any one of their absolute merit in slaying the slayer of their country's liberty.

ARISTOKRATES, c. 73.

Father of Kasambos (q. v.) a native of Aegina.

ARISTOMACHOS, c. 52.

Father of Aristodemos (q. v.).

ARISTON, cc. 51, 61.

King of Sparta, father of Demaratos (q. v.). During his reign the Spartans were constantly engaged in warfare with the Tegeans [1, 67].

ARISTONYMOS, c. 126.

Father of Kleisthenes of Sikyon (q. v.).

ARISTOPHANTOS, c. 66.

A Delphian, father of Kobon (q. v.).

ARKADIA, cc. 74, 83, 127.

The central district of the Peloponnese. Its strong position had prevented its being occupied by the Dorian invaders. Its inhabitants therefore were Pelasgians or Achaeans, and connected in blood with the subject population of Argos and Lakonia. Hence its people had in old times been in alliance with the rebellious Messenians, and were now inclined to favour the revolting Argive slaves.

ARTAKE, c. 33.

A sea-port town on the west of the peninsula of Kyzikos opposite Priapos, a colony of Miletos [mod. *Erdek*]. See 4, 14. It was not rebuilt after this burning for many centuries.

ARTAPHERNES (1), cc. 1, 2, 4, 30, 42.

A brother of Darios, who was Satrap of Asia Minor, with Sardis as the seat of his government [5, 25, 30]. He had to cope with the Ionian revolt, and did so with success, holding the citadel of Sardis when the town was taken, and gradually reducing the other rebellious states [5, 100, 123]. His policy of dealing with the Ionian states was to support absolute government in them, the several tyrants being dependent for their power on the support of Persia. This policy seems to have at length been regarded by the Court as a mistake, and in 493 B.C. he was superseded by Mardonios who was sent down to institute a new policy of supporting and maintaining popular governments.

(2) cc. 91, 119.

After the failure of Mardonios in B.C. 492, the son of Artaphernes (1), also named Artaphernes, was sent in conjunction with Datis to take Eretria and Athens. His failure at Marathon does not appear to have altogether discredited him; for we find him in command of the Lydians and Mysians in the great army of Xerxes [7, 74].

ARTAXERXES, c. 98.

Son of Xerxes, king of Persia B.C. 465—425. He was called Longimanus (μακρόχειρ) because his right hand was longer than his left. In his reign a revolt in Egypt took place, supported by help sent by the Athenians, in the course of which they suffered many disasters [Thucyd. 1, 104—110], to which Herodotos may here refer [B.C. 460—455]. On the other hand, during his reign, the supremacy of Greece in the Aegean was firmly established. and the freedom of the Greek towns secured by the so-called 'peace of Kimon' abt. B.C. 449—8 [Grote vol. 5, p. 191 sq.]. See also Thucyd. 4, 50; Diodor. Sic. 11, 69; 12, 64.

ARTAZOSTRA, c. 43.

Daughter of Darios, and wife of Mardonios.

ARTEMIS, c. 138.

According to the prevailing Greek legend she was the daughter of Zeus and Leto; but like other divinities the conceptions entertained of her, and the worship paid to her, differed in different places. In Arkadia she was the 'virgin huntress', and her temples were generally by the side of lakes or rivers. The worship of Artemis at Brauron on the other hand was mystic, and connected originally with human sacrifice. The chief seat of this worship had been, according to tradition, the Tauric Chersonese (Crimea), from which Orestes and Iphigenia were said to have brought the image of the goddess.

ASIA, cc. 24, 43, 45, 58, 70, 116, 119.

According to some ancient Geographers the world was divided into two continents, Europe and Asia, the former embracing so much of Libya as was west of the Nile, the latter all that was east of it. Herodotos, however, conceives the world as divided in three, Europe, Asia, Libya. The boundary between Asia and Libya is the Sinus Arabicus (Red Sea), though he says that most Greeks still regarded the Nile as the boundary. In Homer (*Il.* 2, 461) Asia is only a district of Lydia, the basin of the Kayster: and after Herodotos it sometimes was confined to what we call Asia Minor. The word is said to mean 'the land of the Sun', though as usual a derivation was sought from an individual. [See 4, 36—45.]

ASSYRIAN, c. 54.

An inhabitant of Assyria. Though the Assyrian monarchy included, before breaking up into various nationalities, the greater part of upper Asia, the district of Assyria proper—which seems to be what is meant here—was a strip of territory separated from Armenia on the N. by Niphates, on the W. from Mesopotamia by the Tigris, on the S. E. from Susiana, and on the E. from Media by a range of mountains called Zagros [*Mts of Kurdistán*], with Nineveh for its capital. Herodotos however includes Babylonia, and uses the two as synonymous. [1, 192—3.]

ASTRABAKOS, c. 69.

A hero, of whom a chapel or Heroum existed at Sparta [Paus. 3, 16, 6—9]. But little seems to be known about him, the story told by Pausanias not giving any indication of such achievement as usually served to raise a man to such semi-divine honours. The name seems derived from ἀστραβή a mule, or rather a mule's saddle, whence the scandalous story.

ATARNEOS, cc. 4, 28—9.

A tract of land in Mysia opposite Lesbos, which was placed under the government of the Chians, by the Persian Harpagos [1, 160; 7, 42].

ATHENIANS, cc. 21, 34, 35—6, 49, 73, 75, 86—94, 100—5, 108—9, 111—3, 115—7, 120—2, 125, 130—2, 135—140.

ATHENS, cc. 35, 39—41, 45, 85—6, 103, 107, 121.

At the time of the Ionian revolt Athens had not long got rid of her rulers the Peisistratidae, and received those democratical institutions, which starting with the reforms of Kleisthenes developed into the government with which we are afterwards familiar. She was rapidly rising in wealth and importance, but was not yet in a leading position in Greece. The constantly recurring quarrels with Aegina, however, indicate that her commerce was becoming considerable; though she had few ships of war [5, 85], for she had to borrow or hire them from Korinth [B.C. 492—1] when she wished to attack Aegina. The repulse of the Persians at Marathon, by the almost unaided efforts of Athens, gave her a prestige in Greece, which the subsequent policy of Themistokles, Aristeides, Kimon and Perikles confirmed and enhanced.

ATHOS, cc. 44—5, 95.

The most easterly of the then projecting peninsula of the Chalkidic Chersonese, ending in a lofty mountain. The peninsula was properly called Acte (about 40 miles long), but it was often called by the name Athos which belonged to the mountain; hence perhaps the incredulity of subsequent writers as to the cutting of the canal, which they imagined to be through the mountain, but which was really cut across the narrow neck of the peninsula (2500 yards). Distinct traces still remain. Aeschin. *in Ctes.* 132, Isocr. *Paneg.* 89. The headland, where the mountain rises 6350 ft., was a dangerous one from the winds and high seas which sweep round it. It is now called *Monte Santo*, from the number of monasteries on it.

ATTICA, cc. 102, 137—140.

Derived probably from ἀκτη headland or coast-land, is a peninsula about 50 miles long and 30 broad, at its longest and broadest, and contains 700 sq. miles. Its position, and the poorness of its soil, had prevented those frequent changes of inhabitants which were common in other parts of Greece. Its people therefore were regarded as of great antiquity, and even as members of the Pelasgic, or prehellenic stock [1, 56]. They regarded themselves as autochthonous or native to the soil, and declined to be called Ionians (1, 443): still they were generally regarded as the leading Ionian state, if not the origin of all the Ionian settlements (1, 143). Attica was originally subdivided into 12 small independent cantons; the combination of them all under one government was generally attributed to Theseus. [See also cc. 129, 138.]

AUTESION, c. 52.

Father of Argeia (q. v.) and of Theras, reputed colonizer of the island of Thera. According to Herodotos he was great grandson of Polyneikes (Pollux) and son of Tisamenos [4, 147]. He migrated from Thebes to Sparta, where there was a statue of him [Pausan. 3, 1, 7; 3, 15, 6; 4, 3, 4; 9, 5, 15].

AZANIA, c. 127.

A district of Arkadia, the name of which legend derived from Azan, son of Kallistos, son of Arkas, who divided Arkadia with his two brothers Apheidas and Elatos [Pausan. 8, 4, 2].

BAKTRA, c. 9.

The chief city of Baktriana, a large district in Asia corresponding to the modern Bokhára. It was a very ancient city, on the site, it is believed, of the modern Balkh. The remoteness and wildness of the district made it a kind of Siberia of the Persian empire [9, 113].

BISALTES, c. 26.

Son of Apollophanes, a native of Abydos.

BOEOTIANS, cc. 34, 108.

The inhabitants of Boeotia. According to Thucydides [1, 12] they were an Aeolian people originally living in Phthiotis in Thessaly, who being expelled by the Thessalians migrated into the land then called Kadmeïs, to which they gave their name. They were a rather loose confederacy of towns, of which the most powerful had originally been Orchomenos, but in these days was Thebes. Their enmity to Athens was probably caused originally by frontier disputes, especially in regard to the possession of Oropos. They had joined the Chalkidians in ravaging Attica in B.C. 505 [5, 74], and in the subsequent struggle with Persia were for the most part conspicuous medizers [8, 34; 9, 68]. The only exceptions were the towns of Plataea and Thespiae; the former following the lead of Athens.

BRAURON, c. 138.

One of the 12 ancient Cantons of Attica. In historical times it was chiefly known for its temple of Artemis (q. v.), in which the ancient image was believed to have been deposited when brought by Orestes and Iphigeneia from the Tauric Chersonese, though according to other legends this statue was placed at Halae [Eurip. *I. T.* 1452]. It was on the Eastern coast, south of the river Erasinos, and its name is preserved in the village *Paleó Vraóna*. There was a festival every four years held there, at which Attic girls were initiated previous to marriage

[4, 145]. The origin of the festival and the initiations of the young girls, to which the term ἀρκτεύεσθαι was applied, was ascribed by legend to the killing of a bear by the brothers of girls who had been torn by it [Schol. on Aristoph. *Lysist.* 646]. It seems impossible to decide what the true origin of the festival was: it may have been connected with some hunting episode, or with some substitute for human sacrifice.

BRYGI, c. 45.

A Thrakian tribe living above the Chalkidian Chersonesos. They were supposed to be of the same race as the Phrygians [7, 73].

BYZANTIUM, cc. 5, 26, 33. The BYZANTINES, c. 33.

An Hellenic colony from Megara on the site of Constantinople founded B.C. 657 [4, 144]. It was reduced to Persian dependence by Otanes about B.C. 514 [5, 26], but was forced in B.C. 501 to join in the Ionian revolt [5, 103]. The advantages of its situation, as commanding the commerce of the Black Sea were early recognised [4, 144]; and having become a member or subject of the confederacy of Delos, after its rescue from Persian hands by Pausanias in B.C. 478 [Thucyd. 1, 94], it was at various times an object of contention. Thus it revolted from the confederacy in B.C. 440 during the Samian war, but submitted again at the fall of Samos [id. 1, 115—7]. In B.C. 410 the Peloponnesian fleet closed it against the Athenians, and placed in it a Spartan garrison and harmost. Taken again by Alcibiades in B.C. 408, it was retaken by Lysander in B.C. 405 [Xen. *Hell.* 1, 1, 36; 1, 3, 14—20; 2, 21]. Though it suffered much from Philip of Makedon and the Gauls, we find it in the 2nd century B.C. occupying an important position in the commerce of the Levant, and even venturing to resist the Rhodians, who at that time were masters of the sea, and had the most powerful fleet of any Greek people. But though its position was so favourable on the sea side, its prosperity was much checked by the constant hostility of the Gallic tribes that had migrated to its frontier [Polyb. 4, 39, 43—52]; and it was not until it was refounded by Constantine in A.D. 330 as New Rome that it became the powerful and important place that it was in the Middle Ages.

CHALKEDONIANS, the, c. 33.

The inhabitants of Chalkedon, a colony of Megara, in Bithynia, near the entrance of the Pontus [Thucyd. 4, 75]. Megabazos remarked that the settlers must have been blind not to have chosen the better position opposite, where 17 years later Byzantium was founded [4, 144].

CHALKIS, c. 118. CHALKIDIANS, c. 106.

Chalkis (mod. *Egripo*) was on the coast of Euboea, just where the channel, the Euripos, is narrowest, and is now spanned by a bridge, resting on a rock in mid-channel. It was a very ancient town, and had sent out colonies into Italy and Sicily. It was from old times in hostility with Athens, and in B.C. 506 the Athenians had conquered it and divided its territory among 4000 kleruchs or allotment-holders [5, 77].

CHERSONESOS, cc. 33—4, 36—41, 103, 140.

CHERSONESIANS, cc. 38—39.

The Thrakian, or, as Herodotos calls it, the Hellespontine Chersonese, is the Peninsula extending along the western side of the Hellespont. At the narrowest part, near Agora, it was protected by a wall about four miles long. It had in very early times been colonised by Greeks, especially by Athenians, the principal cities being Kardia, Paktya, Kallipolis, Alopeconnesos, Sestos, Medytos, Elaeos. The supremacy of the family which had established their dominion there came to an end by the flight of Miltiades [B.C. 493]; after which it was subject to the Persians until after B.C. 479, when it became again nominally independent, but was an object of contention between Sparta and Athens, until it fell under the power of the Makedonians.

CHIOS, cc. 2, 5, 8, 15—6, 26—7.

CHIANS, cc. 2, 5, 26, 31.

An island off the coast of Asia Minor, separated from the peninsula of Erythrae by a strait 5 miles broad at the narrowest point. It is about 32 miles long, and 18 miles wide at the broadest and 8 miles at the narrowest part. It was celebrated for its wine, which was the best in Greece, and for its pottery; and its inhabitants are said by Thucydides to have been the most wealthy of the Greeks [Thuc. 8, 24—5]. It therefore soon recovered from the effects of Persian severity. It revolted from Athens as head of the Confederacy of Delos in B.C. 412 [Thucyd. 8, 14 sq.], and managed to hold its own. The island maintained an independent position for some centuries; but suffered another devastation by Zenobris, a general of Mithridates, in B.C. 86. It was then restored to freedom under Roman protection by Sulla, B.C. 84; but in the time of Vespasian was included in the *Provincia Insularum*. In A.D. 1300—5 it suffered a massacre by the Turks, in A.D. 1346 was taken by the Genoese, and about A.D. 1600 was retaken by the Turks, and remained fairly prosperous until 1822, when the Turks punished a rebellion in the island by a savage massacre.

CHILON, c. 65.

Son of Demarmenos and father of Perkalos the wife of King Demaratos. He was regarded as the wisest man in Sparta [7, 235; 1, 59]: and as one of the Seven wise men of Greece, and to him is attributed the famous saying γνῶθι σεαυτόν. He is said to have died for joy at his son's winning an Olympic victory [Pliny *N. H.* 7, 32].

CHOEREAE, c. 101.

A place in Euboea in the territory of Eretria. Some suppose that the name applies to some small islets off the main islands now called *Kavalleri*.

DARIOS, cc. 1—3, 9, 20, 24—5, 30, 40—1, 45, 48—9, 70, 94—5, 98.

Darios, son of Hystaspes, was king of Persia from B.C. 521 to B.C. 485. He was of Achaemenid stock [7, 11], and had served Kambyses in Egypt in B.C. 525 [3, 38], and, upon the death of that king, joined the other Magi who put to death the false Smerdis, pretending to be a son of Kyros, and then contrived to be made king himself [3, 70—87], and married Atossa and Artystone the daughters of Kyros [3, 88]. As Kyros had gained the Empire, so Darios was its organizer. He distributed the whole into 20 satrapies, fixing the amount of tribute each was to pay, and introduced an elaborate system of checks upon the officers employed in their management [3, 88—96, Xen. *Oecon.* 4, 5—10]. The Persians expressed their view of him by saying that Kyros was the father of his country, Kambyses the master, and Darios the broker (κάπηλος) [3, 89]. Like all men who have obtained power in an irregular manner over a combination of different nationalities, he was forced to secure his position by military activity. Hence his expedition into Skythia [4, 85—144], and his attempts upon Makedonia and Greece. His enmity to Athens and Eretria arose especially from the help given by them to the revolting Ionians in the burning of Sardis [5, 101—105], after which event he is said to have instructed a slave to say three times each day to him when at supper, 'Sire, remember the Athenians' [5, 105]. He is said to have been about 20 in B.C. 538-7. If that be so, he would be approaching 70 at the time of Marathon [1, 209].

DASKYLEION, c. 33.

Daskyleion was the capital of the Northern Satrapy [the 3rd] of Asia Minor, which included the Phrygians, Asiatic Thrakians, Paphlagonians, Manandynians and Kappadokians [3, 90]. It was in Bithynia on the Propontis, but its exact site is somewhat uncertain. Xenophon [*Hell.* 4, 1, 15] describes it as beautifully situated amidst hamlets and parks well stocked with game, and on a river filled with various kinds of fish.

DATIS, cc. 94, 98, 118, 119.

A Mede by birth, of whom Herodotos tells us nothing except what we read in this book. From Suidas, however, we learn of him that he learnt to speak Greek, but spoke it so imperfectly that Δάτιδος μέλος (Arist. *Pax* 290) or δατισμὸς became a proverb for barbarous Greek. A son of his is mentioned in 7, 88.

ΔΕΚΕΛΕΥΣ, c. 92.

Of Dekelea, 12 miles to the N.E. of Athens, commanding the Eastern pass over Parnes. It was one of the 12 original Cantons said to have been united by Theseus. The modern village of *Tatoi* is on its site.

DELOS, cc. 97—9, 110.

DELIANS, the, cc. 97—8, 118.

Delos, occupying a central situation among the Cyclades, had been to island Greece and Ionia much what Delphi was to continental

Greece or Greece generally. Legend said that it had once been a floating island and had become fixed when Latona brought forth Apollo in it. Its temple accordingly was for a long time the centre of the worship of Apollo, and the meeting place of the Panionian congress. This yearly gathering is described in the Homeric Hymn to Apollo (146—155); and though when the meeting place of the Ionian League was removed to Ephesos by Polykrates of Samos, about B.C. 530—20, this yearly festival lost something in character and in the number of those frequenting it, yet it was still attended largely by Ionians from all parts of Greece. The *θεωρίαι* were conducted with considerable splendour [see Plutarch *Nicias* c. 3], and its temple was enriched by numerous offerings from individuals and states. Its peculiar position was shewn by its selection as the treasure-house of the Confederacy of Delos [B.C. 476]; and its yearly assembly continued to the end of the period of Greek independence to be largely attended; for we find in B.C. 178 that when King Perseus wished to circulate a proclamation among all Greeks he sent copies of it to Delos, as well as to Delphi and the temple of Itonian Athene (Polyb. 25, 2). It appears however to have gradually degenerated into a fair (*ἐμπορικόν τι πρᾶγμα* Strabo 10, 5). The island regained some importance after the destruction of Korinth (B.C. 146), because it was made a free port and was regarded as a half-way place between Athens and Ephesos, the ordinary route taken by the Roman ships going to Asia (Cicero *ad Att.* 5, 12). It is the smallest of the Cyclades, lying between Rheneia and Mykonos.

DELIUM, c. 118.

A small town or hamlet in Boeotia, within the territory of Tanagra. It got its name from a temple of the Delian Apollo on the sea coast, and was afterwards rendered famous by the defeat of the Athenians there in B.C. 424 [Thucyd. 4, 90].

DELPHI, cc. 19, 27, 34—5, 52, 57, 66, 70, 76, 135, 139.

The town of Delphi stood in a kind of natural amphitheatre to the S. of the sloping foot of a precipitous two-headed cliff which terminates the range of Parnassos. The valley is watered by the river Pleistos flowing to the S.W. into the Krissaean gulf. The name of the town in the Homeric poems is Pytho (Πυθώ), hence the 'Pythian games', and the 'Pythia', the priestess who delivered the oracles. To its famous temple of Apollo men came from all parts of Greece to consult the oracle on all manner of questions. The oracle was also often consulted on matters of international dispute between the Greek States; it was therefore of great importance to secure its impartiality, and position of independence. Hence the question of the custody of the temple, disputed between the Delphians and the Phokian League, was one which Sparta and Athens thought it worth while to maintain in arms [circ. B.C. 449 Thucyd. 1, 112]. The Amphiktyonic League met there and at Thermopylae alternately, and was specially bound to maintain its inviolability. The restoration of the temple by the Alkmaeonidae, mentioned in this book, was in consequence of its destruction by fire in B.C. 548.

DEMARATOS, cc. 50—1, 61, 63—7, 70—5.

Son of Ariston, who was the 14th of the Spartan kings of the junior house (the Eurypontidae). The story of his deposition is told in this book. But though he was thus forced to take refuge in Persia, he did not lose all concern for Greece; for he took pains in B.C. 481 to warn his countrymen of the intended invasion of Xerxes (7, 239). Still he accompanied the king in his invasion, and occupied a high place in his confidence [7, 101—4, 209, 234—7]: and was rewarded by the possession of the cities of Pergamum, Teuthrania and Alisarna, which were in the possession of his descendants in B.C. 399 [Xenoph. *Hellen.* 3, 1, 6]: though he nearly forfeited the favour of Artaxerxes by imprudently asking to be allowed to wear the tiara [Plutarch *Themist.* 29.].

DEMARMENOS, c. 65.

A Spartan, father of Prinetidas, see 5, 41.

DEMETER, C. 91.

The most venerable of the goddesses in Greek mythology. She was the daughter of Kronos and represented mystically the secret operations of nature. Her worship was mystic, and the initiation in the mysteries was regarded as the most solemn act in a man's life. Her temples were usually in a solitary place some way from a town [see 9, 69, 97, Vergil, *Aen.* 2, 714]; but the most celebrated one was that at Eleusis.

DIAKTORIDES, C. 71.

A Spartan, father of Eurydame, wife of king Leotychides.

DIAKTORIDES, C. 127.

Of Crannon in Thessaly, one of the suitors of Agariste: of the wealthy family of the Scopadae (q. v.). See also *Krannon.*

DIDYMA, C. 19.

A place just outside the gates of Miletos in which there was a very ancient temple and oracle of Apollo, said to be anterior to the settlement of the Ionians, and to have been founded by Herakles [Pausan. 5, 13, 11; 7, 2, 4]. It was also called Branchidae, and was with others presented with rich offerings by Kroesos [1, 46, 92, 157; 5, 36]. Some of the works of art in the temple were taken to Ecbatana [Pausan. 1, 16, 3].

DIONYSIOS, CC. 11, 12, 17.

A man of Phokaea elected by the Ionians to command their combined fleet. The people of Phokaea had long taken the lead in naval enterprise among the Greeks, and had been the first to shew the way to the Adriatic, Italy, and Spain. Rather than yield to Harpagos, the commander sent against them by Kyros, they had migrated almost in a body to Corsica and thence to Rhegium [1, 163—167]: hence the departure of Dionysios, to lead a buccaneering life in and near Sicily, was in the natural course of things.

DIOSCURI (Διὸς κοῦροι), C. 127.

Castor and Pollux, the twin sons of Zeus and Leda. The story of their entertainment by Euphorion is not mentioned elsewhere; but there is another similar story told by Pausanias (3, 16, 2) of their asking to be entertained by one Phormio, and of this man being punished by the loss of his daughter for refusing.

DOLONKI, cc. 34—5, 60.

A Thrakian tribe, supposed to be named from Dolonkos, a son of Kronos and Thrake. They continued to exist and retain their name until after the Christian era, but were never important.

DORIANS, the, cc. 53, 55.

One of the great divisions of the Hellenic race, which settling first in the north afterwards migrated into the Peloponnese and elsewhere. According to the myth they were descended from Dorus, the eldest son of Hellen [1, 56].

EGYPTIANS, cc. 6, 54—5, 60.

The inhabitants of Egypt; to whom Herodotos ascribes the commencement of most of the arts of life: history [2, 77], knowledge of immortality [2, 123], writing [2, 37], the divisions of the year [2, 4], religious worship [2, 4], divination [2, 57]. Herodotos also believed that there had been before the immigration of the Dorians an Egyptian occupation of the Peloponnese led by Danaos [7, 94; 2, 71].

ELAEŪS, c. 140.

A town near the extremity of the Thrakian Chersonese, at the entrance of the Hellespont. It was a colony of Teos in Ionia, and was celebrated for its tomb and temple of Protesilaos [7, 33], the wealth of which was plundered by Artayktes, who deceived Xerxes into granting him permission to do so [9, 116].

ELEUSIS, cc. 64, 75.

Eleusis, situated on a bay to which it gives a name, was about 11 miles W. of Athens. It is opposite Salamis and at the mouth of the western branch of the Attic Kephisos. It was famous for its temple of Demeter, at which mystic initiations were performed yearly; and the road to it from Athens was accordingly called the 'sacred way', along which the citizens went in solemn procession to attend the yearly celebration.

ELEANS, the, c. 127.

ELIS, c. 70.

Elis was the North-Western province of the Peloponnese. Its inhabitants were partly immigrants from Aetolia, and are not prominent in this period of Greek history except as managers of the Olympic festival held in their territory.

ENNEAKROUNOI, c. 137.

The Nine-Springs, the name of a spring in Athens, which had been constructed into a conduit with nine pipes by Peisistratos for the use of the public. It was close to the Odeium and Olympeion, on the S.E. of the Acropolis. It was the only natural spring in Athens, though there were numerous wells [Thucyd. 2, 15; Pausan. 1, 14, 1]. It was, most likely, from this fountain that certain citizens illegally drew off water for private use and were prosecuted by Themistokles as ὑδάτων ἐπιστάτης [Plutarch *Them.* 31].

EPHESOS, c. 16, territory of, c. 16.

A town on the coast of Lydia at the mouth of the Kayster. It had a good harbour called Panormos, which has now been silted up. It was one of the twelve Ionian cities which joined in the yearly festival at the Panionium, a temple of Poseidon on Mykale [1, 142, 148]. This festival was afterwards called the Ephesia, indicating that Ephesos was the chief city of the Ionian league [Thucyd. 3, 104]. Its celebrated temple of Artemis was believed to be far older than the Ionian settlement; that is to say, that, when the Ionians came there, they found a frequented temple of some Goddess, to whom they gave the name of Artemis [Pausan. 7, 2, 6—7].

EPIDAMNOS (or -um), c. 127.

A colony of Korkyra on the Illyrian coast between Apollonia and Lissos [Thucyd. 1, 24]; its name was changed by the Romans, from an idea of its evil omen, to Dyrrachium (*Durazzo*), Pliny *N. H.* 3, 145: and it became the usual port for ships starting from Brundisium.

EPIKYDES, c. 86.

A Spartan, father of Glaucos.

EPIKYDEIDES, c. 86, = S. of Epikydes, see Glaukos.

EPISTROPHOS, c. 127.

Of Epidamnos, father of Amphimnestos, one of the suitors of Agariste.

EPIZELOS, c. 117.

An Athenian, who distinguished himself at Marathon, and was represented in the picture in the Stoa-Poekile.

ERASINOS, c. 76.

A river rising at the foot of Mt Chaon and flowing across the plain of Argos. Its modern name is *Kefalari*, and it was believed to be united by an underground stream with the river of Stymphalos in Arkadia [Pausan. 2, 24, 6].

ERETRIA, cc. 43, 94, 102, 105, 119.

ERETRIANS, the, cc. 99, 100, 119, 127.

In Euboea nearly opposite Oropos. Its inhabitants were mostly Ionians [8, 46], and it had long been the seat of considerable commercial activity and a rival of Chalkis. The Eretrians sent a contingent of five ships to help the Ionians in their revolt, in gratitude for help long ago given them by the Milesians, in their struggles with Chalkis [5, 99; Thucyd. 1, 15].

ERYTHRAEANS, the, c. 8.

The inhabitants of Erythrae, a town in Ionia (mod. *Ritri*), standing on a peninsula which juts out into the bay of Erythrae. It was one of the 12 Ionian cities [1, 142]: and is grouped by Herodotos with Chios as using the same dialect of Ionic Greek [1, 142], though the two states were at times at war with each other [1, 18]. According to the legend given by Pausanias, it was founded by Erythros from Krete, who brought beside Kretan settlers, Lykians, Karians and Pamphylians, and was afterwards strengthened with Ionians under Kleopos son of Kodros [Paus. 7, 3, 7].

ERYTHRAEAN SEA (ἡ Ἐρυθρὴ θάλασσα), c. 20.

The 'Red Sea' of Herodotos includes the Indian Ocean with its gulfs, the Red Sea and Persian Gulf. What we call the Red Sea he calls the Arabian gulf [ὁ Ἀράβιος κόλπος 2, 11; 4, 39].

ETESIAN WINDS, the, c. 140.

The periodical N.-West winds which in the Aegean blow for forty days from the rising of the Dog-star (26 July), cp. 7, 168.

EUBOEA, cc. 100, 127.

A long island, not, as H. says, as large as Kypros [5, 31], extending from the Malian gulf as far south as about half the length of Attica. The channel between it and the mainland was very narrow at one point, but was not bridged until B.C. 410. Down the centre went a long range of mountains dividing it into only three plains of any extent, that of Histiaea (Oreos) in the north, of Chalkis and Eretria in the centre of the west side, and of Karystos in the south. It had long been the seat of a flourishing commerce, and had been fruitful in colonies. Its country folk seem to have been a hardy race of mountain shepherds, who even in Pausanias' time were clothed in coats of pigs' skin [Paus. 8, 1, 5]. These were of three different races inhabiting the three districts mentioned above, Ellopians (Ionians), Thrakians from Abae, and Dryopians (Pelasgi).

EUPHORBOS, c. 101.

Of Eretria, son of Alkimachos. With Philagros he betrayed his native city to the Persians, and was rewarded by the king with a gift of territory [Plutarch *de Garrul.* 15. Pausan. 7, 10, 2].

EUPHORION, c. 114.

The father of the poet Aeschylos and of Kynageiros [2, 156]. He seems to have lived at Eleusis and been connected with the worship of Demeter.

EUROPE, cc. 33, 43.

According to Herodotos, Europe ended at the Colchian Phasis (*Rioni*); though others looked upon the Tanais (the *Don*) as the boundary to the N.-East. The question of whether a sea washed its northern shores was looked upon as entirely insoluble [4, 45], and the western shores of the ocean were almost as unknown. He derived the name from εὐρύς and ὄψ, because of its lengthened frontier stretching opposite to Asia and Libya.

EURYBATES, c. 92.

An Argive athlete, who commanded a thousand volunteers in the defence of Aegina [9, 75; Pausan. 1, 29, 5].

EURYDAME, c. 71.

Second wife of king Leotychides of Sparta, daughter of Diaktorides.

EURYSTHENES, cc. 51—2.

The descendant of Herakles with whom began the elder family of the kings of Sparta [8, 131].

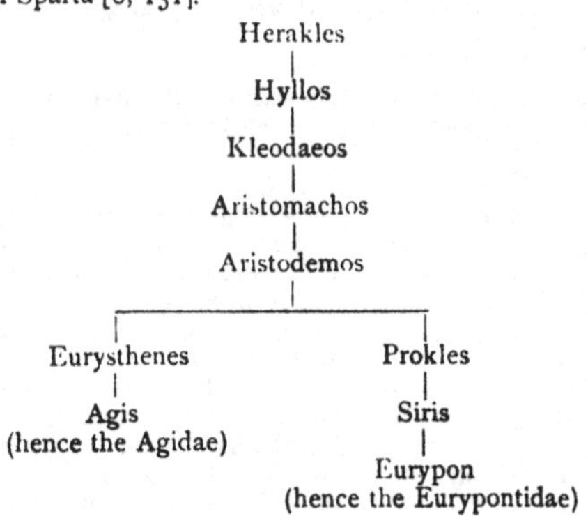

EUXINE, the, c. 33.

Called also by Herodotos the North Sea (ἡ βορηΐη θάλασσα, 4, 37). It was earlier called the inhospitable sea (πόντος ἄξενος Pind. *Pyth.* 4, 362). The English name of Black Sea appears to be a translation of the modern Greek *Maurothalassa*.

EVAGORAS, c. 103.

Of Sparta, the owner of some famous racing mares.

GELA, c. 23.

On a river of the same name in Sicily. It is on the S. coast between Agrigentum and Kamarina. It was a joint colony from

Rhodes and Krete formed B.C. 690 [7, 153. Thucyd. 6, 4, 3]. Its name was believed to be from an African word meaning 'white-frost' (gelu), from the appearance along the banks of the stream (the *Fiume di Terra-Nuova*): but others have derived it from the notion of brightness contained in γελᾶν to laugh [cp. Hom. *Il.* 19, 362 αἴγλη δ' οὐρανὸν ἷκε, γέλασσε δὲ πᾶσα περὶ χθὼν Χαλκοῦ ὑπὸ στεροπῆς]. For a while it was independent and flourishing; but when one of its citizens Gelo became tyrant of Syracuse (B.C. 485) he took away a great part of its citizens and settled them in Syracuse, giving over Gela to his brother Hiero, who in his turn became tyrant of Syracuse and continued the same policy. After the expulsion of the Syracusan tyrants however (B.C. 466) it again rose to importance, until it was laid waste by the Carthaginians in B.C. 405. It partially recovered this, but was again destroyed by Phintias, tyrant of Agrigentum, some time between the 1st and 2nd Punic wars, and from this it never thoroughly recovered [Diodor. Sic. 22, 2].

GLAUKOS, c. 86.

A Spartan, son of Epikydes, a man of high position and character, whose attempted dishonesty was denounced by the Pythia [Pausan. 2, 8, 2; 8, 7, 8].

GOBRYAS, c. 43.

The father of Mardonios [7, 82]. He was one of the Persian nobles who killed the false Smerdis [3, 70, 73, 78]. He accompanied Darios into Skythia and interpreted the meaning of the Skythian emblematical missive [3, 132]. He was married to a daughter of Darios, and Darios married his daughter [7, 2, 5].

HARMODIOS, cc. 109, 122.

One of the two conspirators who killed Hipparchos, son of Peisistratos [5, 55]. See *Aristogeiton.*

HARPAGOS, cc. 28, 30.

A Persian in command of an army of the Persian king in Mysia.

HEGESANDER, c. 137.

Of Miletos, father of Hecataeos (q. v.).

HEGESIPYLE, c. 39.

Wife of Miltiades son of Kimon, and daughter of the Thrakian king Oloros.

HEKATAEOS, c. 137.

A native of Miletos [circ. B.C. 550—476]. One of the earliest Greek historians and geographers; and the first to travel personally to various countries to collect his information, afterwards set forth in two works, Περίοδος γῆς or περιήγησις, and Ἱστορίαι or γενεαλογίαι. Herodotos knew and used his writings. He was the son of Hegesander, and of an illustrious family [2, 143]. He displayed great wisdom and moderation at the time of the Ionian revolt. He dissuaded Aristagoras from beginning it without preparation [5, 36]; and on his failure again offered him salutary advice as to the best way of securing himself [5, 125]. After the collapse of the revolt he went as ambassador to Artaphernes and obtained mild terms for his countrymen.

HELEN, c. 61.

Wife of Menelaos, and daughter of Leda and Tyndareos. Her temple at Sparta was in the Phoebaeum, a part of the Platanistas, a broad level south of the Dromos planted with plane trees [Pausanias 3, 15, 3].

HELLAS, cc. 24, 48—9, 61, 85, 94, 98, 106, 109.

HELLENES, the, cc. 17, 29, 43, 53—4, 106, 112, 126—7, 134, 137.

Herodotos uses Hellas and the Hellenes in the widest sense, including all those who were united by common descent (the mythical ancestor of all being Hellen), common language, and religion. Thus in c. 24 we see that the Greek cities in Sicily were included, as well as those in Asia Minor (c. 29).

HELLESPONT, the, cc. 33, 43, 95, 140.

The narrow strait (varying from 1 to 3 miles) now called the Dardanelles, between the Thrakian Chersonese and the coast of Asia. It was lined with Greek colonies—the list of which on the European side is given in c. 33—attracted thither by the facilities for trade in the Black Sea (c. 26).

HEPHAESTIA (or -ias), c. 140.

A town on the north of Lemnos, which like the island was sacred to Hephaestos, who was said to have fallen upon it when thrown from heaven [Hom. *Il.* 1, 594]: a legend accounted for by the evident marks throughout of violent volcanic action; on which account also it was sometimes called Aethaleia.

HERAEON, cc. 81—2.

Temple of Herè, sister and wife of Zeus. Her worship was especially common among the Dorians; but the most celebrated and frequented temple of all was this between Mykenae and Argos. There was held a solemn yearly festival [1, 31]; and the names of the priestesses, appointed annually, served, like those of the Archons at Athens, to mark the years in the calendar [Polyb. 12, 11]. Pausanias says that the temple was 15 stades (or 1⅞ miles) from Mykenae, on a stream called Eleutherium [2, 17, 1]. It was burnt down in B.C. 423 by the carelessness of the priestess Chrysis [Thucyd. 4, 133], and rebuilt on a somewhat lower site [Pausan. 1, 17, 7].

HERAKLES, cc. 53, 108, 115.

Son of Zeus and Alkmena, according to the commonest legend.

His worship was very widely spread. Herodotos says that he found it going on in Egypt [2, 43, 145], at Tyre [2, 44] and in Skythia [4, 59]: though the Egyptians did not know the Greek myths in regard to him. The worship of Heroes was usually of a different character to that of Gods [ὡς ἥρωϊ ἐναγίζουσι ὡς θεῷ θύουσι, Pausan. 2, 11, 7]; but the peculiarity of the worship of Herakles was that both sorts were paid to him [2, 44]. He was the mythical ancestor of the royal families of Sparta and of Lydia [1, 7; 7, 204; 8, 131; 9, 26], and temples to him existed in many parts of Greece.

HERAKLEION, c. 116.

A temple and precinct of Herakles at Marathon (q. v.) and Kynosarges (q. v.).

HERMIPPOS, c. 4.

A native of Atarneos in Mysia, an agent of Histiaeos.

HESIPEIA, c. 77.

The MSS. vary between ἡ Σίπεια, Ἡσίπεια and σήπεια. This place was apparently between Argos and Tiryns, but it is not mentioned elsewhere.

HIMERA, c. 24.

Himera (mod. *Termini*) was the only independent Greek city on the N. coast of Sicily. It was colonised from Zankle (about B.C. 648), mostly with Chalkidians, but also with some exiles from Syrakuse [Thucyd. 6, 5]. Its modern name is derived from Thermae, which was built not far from the site of Himera, after its destruction in B.C. 408, and was so called from its hot baths or springs [Cicero *Verr.* 2, 35].

HIPPARCHOS, c. 123.

The younger son of Peisistratos, according to Thucydides [6, 54—55], whose elder brother Hippias (q. v.) succeeded his father as tyrannus. Thucydides says that he had made special inquiries and ascertained this to be the case; although there was a common notion that Hipparchos was the elder, and that Hippias succeeded at his death. This popular view is assumed in the dialogue *Hipparchos* [228 B] falsely attributed to Plato, but not much subsequent in date to him. Herodotos [5, 55] confirms the statement of Thucydides that Hippias and not Hipparchos was the tyrannus: yet Thucydides seems to attribute some ἀρχή to Hipparchos; and the explanation may be that they succeeded jointly to the functions of their father, but that Hippias, as the elder, had the chief dignity and authority. The story of the assassination of Hipparchos told by Thucydides attributes the action to jealousy on the part of Aristogeiton of Hipparchos' affection for Harmodios.

HIPPIAS, cc. 102, 107—9, 121.

The elder brother of Hipparchos (q. v.) and son of Peisistratos. The assassination of his brother [B.C. 514] embittered Hippias, and caused his rule, which had been mild and beneficent, to degenerate into a harsh tyranny [Thucyd. 6, 59]; and he was expelled in B.C. 510 [5, 63]. He went first to Sigeium, of which his half-brother Hegesistratos was despot, and thence to the court of Darios, to request his help to regain his power [5, 96], after a vain attempt of the Lakedaemonians to restore him [5, 93]. He was doubtless influential in prompting the invasion of Attica, and suggested Marathon as a place of landing; for it was there that his father Peisistratos had landed in B.C. 537, accompanied by Hippias, when successfully endeavouring to regain his power [1, 62].

HIPPOKLEIDES, cc. 127—9.

An Athenian, son of Tisander, one of the suitors of Agariste.

HIPPOKRATES.

(1) c. 103.

The father of Peisistratos. When a private citizen he was warned by an omen at the Olympic games that his son was destined to work some mischief [1, 59].

(2) c. 23.

Tyrant of Gela (q. v.) in Sicily, succeeding his brother Kleander in B.C. 498. He conquered the Syrakusans in a decisive battle, and much extended his territory. In B.C. 491 he died while besieging Hybla in the course of a war with the native Sikels [7, 154—5. Thucyd. 614].

(3) c. 127.

Of Sybaris, father of Smindyrides who was one of the suitors of Agariste.

(4) c. 131.

Son of Megakles and Agariste, and brother of Kleisthenes the reformer.

HIPPONIKOS, c. 121.

Son of Kallias. The two names were taken in alternate generations by the eldest sons [7, 151]. See Arist. *Av.* 283 ὥσπερ εἰ λέγοις Ἱππόνικος Καλλίου κἀξ Ἱππονίκου Καλλίας. They were the head of a very ancient and wealthy family at Athens.

HISTIAIOS, cc. 1—5, 26—9, 46.

Histiaios, son of Lysagoras, was tyrant of Miletos, at the time of Darius' expedition into Skythia: and was one of the Greek tyrants who resisted the proposal to cut the bridge over the Danube, and so prevent the king's retreat [4, 137—8]. He was rewarded by the gift of Myrkinos and its territory in the country of the Edones [5, 11]. Leaving Miletos in the care of his son-in-law Aristagoras, he proceeded to fortify Myrkinos. This roused the suspicions of Megabazos, who warned Darios that he was meditating some act of insubordination. He was consequently summoned to Sardis under the pretext of his advice being needed by the king, and was kept there and at Susa in a sort of honourable captivity [5, 23—4]. Getting tired of this he determined to stir up a movement among the Ionians, with the hope of being sent down himself to repress it. He did this by sending a slave down to Aristagoras with the words 'Raise the Ionians' [ἀποστῆσον τοὺς Ἴωνας] branded on his head, with only a message to Aristagoras that he was to shave his head and read [5, 35]. The device succeeded; and though Darios blamed Histiaios for the revolt he sent him down to suppress it [5, 105—7]. But on arriving at Sardis he found that Artaphernes was aware of his secret practices; he therefore tried to return to Miletos; but the citizens refusing to receive him he went to Chios, and after being imprisoned and released there, to Byzantium [6, 1—5]. There for a time he supported himself by acts of piracy, and levying black mail on the ships coming out of the Hellespont, and retaliated on the Chians by seizing their Island [6, 26—7]. He then besieged Thasos, for the sake of its gold mines. Hearing that the Phoenikian fleet of the king was about to leave Miletos he removed his army southward to Lesbos. But not finding sufficient provisions he crossed to the mainland, where he was captured and taken to Sardis. There Artaphernes without waiting to refer to the king, crucified him [6, 28, 30]. He is described as a man of great subtilty and acuteness [δεινὸς καὶ σοφός, 5, 23].

HYDARNES, c. 133.

One of the assassins of the false Smerdis [3, 70]. He was afterwards governor of the maritime district of Asia Minor [7, 135], and his son commanded the Immortals in the Army of Xerxes [7, 83, 211]. His descendants became kings of Armenia [Strabo 11, 14, 15].

HYLLOS, c. 52.

Son of Herakles, see *Eurysthenes*. He it was who, according to the legend, led the Herakleidae back to the Peloponnese. According to one story his mother was Melite (Apoll. Rhod. 4, 538), according to another Omphale (Paus. 1, 35, 8), to another Deianeira [Soph. *Trach.* 56]. He was killed in single combat by Echemos of Tegea [9, 26].

HYMETTOS, c. 137.

A double range of mountains to the S.E. of the plain of Athens, divided by a deep depression. Its greatest height is 3506 feet. It was celebrated for its bees, as Pentelicos for its marble quarries, and Parnes for its hunting [Pausan. 1, 32, 1].

HYSIANS, the, c. 108.

The people of Hysiae, a small town on the N. slope of Kithaeron, on the road from Athens to Thebes [9, 15, 25].

HYSTASPES, c. 98.

Father of Darios, and son of Arsames, of the family of the royal Achaemenidae, of which Kyros was a descendant by an elder branch [7, 11]. He accompanied Kyros in his expedition against the Massagetae, but was sent back to Persia to prevent a movement of his son

Darios, which Kyros wrongly fancied was foretold in a dream [1, 209—210]. At the time that his son succeeded to the throne he was governor (ὕπαρχος) of the Persians [3, 70]. And from inscriptions it appears that he acted as his son's general against the Parthians.

IAS, cc. 9, 31.

Ionian (f. adj.) see *Ionians*.

ICARIAN SEA, the, cc. 95—6.

The S.E. part of the Aegean, along the coasts of Caria and Ionia, deriving its name from the island of Icaria. The name was as old as Homer [*Il.* 2, 145].

IMBROS, cc. 41, 104.

An island off the Thracian Chersonese, about 35 miles in circumference. It is mountainous and well-wooded, with fertile valleys. It was first inhabited by Pelasgians [5, 26], and then colonised from Athens, of which it was always regarded as a possession.

INYCUM, cc. 23—4.

A small town in the S.W. of Sicily, on the river Hypsas [mod. *Belici*]. In mythology it was connected with Daedalos, who took refuge there when flying from Minos of Crete.

IONIA, cc. 1, 7, 13, 17, 22, 28, 43, 86, 95.

IONIANS, the, cc. 1—3, 7—14, 17, 22, 26, 28, 31—2, 41—3, 98.

Ionia was the maritime district of Asia Minor, extending from the River Hermus on the North to a short distance S. of Miletus. 'The 'region', says Herodotos, 'is the fairest in the whole world: for no 'other is so blessed as Ionia, either North or South or East or West of 'it. For elsewhere the climate is either too cold and damp, or else the 'heat and drought are oppressive' [1, 142]. Pausanias [7, 5, 2] also says that, 'The Ionians have the very best possible temperature of the seasons'. In this pleasant region the cities were colonized by the Ionians, which together with the islands of Samos and Chios established an Amphiktyone, or community for religious worship, the centre of which was the Panionium, or temple of Poseidon, on the promontory of Mykale; though for political purposes the states were independent. Ephesos and Miletos, the two principal cities, were believed to have been founded by the two sons of Kodros, king of Athens, Androkles and Neileos; and therefore Athens was regarded as in a sense the Metropolis or mother city of Ionia. But the Athenians and other Ionic states in Greece became ashamed of the title 'Ionian', and did their best to renounce it. The Asiatic cities on the other hand were proud of it, and they became to all intents and purposes 'Ionia'. The cities were Miletos, Myūs and Priene, *in Karia*: Ephesos, Colophon, Lebedos, Teos, Clazomenae, Erythrae and Phokaea, *in Lydia*. They were not purely Ionic; for the Ionian settlers had intermarried with the original Karians, Lydians, and other inhabitants, and they did not all speak the same dialect; but the predominant element in them all was Ionian. By B.C. 528 all these towns had been reduced to subjection to Persia, by Harpagos the general of Kyros [1, 162—170]. The only member of the confederacy still remaining free being Samos, which under Polykrates for a time maintained great power and prosperity. But by B.C. 519 Samos too submitted to become tributary [3, 39—47; 54—6; 120—5]. It is the rising of these tributary states, beginning by the expulsion of the despots established by Persian influence, and its suppression as related in the first part of this book, that served to bring about the collision between Persia and the continental Greeks. See *Introduction*.

IONIAN GULF, the, c. 127.

By the Ionian Gulf (κόλπος Ἰόνιος) Herodotos means the Adriatic [9, 92]. The name must have arisen from 'Ionians' being from their extensive colonies regarded as equivalent to 'Greeks'.

ITALIA, c. 127.

By Italy Herodotos seems always to mean the Greek cities in Italy. He does not mention Rome, but the power in central and northern Italy known to him is the Etruscan empire (Tyrrhenia) 1, 94, 163.

KAIKOS, c. 28.

The Kaikos flows through the S. of Mysia, past Pergamum, and discharges itself into the Sinus Eleaticus between Pitane and Elaea. By the plain of Kaikos Herodotos apparently means the part near Pergamum. It was very rich and fertile.

καλὴ Ἀκτή, cc. 22—3.

The Fair-Strand, on the N. coast of Sicily, between Helaesa and Haluntium, was in Roman times called Calacta [Cic. II *Verr.* 3 § 43]. It was noted for the fisheries in the sea near it [Silius Ital. 14, 251]. There was another place of the same name on the W. coast of Krete.

KALLIAS, cc. 121, 122.

Son of Phaenippos, a wealthy Athenian, and violent opponent of the Peisistratidae. See *Hipponikos*.

KALLIMACHOS, cc. 109—111, 114.

An Athenian of the deme Aphidna, who was Archon Polemarchos in B.C. 470 and fell at Marathon. He and Miltiades were the most conspicuous figures in the picture of the battle of Marathon in the Stoa Poikile [Paus. 1, 15, 3].

KARIA, c. 25. KAR, c. 20.

Karia was the district in Asia immediately South of Ionia. The boundaries between the two differed at different times, sometimes being reckoned as formed by the Messogis range, and sometimes by the Maeander. It had in the last century been subdued and annexed by Kroesos [1, 28], and afterwards by the Persians under Harpagos [1, 174], and joining in the Ionian revolt was again subjugated by the

Persians [5, 117—120]. The original inhabitants were Leleges, and had once been widely spread in Island Greece or, as Herodotos says, had come from the Islands, though they denied this themselves [1, 171], but were reduced to subjection by Minos of Crete [Thucyd. 1, 8]; and either their inferior position among the Ionian settlers, or the fact of their practice of serving as hired troops, brought them into such contempt, that 'the value of a Karian' and 'to risk a Karian' came to be proverbs for what was worthless. [Plato *Laches* 187 B. Eurip. *Cycl.* 654. Polyb. 10, 32]. The Greeks regarded them as βάρβαροι, but though their language was not Hellenic, it was largely mixed with Hellenic words [1, 171].

KARCHEDONIANS, the, c. 17.

The inhabitants of Carthage [Καρχηδών], who in the time of Herodotos were a great mercantile people, and were encroaching on Sicily, though resisted by successive rulers of Syrakuse [7, 165], as they had done on Corsica [1, 166].

KARDIA, cc. 33, 36, 40.

A town in the Northern part of the Thrakian Chersonese, at the head of the Black Gulf (μέλας Κόλπος q.v.). It was originally a joint colony from Miletos and Clazomenae; but was strengthened afterwards from Athens. It was destroyed (about B.C. 300) by Lysimachos, and partially restored with the name of Lysimachia.

KARYSTOS, c. 99.

A town on the South of Euboea near Mt Ocha, still called *Karysto*, though now a mere village. It was celebrated for its marble quarries. Like Styra in the same district its inhabitants were Dryopians, a Pelasgic race from a district near Malis [4, 33; 8, 66]. In B.C. 469—7 they waged a not unsuccessful war with the Athenians [9, 105. Thucyd. 1, 98].

KASAMBOS, c. 73.

An Aeginetan, son of Aristocrates. He was one of the leading men selected by the kings of Sparta to be delivered, as a hostage for the loyalty of Aegina, to the care of the Athenians.

KILIKIA, c. 95. KILIKIANS, c. 7.

The South-Eastern district of Asia Minor, bordering on the Mediterranean opposite Kypros [5, 49], bounded on the W. by Pamphylia and Pisidia, and on the East by Mt Amanus, though Herodotos [5, 52] extends it to the Euphrates. Its Northern frontier is Mt Taurus, it therefore counted as ἔξω τοῦ Ταύρου as opposed to that part of Asia which was spoken of as ἡ ἐπὶ τάδε τοῦ Ταύρου, a division which may be taken to correspond to that of Asia Minor and Palestine, Kilikia being on the borders of the two. The Kilikians, originally called Hypachaei [7, 91] were probably of Aramaic origin, and connected with the Phœnikians.

KIMON.

(1) cc. 34, 38—40, 103, 140.

The father of Miltiades, an Athenian of wealth and influence, but according to Plutarch [*Cimon* 4] of bad character, and of such mean intellect as to get the nickname of Κοάλεμος 'the fool'. For his family and descendants, see *Miltiades*.

(2) c. 136.

Son of the great Miltiades and Hegesipyle. His youth, according to the gossip in Plutarch's Life (which however is probably derived from hostile and untrustworthy sources) was discreditable; but his after career was of great service to Athens. In conjunction with Aristeides he was the founder of the confederacy of Delos [B.C. 478], and his victory over the Persians at the mouth of the Eurymedon in Pamphylia [B.C. 466] finally decided the freedom of the Greek states from Persian influence: though the actual conclusion of the treaty after it, mentioned by Plutarch, has been denied. The rising power of Perikles, combined with the unfriendly attitude of Sparta, of which Kimon had been always a partisan, contributed to secure his ostracism in B.C. 461. But on his recal in B.C. 456 he again did good service to his country in Egypt and Kypros, in which latter island he died while engaged in the siege of Kitium [B.C. 449]. Taken as a whole the policy of Kimon was for the Athenians to keep on friendly terms with Sparta, and in conjunction with her to maintain the *status quo* in Greece, while he pushed the Athenian Empire abroad: that of Perikles was to make Athens supreme in Greece, and in order to do so, to depress Sparta.

KISSIA, c. 119.

A district in Asia [mod. *Khuzistán*] of which the Capital was Susa.

KLEANDROS, c. 83.

A native of the town Phigaleia in Arcadia, a mantis or soothsayer.

KLEISTHENES.

(1) cc. 126, 128—130.

The last tyrannus of Sikyon, of a dynasty established by Orthagoras, which lasted 100 years, owing, says Aristotle (*Pol.* 5, 11), to the moderate and law-abiding character of the sovereigns. In B.C. 595 he commanded the joint army of the Amphiktyons in the Sacred War, which after 10 years ended in the destruction of Kirrha. The year of his death and of the end of his dynasty is not known, but it was subsequent to B.C. 582

(2) c. 131.

Son of Megakles and Agariste the daughter of Kleisthenes tyrant of Sikyon. His reforms on the Solonian constitution of Athens were introduced shortly after the expulsion of Hippias and his family [B.C. 510]. They were in a democratic direction; the foundation of them being the division for political purposes of the people into 10 local tribes instead of the four ancient tribes, which were combinations of φρατρίαι or clans. His rival Isagoras applied for help from Sparta, and in the course of the ten years [510—500 B.C.] Kleisthenes with the rest of the Alkmaeonidae had to leave Athens; but was soon afterwards recalled. Of the end of his life we know nothing. [See 5, 63, 66, 69—73.]

KLEODAEOS, c. 52.

Son of Hyllos (q. v.), and grandson of Herakles [see pedigree given in article *Eurysthenes*, and comp. 7, 204; 8, 131]. He had an Heroum or Chapel at Sparta near the Theatre [Pausan. 3, 15, 10].

KLEOMENES, cc. 50—1, 61, 64—6, 73—5, 78—81, 84—5, 92, 108.

The 16th king of Sparta of the elder line, the Agidae, from B.C. 520 to B.C. 491. He was the son of Anaxandridas by his second wife [5, 41—2]. He was all his life strange and eccentric, and finally died by his own hand in a state of absolute madness. His career as king of Sparta was, as might be expected from his character, a strange and sometimes discreditable one: though it does not appear to have been stained by the usual corruption of Spartan kings in regard to taking bribes [3, 148; 5, 51; 6, 82]. In B.C. 511—10 he led an army against Athens, and compelled Hippias and his family to quit [5, 64—5], and two or three years later supported Isagoras and expelled Kleisthenes with the other Alkmaeonidae from Athens [5, 70], but was shortly afterwards forced to give up the Akropolis which he had seized [5, 72—3, 90]. He however made another attempt to set up Isagoras as tyrant of Athens, which was frustrated by a quarrel with the other king Demaratos [5, 74—6]. While on the Akropolis he had discovered the oracles which the Alkmaeonidae had fraudulently obtained from Delphi ordering the expulsion of the Peisistratidae, and had taken them to Sparta, and this induced the Spartans to make an attempt to bring back Hippias [5, 72]. After refusing to take part in assisting the Ionian revolt [5, 49—51], his next enterprise was that against Aegina recorded in chs. 49—51 of this book. His most outrageous proceeding was his invasion of Argos, and the severity with which he treated the inhabitants [cc. 76—80, cp. 7, 148], which so embittered the Argives against the Spartans, that they readily entered into terms with the Persians, and all through the subsequent Persian war consistently *medized*. His final loss of reason seems to have been brought to a crisis by intemperate habits [6, 76].

KNOETHOS, c. 88.

An Aeginetan, father of Nikodromos.

KOBON, c. 66.

A man of high position at Delphi, son of Aristophantos, who acted in collusion with Kleomenes in the matter of Demaratos.

KOENYRA, c. 47.

A place in the island of Thasos, near which were some silver mines once worked by the Phoenikians.

Κοῖλα, τά, c. 26.

A place in the Island of Chios, 'the Hollows,' but whether so called from a depression in the land, or from the configuration of the shore is not known. The same name in Euboea refers to the lie of the coast [8, 13].

Κοίλη ὁδός, ἡ, c. 103.

The 'Hollow Road' or the 'Valley Road,' the name of a road outside Athens, on the north of the town. It leads through a place called Κοίλη, in which was the tomb of Thucydides.

KORINTHOS, c. 128. KORINTHIANS, the, cc. 89, 108.

The territory of Korinthos was separated from the Megarid on the north by the range of Geraneia, and from Argolis on the south by Oneum. Korinth itself consisted of a citadel, the Akrokorinthus (1900 ft.), with a town round it enclosed with wall, and joined to its western port, Lechaeum, by long walls. Its eastern port, Kenchreae, was more than eight miles distant. Its position on a narrow isthmos commanding the road from Northern Greece to the Peloponnese, and with easy access to both seas, made it early important; and in it the first triremes were built [Thucyd. 1, 13]. Though the inhabitants were for the most part Dorians, they were at this time on friendly terms with Athens rather than with Sparta, and refused to join in the attempt to re-establish the tyranny there [5, 75, 92].

KOUPHAGORAS, c. 117.

An Athenian, father of Epizelos (q. v.).

KRANNONIAN, a, c. 127.

Of Krannon, also called Ephyra [Pind. *Pyth.* 10, 85]. It was a town in the district of Thessaly called Pelasgiotis, in which the most wealthy and important family was that of the Scopadae, whose great flocks feeding on the fertile plain round it are mentioned by Theocritus *Id.* 16, 36:

> *πολλοὶ δὲ Σκοπάδαισιν ἐλαυνόμενοι ποτὶ σακοὺς*
> *μόσχοι σὺν κεραῇσιν ἐμυκήσαντο βόεσσιν·*
> *μυρία δ' ἀμπέδιον Κραννώνιον ἐνδιάασκον*
> *ποιμένες ἔκκριτα μῆλα φιλοξείνοισι Κρεώνδαις.*

KRIOS, cc. 50, 73.

A man of high position in Aegina. His son Polykritos distinguished himself at the battle of Salamis (8, 92—3). Krios seems, from the Scholiast on Aristoph. *Nubes* 1356, to have been an athlete.

KROISOS, cc. 37—8, 125.

Kroisos was king of Lydia from B.C. 560 to B.C. 546, famous for his wealth and liberality. He conquered the Asiatic Greeks [1, 26], and attempted to do the same, though unsuccessfully, to the Islanders [1, 27]. The consultation of the oracles mentioned in c. 125 refers to his sending round to the Greek Oracles owing to his terror at the growing power of Kyros [1, 46 sqq.]. He was conquered near Sardis in B.C. 546, and was afterwards kept in honourable arrest at the Persian court [1, 79—88, 155—7, 3, 34—6].

KROTONIANS, the, c. 21.

The inhabitants of Kroton, a Greek colony on the E. coast of Bruttium, about six miles N. of the Lacinian promontory, founded about B.C. 710 by a band of Achaeans led by Myskellos of Rhypae. It was long the residence of Pythagoras, and his club of 300 for some time exercised supreme political influence there; and during this influence (about B.C. 510) occurred the war with Sybaris referred to in this passage, which ended in the total destruction of that town [5, 44].

KYNEGEIROS, c. 114.

Son of Euphorion, and brother of Aeschylos. According to Trogos (Justin. 2, 9) his right hand with which he was holding on to a Persian ship being cut off, he laid hold with his left, and when that too was cut off, with his teeth.

KYNEAS, c. 101.

An Eretrian, father of the traitor Philagros (q. v.).

KYNISKOS, c. 71.

A name ('the Whelp') given by the Spartans to Zeuxidamos son of Leotychides.

KYNOSARGOS, c. 116.

A precinct of Herakles, and gymnasium outside Athens, on the road to Marathon, near the Dromeian gate. The gymnasium was frequented by half-bred Athenians and illegitimately born boys,—Herakles, half man and half god, being the patron of such [Plutarch *Themist.* 1]. It was on high ground commanding a view of the sea.

KYPRIANS, the, c. 6.

The inhabitants of Kypros, an island opposite the coast of Kilikia. It was valuable to the Persians, both as possessing a navy of its own, and as being in the way for the Phoenikian ships crossing to Asia Minor. It had been under the power of Amasis king of Egypt [2, 182]; then, with Asia and Egypt, tributary to the king of Persia [3, 91]; and joining in the Ionian revolt had again been subdued. The Kyprians were originally Phoenikians; but Greek colonies had been settled in it from Salamis, Athens, Arkadia and Kythnos, and some from Aethiopia [7, 90]. It is 150 miles long, with a maximum breadth of 40 miles.

KYPSELOS, c. 34.

The father of Miltiades, the first tyrant of the Chersonese. See *Miltiades*. He was supposed to be descended from the Aeakidae, and was a man of great wealth.

KYPSELIDAE, c. 128.

The descendants of Kypselos, for 30 years tyrant of Corinth, having deposed and decimated the oligarchical clan of the Bacchiadae [about B.C. 655]. His name means a 'chest' or 'corn-bin,' in which the infant Kypselos was concealed from the emissaries of the Bacchiadae, who in consequence of an oracle wished to kill him [5, 92]. This chest was preserved in the family, and, being covered with cedar and gold, was dedicated in the temple at Olympia. [Pausan. 5, 17, 5]. He was succeeded by his son Periander who died after a reign of forty years [about B.C. 585], and was succeeded by a relative, Psammētichos, after a few years of whose reign the dynasty was suppressed [Aristot. *Pol.* 5, 9].

KYZIKOS, c. 33.

A city in Mysia on the extremity of a peninsula (or island joined to the mainland artificially, which has now become a peninsula) projecting into the Propontis. The ruins of it are now called *Bal Kiz*. The Kyzikenes were apparently of Greek origin, but it is not known whence they came or when the town was first formed. It was included under the supremacy of Athens after the Persian wars; revolted and was recovered in B.C. 411 [Thucyd. 8, 107]; was taken by the Spartans in B.C. 410, and recovered by Alkibiades [Xen. *Hell.* 1, 1, 16], and by the peace of Antalkidas (B.C. 387) reverted to the Persians. It was a wealthy mercantile town, and its gold staters were widely current in Greece.

LADE, cc. 7, 11.

The largest of a group of islands off Miletos. It is now part of the continent, and represented by a hill in the plain of the Maeander, the deposits of the river having extended the land beyond the scene of the battle.

LAKEDAEMONIANS, the, cc. 52—3, 58, 60, 70, 75, 77, 85, 92, 106, 108, 120, 123.

The inhabitants of the district of Lakonia, over which, as well as over Messenia, the citizens of Sparta (which contained 8000 men of military age, 7, 234) were supreme. At this time their military reputation gave them an informal supremacy throughout Greece, as is shown by the Athenians accusing the Aeginetans to them, as though they were the natural judges in a matter of international dispute (c. 49).

LAKONIAN DANCES, c. 129.

Probably war dances, or steps like a march to music; for we learn that 'the Lakedaemonians used to march out to battle, not to the sound of trumpets, but to the strains of pipes, lyre, and kithara' [Pausan. 3, 17, 5]. See also note to this passage.

LAMPITO, c. 71.

Daughter of king Leotychides, and wife of Archidamos. She was the mother of king Agis, and was alive in B.C. 427 [Plato, *Alcib.* I. 124 A].

LAMPSAKENES, the, cc. 37—8.

The inhabitants of Lampsakos, a city of Mysia on the Hellespont. It was originally called Pityusa, but received the name of Lampsakos after being colonised from Phokaea and Miletos. It had an excellent harbour, and its territory produced good wine. It had been captured by the Persians during the Ionian revolt [5, 117]. Its tyrant was Hippokles, whose son Aeantides married Archedike, a daughter of Peisistratos [Thucyd. 6, 59].

LAOS, c. 21.

Laos, 35 miles from Sybaris, on the west coast of Italy and the mouth of the river Laos (mod. *Lao*).

LAPHANES, c. 127.

An Arkadian of Paeos, in the district of Azania (q. v.), one of the suitors of Agariste; he was son of Euphorion (q. v.).

LEMNOS, cc. 137—140.

An island lying between Mt Athos and the Hellespont, about 150 square miles in area. Its earliest known inhabitants were a Thrakian tribe called Sinties, which perhaps means robbers or pirates (*σίνομαι*); it then received the Pelasgi expelled from Attica. It was conquered by the Persian Otanes (5, 26), but after its deliverance by Miltiades it remained a possession of Athens till late times [Thucyd. 4, 28. Polyb. 30, 18].

LEMNIAN DEEDS, c. 139.

The *Λήμνια ἔργα* referred to, besides the slaughter of these sons of the Attic women, were as follows. The Sintians of Lemnos loathed their wives, on whom Aphrodite had sent a curse (*δυσοσμία*), and married others from Thrace. In revenge all the women murdered their husbands and fathers, except Hypsyle who concealed her father Thoas [Apollodor. 1, 9, 17].

LEOPREPES, c. 85.

A Spartan, father of Theasides.

LEOTYCHIDES, cc. 65—7, 69, 71—3, 85—86.

The sixteenth king of Sparta of the junior branch (the Eurypontidae). There is some confusion as to his pedigree, for in 8, 131 Herodotos calls his grandfather Agesilaos, not Agis, as here. His and Demaratos' families seem to have parted 8 generations earlier, both being descended from sons of Theopompos, king circ. B.C. 772—713. See *Agis*. He commanded the Greek fleet at Mykale [9, 98].

LESBOS, cc. 8, 14, 28, 31. THE LESBIANS, cc. 5, 8, 26—7.

An island about 7 miles off the coast of Mysia. It had two excellent harbours, and was celebrated for the healthiness of its climate, and the excellence of its wine. It had been originally divided into six territories; but four of them became absorbed in the dominions of the two cities of Methymna and Mytilene, the latter of which gave its name afterwards to the whole island. The inhabitants were principally Aeolians; and the island was regarded as the central seat or *μητρόπολις* of the Aeolians. It never became part of the Lydian kingdom, but submitted for a time to the Persians [1, 169]. After the Persian wars it joined the confederacy of Delos, and was attached to the Athenian interest; but in the course of the Peloponnesian war revolted from Athens [Thucyd. 3, 15], and was punished by the division of the territory of Mitylene among Athenian cleruchs [Thucyd. 3, 50].

LOKRI EPIZEPHYRII, c. 23.

The 'Western Locrians,' that is, the Lokrians in Italy as opposed to the Lokrians in Greece. The town of Lokri was on the S.-E. coast of Bruttium, 5 miles from the modern *Gerace;* but even its ruins have almost entirely disappeared. There seems a doubt as to whether they were colonists from the Opuntian or Ozolian Lokrians, Pausanias (3, 19, 11) and Vergil (*Aen.* 3, 399) favouring the former; and Strabo and others the latter. Aristotle said that the colony was formed by certain slaves of the Lokrians, who, in the absence of their masters on an expedition, intrigued with the Lokrian ladies, and when their masters were about to return, fled with the women to Italy. This account however was strenuously denied by Timaeos, who asserted that Italian Lokri was a colony from Greek Lokris, sent out under the usual honourable circumstances [Polyb. 12, 5—9]. Some derive the appellation of Epizephyrii, not from the distinction between western and eastern Lokrians, but from the promontory Epizephyrium in their territory.

LYDIANS, the, cc. 32, 37, 125.

Lydia was properly the district in Asia Minor separated from Mysia on the N. by the range of Temnos, and from Karia on the south by the Messogis Mts or by the R. Maeander. In Homer [*Il.* 2, 865 etc.] the inhabitants of this country are called Meiones (*Μήονες*), who appear to have been of Pelasgic origin. At some period before B.C. 700 these Meiones were conquered by the Lydi, whose place of origin is quite unknown, but who appear to have been connected ethnologically with the Karians. In the reign of Kroesos [B.C. 560—546], of the 3rd dynasty established by Gyges [1, 18—13], the kingdom of Lydia included all Asia Minor except Lykia and Kilikia. This kingdom was annexed to Persia by Kyros [B.C. 546], and by Darios Lydia and Mysia were formed into a Satrapy, the seat of government being Sardis [3, 90]. Herodotos describes them as an active and warlike race, [1, 7], the first to engage in commerce and coin money [1, 94], with institutions similar to the Greeks [1, 35, 74, 94]. Yet their blood relationship to the Karians was commemorated by a joint worship of the Karian Zeus [1, 171].

LYKURGOS, c. 127.

An Arcadian, father of Amiantūs, of the city Trapezūs (q. v.), who was one of the suitors of Agariste.

LYSAGORAS, c. 133.

Son of Tisias, of the island of Paros.

LYSANIAS, c. 127.

Of Eretria, in Euboea, one of the suitors of Agariste.

MAKEDONIA, cc. 44—5.

Makedonia proper consisted of two parts: Upper Makedonia, an inland district on the east of the chain of mountains which form a continuation of Pindos, and watered by the confluents of the Upper Haliakmon: Lower Makedonia consisting of the basin of the Axius, but not reaching to the sea. In this latter district of Lower Makedonia, a dynasty had established itself with Edessa as its capital, founded by Perdikkas, who claimed to be a grandson of Têmenos of Argos [8, 127; Thucyd. 2, 99]. Under the reign of Amyntas I., fifth in descent from Perdikkas, and his successor Alexander [5, 22; 8, 34], Makedonia, having been gradually organised by them and their predecessors, began to be mixed up in Hellenic politics.

MALENE, c. 29.

A town in the district of Atarneos (q. v.).

MALES, c. 127.

An Aetolian, one of the suitors of Agariste.

MARATHON, cc. 102—3, 107, 111, 113, 116, 120, 132.

A small plain on the E. coast of Attica, 6 miles long and varying from 3 to 5 miles broad between the mountains and the sea. In it were originally four townships, Marathon, Probalinthos, Trikorythos and Oenoe, which together under the name Tetrapolis formed one of the 12 original cantons of Attica. On the union of all Attica, attributed to Theseus, all four of these places became Attic demes. Its northern end is filled up by a marsh which leaves only a narrow pass between it and Mt Koráki: on the south the hills approach close to the sea. Four roads lead from this plain over the hills into the interior: the southern between Pentelikos and Hymettos; the next more northwards to Athens by Kephisia; the next through Aphidna and Dekelea to Athens; the next between Mt Koráki and the large marsh to Rhamnos and so along the coast to Oropos. It had long been known as a place of landing in Attica [Homer, *Odyss.* 7, 80], and had been used by Peisistratos and Hippias in B.C. 537 [1, 62], see *Hippias*. For the battle of Marathon, see Introduction.

MARDONIOS, cc. 43—4, 94.

Son of Gobryas by a sister of Darios [7, 5], whose daughter Artazostra he married. After his failures recorded in this book and his removal from command, he still seems to have retained influence at court. For he is represented as being the chief means of persuading Xerxes to undertake his expedition against Greece [7, 5], and was commander in-chief of the land forces [7, 82]. After the battle of

Salamis he persuaded Xerxes to go home, and leave him with 300,000 men to complete the conquest of the Peloponnese [8, 100—6]. After wintering in Thessaly [8, 113] and vainly trying to win over Athens by negociation [8, 133—141], he occupied it once more [9, 1—4], and after retreating into Boeotia [9, 12—15] finally fell at the battle of Plataea [9, 63].

MEDES, cc. 9, 22, 24, 67, 109, 120. MEDIKE, c. 84. A MEDE, c. 94. MEDIAN ARMY, c. 111. MEDIAN DRESS, c. 112.

The Medes were an Aryan people [7, 62] who when first heard of inhabited a district south of the Caspian now called Khorassan. About the middle of the 7th century B.C., having removed to Media,—a country which with its capital Ecbatana (*Hamadân*) is included in the modern *Irâk Ajem*,—they fell under the power of the Assyrian monarchy. An independent Median kingdom seems to have been established by Kyaxares about B.C. 635. In B.C. 624 he took Nineveh, and before his death extended his dominions up to the river Halys, and threatened Asia Minor [1, 103]. His successor Astyages was conquered by Kyros and the Persians. The new monarchy thus formed was called the Persian Empire or the Medo-Persian; the official title being often the 'Medes and Persians', as in Daniel. The Greeks spoke of them as Medes or Persians indifferently, as does Herodotos; though when necessary he clearly distinguishes them. The official class seems almost always to consist of Persians, and therefore Herodotos (c. 94) is careful to note that Datis was a Mede: cp. 1, 156; 1, 62; 7, 88 for other cases.

MEGABAZOS, c. 33.

Megabazos, or Megabyzos [4, 143] was one of the conspirators against the false Smerdis [3, 81], who wished to establish an oligarchy in place of a king. When Darios obtained the crown, however, he held office under him. He was left in command of the forces in Europe after the Skythian campaign, and subdued the Hellespontine cities [4, 143—4] and Thrace [5, 1], transferred the Paeones to Asia [5, 14], and warned the king of the ambitious designs of Histiaeos [5, 23].

MEGAKLES,

(1) cc. 127, 130—1.

The son of Alkmaeon and husband of Agariste. He was at first a great opponent of Peisistratos, and managed to expel him from his tyranny soon after he had first obtained it [560 B.C.], but presently assented to his return on the condition of his marrying his daughter [1, 59—60]. Peisistratos however neglected his new wife, and Megakles again managed to expel him. But when after 11 years' exile Peisistratos returned [about B.C. 547] Megakles with the other Alkmaeonidae retired from Athens [1, 61—4].

(2) c. 125.

Son of Alkmaeon, and grandfather of the subject of the last article, the successful suitor of Agariste.

(3) c. 131.

Son of Hippocrates, grandson of Megakles and Agariste. See *Alkmaeonidae*.

μέλας κόλπος, c. 41.

The gulf on the west side of the Thracian Chersonese, mod. *Xeros*, into which a river also called Μέλας (mod. *Saldatti*) flows.

MENARES, cc. 65, 71.

A member of the junior royal family of Sparta, father of Leotychides (q. v.).

MENIOS, c. 71.

A Spartan, son of Diaktorides and brother of Eurydame, second wife of Leotychides.

MESAMBRIA, c. 33.

One of the five important Greek cities in Thrace on the shores of the Euxine. It had apparently existed before the Greeks came there, and was first colonized by Megarians; and afterwards, as here stated, reinforced by fugitives from Byzantium and Chalkedon. The orthography of the name appears to vary between Mesambria, Mesembria, and Melsembria, the latter being derived from a founder Melsas [Steph. B.].

MESSENIAN, a, c. 53.

That is, a native of Messene, the S.W. province of the Peloponnese, which since about B.C. 668 had been completely subject to Sparta.

METIOCHOS, c. 41.

A son of Miltiades, who was taken prisoner by the Persians, and settled in Persia.

MILETOS, cc. 1, 5—7, 9—10, 13, 18—9, 21—2, 24—6, 28—9, 31, 86.
    " inhabitants of, cc. 5, 7—8, 19—22, 29, 77, 86.
    " territory of, c. 9.

Miletos, an Ionian city in Karia, stood on a peninsula on the Southwest of the Latmian bay, opposite the mouth of the Maeander, which was at this time about 10 miles distant. The deposits of the Maeander have now filled up the Latmian bay, and covered the ancient site of the city. It was formerly inhabited by Karians, whose husbands and other male relations were massacred by the Ionians when under their leader Neleus they occupied the town [1, 146. Homer *Il.* 2, 867], though some authorities speak also of Leleges and Kretans as forming part of the inhabitants. Between the time of its settlement by Ionians and its capture by the Persians, in B.C. 494, it had risen, greatly owing to its favourable situation and excellent harbour, to a position of high prosperity and power: though frequently in the hands of tyrants [1, 20—2], and torn by violent civil strife [5, 28—9]. It had offered a firm resistance to the encroachment of the Lydian kings, and had made a treaty on favourable terms with them [1, 17—22]; as also with Kyros [1, 143, 169]. Induced by Aristagoras to join the Ionic revolt, it had at the period at which this book opens sustained a crushing defeat [5, 120]. After its depopulation, here narrated, it was restored on the defeat of the Persians at Mykale [B.C. 479] and joined the confederacy of Delos, but revolted from the Athenian supremacy in B.C. 445 [Thucyd. 1, 115], and maintained its independence for some time [Thucyd. 8, 25, 84]. It never however quite recovered its old position as the chief city of the Ionians.

MILTIADES,

(1) Son of Kypselos, cc. 34—7, 41, 103.

Miltiades was a wealthy Athenian, who traced his descent to Philaeos, son of Ajax. He was the first tyrant of the Chersonese, invited there by the Dolonki, and has been confounded with his kinsman the son of Kimon (q. v.). The connexion will be best seen by the following table:

Kypselos = *Mother* = Stesagoras I.
  Miltiades I.   Kimon
    Stesagoras II.   Miltiades II. = Hegespyle
        Metiochos   Kimon II.

Though other authorities make the pedigree thus

Kypselos = (*Mother*) = Kimon I.
  Miltiades I.   Stesagoras   Miltiades II.

(2) Miltiades, son of Kimon, cc. 34, 38—40, 103, 108—9, 133—7, 140.

The career of Miltiades, the victor at Marathon, is contained for the most part in this book. His advice to the Ionian princes to cut off Darios in Skythia by breaking the bridge over the Danube is narrated in 4, 137, cp. Nep. 3. Nepos (*Milt.* 1) appears to confuse the circumstances of his going to the Chersonese with those of the elder Miltiades; but his narrative, like that of Herodotos, seems to assume that Miltiades was not 'tyrant of the Chersonese' in the ordinary sense, that is, he was not an independent prince, but held the Chersonese for Athens.

The affair at Paros, which cost Miltiades his reputation and life, is very differently related by Nepos. According to him Miltiades was despatched with the regular Athenian fleet of 70 sail to punish the islanders who had helped the foreigners, and was very largely successful in reducing them to obedience; only landed troops on Paros upon the refusal of the Parians to submit; and retreated from a false alarm of the coming of the Persian fleet. Nepos' authority was apparently Ephoros, and it may be urged that he indicates relationship between Athens and the islands which did not exist at this time; and confuses the proceedings of Miltiades with those of Themistokles after the battle of Salamis, ten years later, who even then acted not as an Athenian, but in the name of all Greece. The account in Herodotos is not so intelligible, though it may be the truer, and is not less discreditable to the Athenian Demos than to Miltiades. For the former, after giving Miltiades free licence to maraud, the 70 ships being necessarily for that purpose, and after taking no step to recall him during his 26 days' siege of Paros, condemned him not really for his attempt upon Paros, but for its failure, as Pausanias (1, 32, 4) also seems to think when he says 'Miltiades came to his end for having *failed to take* Paros '(Πάρου ἁμαρτόντι) and being thereupon brought to trial by the Athenians.'

MOLOSSI, c. 127.

A tribe of Epeiros, inhabiting the district between the R. Aoos and the Ambrakian Gulf.

MYKALE, c. 16.

A high headland in Karia, opposite Samos, and between Ephesos and Miletos; on it was a temple and precinct of Poseidon common to the Ionians, and called the Panionium [1, 148]. It is a high ridge terminating Mt Mesogis; its extreme point was called Trogylium (*S. Maria*). In the narrow channel between it and Samos, and on the shore at its foot, was fought the final battle in B.C. 479 between the Greeks and Persians [9, 98—104].

MYKONOS, c. 118.

A small island of the Cyclades only two miles E. of Delos, about 10 m. long by 6 m. broad. It was colonised from Athens, but was in great part a barren rock.

MYRINAEANS, c. 140.

Inhabitants of Myrina (or -inna) a town on the Western coast of Lemnos (q. v.). Mod. *Castro*. There was another town of the same name in Mysia.

MYSON, c. 126.

King of Sikyon and grandfather of Kleisthenes (q. v.). About B.C. 648 he won the chariot race at Olympia, and built a treasure-house in the Altis in commemoration of it [Pausan. 6, 19, 1].

MYSIANS, the, c. 28.

The inhabitants of Mysia, the N.W. district of Asia Minor on the Propontis and Aegean. The part bordering on the Propontis was called Mysia Minor, and its eastern boundary separating it from Bithynia seems to have varied, for Xenophon speaks of Kios as in Mysia [*Hell.* 1, 4, 7]. The southern part bordering on Lydia with its capital Pergamum was called Lydia Major; while the N.W. district bordering on the Hellespont and Aegean was the Troas. It had numerous Hellenic colonies; but the native Mysians, a simple pastoral people, were connected in race with the Lydians and Karians [1, 171; 7, 74]. They were conquered by Kroesos [1, 28], and afterwards included by Darios in the second Satrapy [3, 90].

MYTILENEANS, cc. 5, 6.

The inhabitants of Mitylene, the chief town of the Island of Lesbos (q. v.). Its inhabitants were Aeolians [2, 178], and had been at war with the Athenians for the possession of Sigeium [5, 94].

MYUSIANS, c. 8.

The inhabitants of Myūs in Karia, one of the 12 Ionian cities [1, 142] situated on the S. of the Maeander, about 4 miles from the mouth. Its inhabitants were afterwards transferred to Miletos [Pausan. 7, 2, 7].

NAUPLIA, c. 77.

In Argolis, built on a rocky peninsula in the Argolic bay, connected by a narrow isthmus with the mainland. It was now the port of Argos, but had been once an independent town, said to have been originally founded by Egyptians [Paus. 4, 35, 2]. It came into the hands of Argos about the time of the 2nd Messenian War [B.C. 685—668].

NAXOS, c. 96. NAXIANS, the, c. 96.

The largest and wealthiest of the Cyclades [5, 28] formerly conquered by Peisistratos [1, 64]. It was the resistance made by the Naxians to the attempt of Aristagoras to restore their banished Oligarchs which led to the Ionian revolt [5, 30 sq.]. The inhabitants were Ionians, and afterwards revolted from Athens [Thucyd. 1, 98].

NIKODROMOS, cc. 88—91.

An Aeginetan, who appears to have been a popular leader in opposition to the oligarchs, who had managed to secure his banishment. His confederacy with Athens was doubtless on the understanding that they would establish democracy in Aegina; but we have no farther information concerning him.

NONAKRIS, c. 74.

A city in Arkadia, the next town west of Pheneos [Paus. 8, 17, 6]. It was one of a confederacy of three towns (τρίπολις) with Kallia and Dipoena [id. 8, 27, 4].

NOTHON, c. 100.

An Eretrian, father of Aeschines.

OEBARES, c. 33.

Satrap of Daskyleion (q. v.), and Son of Megabazus (q. v.).

OLORUS, cc. 39, 41.

A king of the Thracians, whose daughter Hegesipyla married

Miltiades. The father of Thucydides, also called Olorus, is supposed to have been descended from Miltiades, which would seem to be confirmed, if his biographer Marcellinus is right in stating that his mother's name was Hegesipyle.

OLYMPIAD, an, cc. 70, 105, 125.

A victory at the games at Olympia in Elis, held every 5th year. This was the great national festival of all Hellenes, at which none were allowed to contend except those of Hellenic descent.

ONOMASTOS, c. 127.

Son of Agaeos of Elis, one of the suitors of Agariste.

ORKOS, c. 86.

'An Oath', personified as the God of Oaths: who according to Hesiod [*Theog.* 231, *Op.* 802] was the son of Eris. See also an oracle in the Anthol. xiv. 72.

OROPOS, c. 160.

A town in the maritime plain of the river Asopos, on the borders of Boeotia and Attica. Geographically it belonged to Boeotia, but it was always a subject of contention between the Boeotians and Athenians, the latter of whom were often in possession of it. It exact site appears to be doubtful; and in fact the ancient site was changed by the Thebans, who removed the inhabitants to a distance of a mile farther from the sea [Diodor. Sic. 14, 17]. It was an important place to the Athenians, as, being opposite Euboea, it served as a port for landing corn and other merchandise brought from the North into the Euripos, and thence conveyed through Dekelea to Athens [Thucyd. 7, 28].

OTANES, c. 43.

Son of Pharnaspis. His daughter married the false Smerdis and detected his imposture [3, 67—9]. He then led the conspiracy by which the Pretender was put to death [3, 70—2], and endeavoured to persuade the Magi to establish a democracy instead of the monarchy, [3, 80]. Failing to carry his point he secured independence for himself and family [3, 83]; but still served under Darios in restoring Syloson to Samos, and in inflicting vengeance on that island [3, 141—9].

PACTYA, c. 36.

A city on the Thrakian Chersonese, on the coast of the Propontis, about 4½ miles from Kardia, on the opposite side of the Isthmus. Alkibiades retired there after his second disgrace on the disaster at Notium [B.C. 407]. Nepos, *Alcib.* 7.

PAEOS, c. 127.

An Arkadian town of the district Azenia (q. v.). Its site is unknown; but Pausanias (8, 23, 9), mentions the ruins of Paos, a village in the district of Kleitor.

PAN, cc. 105—6.

An Arkadian pastoral deity, haunting the mountains, and hunting the game in that wild district. But his worship extended to other parts of Greece, and was often connected with that of Dionysos [Pausan. 2, 24, 6]. His temples were generally caves in mountain sides. Thus the Korykian cave at Delphi was sacred to him [Paus. 10, 32, 7] and also a grotto on the north side of the Akropolis at Athens. Pausanias [1, 28, 4] tells us that Pan met Pheidippides on Mt Parthenium, the skirts of which are crossed near Tegea by the road to Sparta, and that at the spot a temple to Pan was built [8, 54, 7].

PANIONIUM, c. 7.

A name given to a sacred enclosure and temple of Poseidon on Mykale, as being the central place of worship for the 12 Ionian cities of Asia Minor. [See *Ionians*.] They wished it also to be regarded as common to all Ionians wherever they might be living; but no other city except Smyrna [Pausan. 7, 5, 1] took advantage of the privilege, the term Ionian conveying some notion of inferiority [1, 143—4, 148, 170].

PANITES, c. 51.

A Messenian, who apparently lived at Sparta.

PAROS, cc. 133—6. **PARIANS**, the, cc. 133—5. **PARIAN**, a, c. 134.

One of the larger of the Cyclades, 6 miles west of Naxos, celebrated for its quarries of white marble. It was said to have been originally inhabited by Kretans and Arkadians, and afterwards to have been colonised by Ionians. Before the Ionian revolt the Parians were in a high state of prosperity, and enjoyed so high a reputation for equity, that they were called in to settle the civil disputes in Miletos [5, 28]. They do not however appear to have been eager to take part against Persia. Besides the case of the single ship, mentioned in c. 133 as having served in the Persian fleet at Marathon, they kept cautiously aloof from Salamis [8, 67], and after the Greek victory secured themselves by promptly sending Themistokles the indemnity he was demanding from the medizers [8, 112].

PEDASA, the people of, c. 20.

Pedasa was a town in Karia, between Miletos and Halikarnassos, some way from the coast. The people for some time resisted Harpagos, the general of Kyros, fortifying themselves on a hill called Lide [1, 175—6]. The Persians sustained a defeat near it in the course of the Ionian revolt [5, 121]. Its exact site is not known.

PEISISTRATOS, cc. 35, 102—7, 121. **PEISISTRATIDAE**, cc. 39, 94, 123.

Peisistratos, son of Hippokrates, and a relation on his mother's side of Solon, became tyrant of Athens in the usual way by taking the lead of the poorer classes against the two factions of the wealthier men which were led respectively by Lykurgos and Megakles. By pretending that his life had been attempted by his oligarchical enemies he obtained a grant of a body-guard, and thus was enabled to possess himself of absolute power. He was born about 612 B.C. His first usurpation took place in B.C. 560 and his death in B.C. 527. Of the 33 years between his usurpation and death he only was in actual possession of power for 17 years, being twice driven out by Megakles and the Alkmaeonidae, and twice restored [1, 59—63]. The dates of these two exiles are not certain; but we are told that the second lasted 10 years [1, 62]. He made no changes in the laws, and ruled well and wisely [1, 59].

By the PEISISTRATIDAE is meant the sons of Peisistratos, and their families. Peisistratos left three legitimate sons, Hippias, Hipparchos, and Thessalos [Thucyd. 1, 20]. Of these Hippias succeeded his father, and Hipparchos was associated with him in some way not clearly defined. [See *Hipparchos. Hippias.*] The third son Thessalos never appears to have had any part in the government; and it is said he was a philosopher, an ardent admirer of equality, and lived as a private person in great repute at Athens [Diodor. Sic. x. fr.]. Hipparchos was assassinated in B.C. 514, and Hippias and all the family of the Peisistratidae expelled in B.C. 510 [5, 65]. Peisistratos had during his lifetime secured possession of Sigeium, to which the people of Mitylene had long laid claim, and had placed his natural son Hegesistratos in command of it [5, 94—5]. The Peisistratidae accordingly retired thither. Grote vol. 4 p. 43.

PELASGI, the, cc. 136—140.

The inhabitants of a great part of Greece before the coming of the Hellenes, to whom however they seem to have been allied. Their name survived in that of Pelasgic Zeus [*Il.* 16, 233] and Pelasgic Argos [*Il.* 2, 681; 24, 437], and in a tribe living near Larissa in Thessaly [*Il.* 2, 840, 843]. Herodotos mentions remnants of them at Kreston, Skylake, and Plakia (in Makedonia and Mysia), and says that their language was barbarous [1, 57] and that it was those of them that remained in Attica that got the credit of being Hellenes [2, 51]. We know nothing of their language except that Larissa and Argos are said to be Pelasgic for 'fortress' and 'plain'. They settled in rich plains and were great builders and reclaimers of land. Thucydides speaks of them as the prevailing race in Greece before the Hellenic name superseded them [1, 3, 2]; and asserts that some of them migrated to Etruria and returned afterwards to Chalkidike, as well as to Lemnos and Athens [5, 109, 3; cp. Pausan. 7, 2, 2]; though Pausanias was told that they came from Sicily [1, 28, 3]. A reminiscence of the Pelasgic building at Athens was the place called τὸ Πελασγικὸν beneath the Akropolis [5, 64; Thucyd. 2, 16]. Their name was also connected with a part of Arkadia [Paus. 8, 1, 6] and with Pylos in Messenia [id. 4, 36, 1].

PELOPONNESE, the, cc. 86, 127.

The Peloponnese ('Island of Pelops') was not a name known in Homeric times. In the *Iliad* the only name given to the whole seems to be Argos, for Ephyra in Elis is spoken of as being ἐν μυχῷ Ἄργεος

(6, 152); in *Odyss.* 4, 173 'Argos' refers to Lakonia; and in 3, 251 the Peloponnese is called Ἄργος Ἀχαιϊκὸν (as opposed to Pelasgic Argos). In *Il.* 1, 269 ἀπίη, 'distant land', has been regarded by some as a territorial name of the Peloponnese. This appellation was subsequent to the Dorian invasion, and was referred by the Greeks to the wealth and power of Pelops son of Tantalos. In Homeric times it appears to have been the most important part of Greece, and Agamemnon king of Mykenae or Argos, is 'king of men' and natural leader of the united Greek army; as we find the same supremacy still acknowledged as belonging not to Argos, but to Sparta. Of the six divisions in it,—Elis, Messenia, Lakonia, Argolis, Achaia, Arkadia, the four first had been occupied by the Dorians; Achaia had been, as its name implies, the place of retreat of the defeated Achaeans; and Arkadia, from its mountainous situation, defied the invading Dorians and retained to a great extent its ancient inhabitants.

PERIALLA, c. 66.

A Pythia, or prophetess at Delphi, who was bribed by Kleomenes.

PERINTHOS, c. 33.

In Thrace on the N. shore of the Propontis. The name of the town was afterwards changed to Heraklea (mod. *Erekli*): it was a colony from Samos. It had been at war with the Paeonians, and only yielded to Megabazos after a gallant struggle [5, 1—2].

PERKALOS, c. 65.

Daughter of Chilo, and wife of Demaratos.

PERSEUS, cc. 53—4.

Son of Zeus and Danae [7, 61], a hero of Argos [7, 150]. The Persians used the similarity of his name with their own to induce the Argives to believe that they were united in descent [7, 150—2]. They, however, at other times asserted that Perseus was not a Greek, but an Assyrian who settled in Greece [c. 54]. Herodotos, on the other hand, found a temple to his honour at Chemmis [Panopolis] in Egypt, where Greek ceremonies were performed and Greek games celebrated; and he was informed by the priests that Perseus was an Egyptian who migrated to Greece [2, 91].

PERSIANS, the, cc. 4, 7, 9—10, 13, 18—9, 21, 24—5, 28—33, 42—3, 45, 49, 54, 59, 98—9, 101, 112—3, 115.

PERSIAN, the (sc. Darios), cc. 94, 100.

PERSIAN LANGUAGE, the, cc. 29, 100.

The Persians were an Aryan mountain tribe led down about B.C. 559 by Kyros to attack Astyages, king of the Medes, in his capital Ecbatana. They next conquered Lydia B.C. 546, and Babylonia B.C. 538. It was the conquest of Lydia that brought them into contact with the Greeks of Asia and the islands, who had become tributary to Kroesos of Lydia, and were now compelled to take the same position towards Kyros. The Medes were not exterminated, but coalesced with their conquerors; though for a long while all the chief posts and commands were held by Persians [see *Medes* and *Datis*], and Herodotos regards the Persian soldiers as by far the best part of the army of the Great King [9, 68].

PHAENIPPOS, c. 121.

Father of Kallias (q. v.), one of a very wealthy family at Athens.

PHALERUM, c. 116.

The ancient harbour of Athens, somewhat nearer to the city than the Peiraeus, on the E. side of the bay of Phalerum. The Peiraeus did not supersede it until the time of Perikles, when a road was made across the salt-marsh which intervened between the city and the peninsula (once it was said an island) of Peiraeus.

PHASIS, c. 84.

A river flowing through Kolchis into the Euxine (mod. *Rion* or *Fachs*), and regarded at one time as the boundary of Europe and Asia.

PHEIDIPPIDES, cc. 105—6.

A hemerodromos, or swift courier of Athens, sent to Sparta to announce the arrival of the Persians at Marathon.

PHEIDON, c. 127.

King of Argos, the 6th (or 10th according to others) of the Temenidae, or branch of the Herakleidae reigning at Argos, descended from Temenos. He made himself despot, instead of constitutional king of Argos, and extended his supremacy over Phlius, Sikyon, Epidauros, Troezen and Aegina, and attempted to embrace in it the whole of the Peloponnese. His ambitious schemes were eventually defeated by the hostility of the Eleans, whom he tried to deprive of their presidency of the Olympic games. The period at which he lived is generally said to be the middle of the 8th century; but if the text of this passage of Herodotos is right, making him the father of a suitor of Agariste, it would bring down his date nearly a century. 'He first coined both 'silver and copper money in Aegina, and first established a scale of 'weights and measures, which, through his influence, became adopted 'throughout Peloponnesus, and acquired ultimately footing both in all 'Dorian states, and in Boeotia, Thessaly, Northern Hellas generally, 'and Makedonia, under the name of the Aeginetan scale.' (Grote.) The other scale, used generally by the Ionians was the Euboic, which stood to the Aeginetan as 5 : 6. Others hold that Pheidon's scale was called Aeginetan, not because he coined money in Aegina, but from the commercial importance of Aegina, in which it would be most frequently used.

PHENEOS, c. 74.

A city in Arkadia, a few miles east of Kleitor. It was in a valley watered by a river formed by a union of two mountain streams called

Olbios and Oroanios [Paus. 8, 14, 3] which frequently caused dangerous inundations. It was said to have been the home of Evander [Verg. *Aen.* 8, 165].

PHIGALEA, c. 83.

A town in the S.W. of Arkadia, close to the Messenian frontier. On Mt Kobilium, about 4 miles to the N.E. of the town, stood the celebrated temple of Apollo Epikurius (ἐπικούριος), built by the same architect as the Parthenon at Athens, and now standing almost entire. It is called the temple of *Bassae*, because of the glen (Doric βᾶσσαι) in which it stands [Paus. 8, 41, 7—8].

PHILAGROS, c. 101.

Son of Kyneas, of Eretria, who betrayed Eretria to the Persians.

PHILAIOS, c. 35.

Son of Ajax, who with his brother Eurysakes was said to have surrendered Salamis to the Athenians, and settled in Brauron in Attica, whence the deme Philaïdae. Plut. *Sol.* 10.

PHOEBEUM, c. 61.

A precinct sacred to Apollo near Therapna (q. v.), in which stood a temple of the Dioscuri [Pausan. 3, 20, 2].

PHOENIKIA, c. 17.

PHOENIKIANS, the, cc. 3, 14, 25, 28, 33, 40—1, 47, 104.

The Phoenikians appear in this book chiefly as supplying ships and sailors to the Great King. They inhabited the North of Palestine, from which they had sent out colonies, originally for commercial purposes, to Kypros, Libya, and Europe. Herodotos says that they came to Palestine from the borders of the Persian Gulf (1, 1). They were a Semitic people, following, like others of the race, the practice of circumcision [2, 104]. They continued in the time of Xerxes and his successors to be the source from which the Persian kings drew their chief naval strength [see 7, 89, Thucyd. 1, 16, 100; 8, 46, 81]. They were also skilful engineers [7, 23, 34].

PHOENIX, c. 47,

Was, according to the myth, son of Hagenor (or Poseidon) and Libya, and gave his name to the Phoenikians. He was also brother of Kadmos, and father of Europa. [*Il.* 14, 321. Pausan. 7, 4, 1.]

PHOKAEANS, the, cc. 8, 11—12, 17. PHOKAEA, c. 17.

The inhabitants of Phokaea, a city of Lydia on the mouth of the Hermos and a colony from Phokis [1, 80, 142]. They were great mariners and explorers, having first opened up to the Greeks the shores of the Adriatic, and reached Spain, passing through the strait of the Pillars of Herakles and visiting Tartessus [1, 163]. When Harpalos began to besiege their city, a large number of them sailed away to

Corsica, and after being defeated in a battle with the Carthaginians, sailed to Rhegium and thence established the colony of Velia [1, 163—7].

PHOKIANS, the, c. 34.

The inhabitants of Phokis, a considerable district bounded on the S. by the Gulf of Korinth, and by Doris and Eastern Lokris on the N. It contained the range of Parnassos and the sacred city of Delphi. The towns in it lay mostly in the valley of the Kephissos, and the people were a mixed race of Achaeans and Aeolians.

PHRYNICHOS, c. 21.

An Athenian tragic poet, about 12 years senior to Aeschylos. He is said to have introduced many improvements both into the composition and representation of tragedies, but we have no means of knowing with accuracy what they were. In the list of his tragedies given by Suidas we find another, the Πέρσαι, which must have been on some incident in the Persian wars; and from Plutarch we learn (*Them.* 5) that in his last tragedy, the *Phoenissae*, which was a play on the repulse of the Persians at Salamis, Themistokles was his Choragus. The play for which he was fined at Athens was the Μιλήτου ἅλωσις.

PLATAEANS, the, cc. 108, 111, 113.

The inhabitants of Plataea, a town in Boeotia on the northern slopes of Kithaeron, its territory being separated from that of Thebes by the river Asopos. Its distance from Thebes is 6½ miles. Naturally it would have been a member of the Boeotian League; but it was always at enmity with Thebes, and had put itself about B.C. 501 under the protection of Athens [3, 108. Thucyd. 3, 68]. The action of the Plataeans, recorded in this book, of sending the 1000 men unasked to Marathon was always remembered gratefully at Athens, and when in after years the Thebans took the town, its inhabitants were admitted to citizenship in Attica [B.C. 429. Thucyd. 5, 32].

POLICHNE, c. 26.

A city in the Island of Chios of unknown site. There were five other towns of the name, in Lakonia, Messenia, Sicily, Krete, and the Troad.

POLYKRITOS, cc. 50, 73,

Of Aegina, father of Krios (q. v.).

POLYNEIKES, c. 52,

Son of Oedipus and Iokaste, one of the seven heroes who fell in their attack upon Thebes. He and his brother Eteokles slew each other in the struggle.

PONTOS, the, cc. 5, 26.

The Euxine (mod. *Black Sea*). See *Euxine*.

Prienians, the, c. 8.

People of Priene, an Ionian town in Karia near Miletos, which, with Myus, used the same dialect as Miletos [1, 142]; it had fallen with the other Ionian towns first under the power of the king of Lydia [1, 15] and then under that of the Persians [1, 161].

Prokles, c. 52.

Fifth in descent from Herakles, and ancestor of the junior Royal family at Sparta, his twin brother Eurysthenes being the ancestor of the other.

Prokonnesos, c. 33.

An island in the Propontis, now called *Marmora*. It had a town of the same name colonised from Miletos [4, 14].

Pythia, the, cc. 34, 36, 52, 66, 75, 77, 86, 123, 135, 139.

The prophetess of the temple of Apollo at Delphi, so called from the ancient name of Delphi (Πυθώ). She was generally a young girl of the lower class, selected by certain families at Delphi. Seated on the sacred tripod, over a hole from which rose a subterranean gas, she gave out the replies which the priest (*προφήτης*) reduced to writing, generally in hexameters. It was important, as all people and states applied to the oracle to settle international as well as private difficulties, that she should be incorrupt; yet there are other instances besides that narrated in c. 66, in which she was bribed [see 5, 63, 90].

Pythia, c. 122.

The Pythian games, held in a hippodrome near Delphi (Πυθώ), originally in the spring at the end of every eighth year, but from B.C. 631 at the end of every fourth year, the third of the Olympiad.

Pythii, c. 57.

Four officers at Sparta, two nominated by each king, to act as envoys (*θεόπροποι*) to Delphi. They lived with the kings at the public charge and are called by Xenophon their *σύσκηνοι*, *R. L.* 15. Cicero *de Div.* 1, 95.

Pythogenes, c. 23.

Brother of Skythes king of Zankle.

Rhegium, c. 23.

A Greek colony of Chalkidians and Messenians at the extreme S. of Italy on the straits of Messina (mod. *Reggio*), founded about B.C. 668.

Rhenaea, c. 97.

Rhenaea (mod. *Megali-Deli*), an island separated from Delos by a very narrow channel, now less than half a mile in breadth, and in ancient times probably much less: for Nikias is said to have brought a wooden bridge cut out at Athens, which he had laid across in a single night, for the members of his chorus to cross by, at the time of the festival [Plut. *Nic.*]. The little rocky island of Delos, on which the temple stood, was almost excluded from human uses, all persons for instance dying in Delos being properly buried in Rhenaea, although at times this law was neglected [Thucyd. 1, 13, 7; 8, 2].

SAKAE, c. 113. A Skythian people in Tibet [1, 153; 3, 93; 9, 31].

SAMOS, cc. 13—4, 25, 95. SAMIANS, the, cc. 8, 13—4, 22—5.

Samos is a considerable island off the coast of Karia. In it were three works, which Herodotos calls the greatest in all Greece, a great tunnel under a hill, seven furlongs long and eight feet high and broad; a mole in the sea going round the harbour; and a temple of Herè larger than any other Greek temple [3, 60]. The island was also rich from great pottery works. From about B.C. 535 to B.C. 522 under Polykrates it obtained a powerful navy, and dominion over several of the neighbouring islands. Polykrates wished to form an Ionian confederacy with Samos as its centre [3, 39—43, 112]. Soon after his death the disputes about the succession made it tributary to Persia [3, 120—5].

SAMOTHRAKE, c. 47.

'Thrakian Samos', so called to distinguish it from the larger Samos, is a small island opposite the mouth of the Hebros, consisting almost entirely of a great volcanic crater, Mt Saŏke (5,500 ft.). Its inhabitants possessed also a tract of land on the opposite shore from Doriskos to Lissos, protected by a line of fortresses [7, 59, 108]. The people were Pelasgians, though Pausanius says that it was also colonized from the larger Samos [7, 4, 3]. They practised a mystic worship called *τὰ καβείρων ὄργια*, connected with the Korybantes [2, 51]; and were apparently also noted for their skill with the javelin [8, 90].

SARDIS, cc. 1, 4, 30, 42, 101, 125.

The capital of Lydia (q. v.). It stood on the northern slope of Mt Tmolos, and on either bank of the river Pactolos. Though almost an open town, it had a strong and almost impregnable citadel. After its capture by Kyros [1, 84], it became the seat of the Satrapy of Asia Minor, and sometimes the residence of the Great King. It was burnt by the Ionians assisted by the Athenians in B.C. 500, who however were obliged to retreat without attempting the citadel [5, 101, 105].

SARDO, c. 2.

Sardinia, which Herodotos calls the 'largest of all islands', is in fact slightly larger than Sicily, and therefore the largest island known at his time. It was in early times occupied by the Karthaginians, who, according to tradition, found inhabitants who had already come from Libya. Though the Greeks often contemplated colonizing it, no Greek settlement was ever made there. Its fertility was often described by subsequent writers, especially by the Sicilian Timaeos [Polyb. 1, 79, sq.].

SELYBRIA, c. 33.

Selybria (or -ymbria) a town on the Thracian Chersonese (mod. *Silivri*), upon the coast of the Propontis; a colony from Megara.

SEPEIA, see **HESIPEIA**.

SIKELI, c. 23.

The native inhabitants of Sicily, as opposed to the Sikeliotae or Greek settlers in Sicily. They were said to have come from Italy, and to have been preceded by Sikani from Spain [Diod. Sic. 5, 2—6. Thucyd. 6, 2].

SIKELIA, cc. 17, 22—4.

Sicily at this time was held by three different nationalities. These were the native Sikeli or Sikani, mostly in the centre; the Greek settlers, generally on the Eastern and South Eastern coasts; and the Karthaginians who were establishing settlements in the West. The earliest of these Greek colonies was Naxos (Tauromenium) founded B.C. 735 from Chalkis: and the most powerful was Syracuse [B.C. 734], which had sent out other colonies in Sicily and claimed and generally obtained a supremacy among the Greek cities, which its leading position in opposing the incroachments of the Karthaginians helped to consolidate.

SIKYON, c. 126. **SIKYONIANS**, the, cc. 92, 129, 131.

A town on the N. of the Peloponnese in the valley of the Asopos. Its inhabitants were Dorians, mixed with the non-Dorian tribe inhabiting it before the Dorian conquest. When the dynasty of the Orthagoridae came to an end with Kleisthenes, mentioned in these chapters, the Dorians appear to have got possession of power, and from that time Sikyon, though remaining independent, acted generally under the direction of Sparta in war [8, 71; 9, 28].

SIRITAN, a, c. 127.

A native of Siris, an Italian town on the river Siris, half way between Sybaris and Tarentum, said to have been founded by Trojans, but colonized in the 7th century B.C. by Ionians from Kolophon. This Ionian colonization appears to be the ground on which Themistokles claimed it as belonging to Athens [8, 62].

SKAPTESYLE, c. 46.

'The dug-out wood' was a name given to a district and town in Thrace [*Scaptensula* Lucr. 6, 810], from the mining excavations in or near it, in the gold-bearing district of Mt Pangaeum. It was the place of Thucydides' exile and death. [For the form of the word see Notes on the Text.]

SKIDROS, c. 21.

An Italian town near Sybaris, of which it was a colony. Its exact site is not known, but some ruins at the village of *Sapri* are conjectured to mark its position.

SKOPADAE, c. 127.

A wealthy family of Krannon (q. v.) in Thessaly. Kritias, the leader of the thirty tyrants, prays in an elegiac poem for 'the wealth of the 'Skopadae, the nobleness of Kimon, and the victories of the Spartan 'Agesilaos' [Plut. *Cim.* 10]. See also *Diaktorides*.

SKYTHAE, cc. 40—1, 84.

The Skythians, who inhabited the country north of the Danube and the Euxine, including the Tauric Chersonese, or Crimea, and extending to an unknown distance to the north. For the names and customs of some of the numerous tribes of these barbarians see 4, 99—110.

SKYTHES, cc. 23—4.

King of Zankle (*Messina*) in Sicily.

SMINDYRIDES, c. 127.

A man of Sybaris, son of Hippokrates, and one of the suitors of Agariste.

SOPHANES, c. 92.

An Athenian of the deme Dekelea, who slew the Eginetan champion Eurybates. His peculiar 'anchor' shield is described afterwards, which he used at the battle of Plataea [9, 74]. Pausanias says that he was one of the two generals of the Athenians in the expedition into the interior of Thrace made from the recent colony of Amphipolis, and fell in battle near Drabeskos [Paus. 1, 29, 6, cp. Thucyd. 1, 100, 3].

SPARTA, cc. 49, 51, 61, 65—6, 70, 72, 74—5, 81, 84—6, 106.

SPARTANS, cc. 50—2, 56, 58—60, 63, 65—6, 71, 74—7, 82, 84—6, 104.

Sparta, at this period, had no fortifications; and its public buildings in the time of Thucydides were insignificant compared with those of Athens [Thucyd. 1, 10]. But its citizens were all trained soldiers, and its situation in the valley of Eurotas, surrounded by mountains, was strong enough to be defended. The Spartans, properly so called, were ὁμοῖοι or peers, the full Spartan citizens, who were at this period about 8000 in number [7, 234]. They governed Lakonia and Messenia, the other inhabitants being either perioeki or helots, and were regarded as the leading Dorian State, and as possessing a kind of hegemony in all Greece. They had long been at enmity with Argos, principally on account of the disputed territory of Kynuria, the chief town of which was Thyrea [1, 82; Thucyd. 5, 41, Pausan. 2, 38, 5; 10, 9, 12].

STESAGORAS, c. 103.

Father of Kimon, and grandfather of Miltiades (q. v.).

STESILÄOS, c. 114.

One of the ten generals of Athens, and killed at Marathon.

STYMPHALIAN LAKE, the, c. 76.

A lake in the north of Arkadia, just south of the range of Kyllene, and near the town of Stymphalos, the waters of which run off into a chasm on the South shore. It is now called *Zaraka*.

STYREANS, the, c. 107.

The inhabitants of Styra on the S.W. coast of Euboea, just north of Karystos, and opposite the promontory of Attica called Kynosoura. They were said to be Dryopeans [8, 46], but themselves claimed an Attic descent. They supplied two ships at Salamis [8, 1], and troops at Plataea [9, 28]. See also *Aegileia*.

STYX, c. 74.

A waterfall in the Aroanian mountains in the N. of Arkadia, near the town of Nonakris. It was and still is regarded with superstitious reverence: its waters were believed to be poisonous, and incapable of being held by any vessel, and to descend into the infernal regions. Herodotos here describes it as nearly dry, and Pausanias describes it as dripping (στάζει) over the highest precipice he ever saw [8, 17, 6]. It is now called *Nauronero* 'the black water', and forms, when it reaches the valley, a stream which joins the Krathis [Homer *Il.* 15, 37; 8, 369].

SUNIUM, cc. 87, 90, 115.

A promontory forming the southern point of Attica. On it was a fortified hamlet or deme, of which the ruins still remain, and a famous temple of Athenè, the columns of which have given the modern name of the promontory, *Kolónnes*. It was once a rich and flourishing place, owing to the neighbouring silver mines at Laurium, but had decayed in the time of Cicero [*ad Att.* 13, 10].

SUSA, cc. 20, 30, 119.

The capital of Susiana, situated on the Choaspes (*Kirkhah*). The province is bounded on the E. by Persis, on the W. by Assyria, on the N. by Media, and on the S. by the Persian Gulf. From the time of Kyros it was one of the principal Royal residences [1, 188; 5, 49].

SYBARIS, cc. 21, 127. SYBARITES, the, cc. 21, 127.

A colony of the Achaeans and Troizenians in S. Italy between the rivers Krathis and Sybaris. In the sixth century B.C. the Troizenians had been driven out, and the town had risen to great power and opulence, the luxury of its citizens having become proverbial. Its fall was brought about by the refusal of the Krotonians to deliver the fugitive oligarchs driven out from Sybaris, when Telys made himself tyrant. The Sybarites proclaimed war with Krotona, were conquered and their city was taken. The Krotonians determined to utterly destroy it, and, in order to do so, turned the river Krathis over the ruins B.C. 510 [5, 44—5. Diodor. Sic. 12, 9—10].

SYLŎSŌN, cc. 13, 25.

Son of Aeakes, and brother of Polykrates of Samos, by whom he was driven into exile [3, 39]. He made a friend of Darios in Egypt, then serving in the army of Cambyses, by the gift of a cloak; and upon Darios becoming king he went to Susa and begged Darios to restore him to Samos. Darios thereupon sent Otanes and an army to accomplish that object [3, 139—141]. Polykrates was now dead [B.C. 522], and had left the government of the island in the hands of Maeandrios; who willingly left it at the bidding of the Persians. But being exasperated by an act of treachery of Charilaos brother of Maeandrios, they devastated the island and then handed it over to Syloson, who ruled there apparently for the rest of his life, as a tributary of the Persian king, and was succeeded by his son Aeakes [c. 13].

TAMYNAE, c. 101.

A town in the territory of Eretria in Euboea, of unknown site, sacred to Apollo [Strab. 10, 1, 10]. The MSS. gives *Temanos*.

TEGEA, cc. 72, 105.

A town on a plain enclosed by mountains in the S.E. of Arkadia, 10 miles S. of Mantinea. It was on the high-road to Sparta, with which it had been once at war and had long resisted its power [1, 65; 9, 35]. But at length about B.C. 500 Tegea submitted, and, though retaining its autonomy, remained closely allied with Sparta.

TEIANS, the, c. 8.

The people of Teos, a town in Lydia, on the isthmus connecting Mt Mimas with the mainland, with two good harbours. Like the other Ionian towns it was captured by the Persians; but a number of its inhabitants, rather than submit to the Persian rule, sailed away to Thrace and founded the city of Abdera [1, 168].

TENEDOS, cc. 31, 40.

A small island about 10 miles in circumference, and about 5 miles from the coast of the Troad. It had been taken by the Persians [c. 31], but at the end of the Persian wars attached itself to Athens. Its inhabitants were Aeolians; and its constitution was much celebrated.

THASOS, c. 47.

Son of Agenor, who led the Phoenikian colony which occupied Thasos, and from whom the island derived its name. Stephanos of Byzantium who writes the name Thassos, says he was ten generations before Herakles. He was brother of Kadmos [4, 147].

THASOS, cc. 28, 44, 47. THASIANS, the, cc. 46—8.

Thasos is an island in the N. of the Aegean between 3 and 4 miles from the plain of the river Nestos in Thrace. It had valuable gold mines, from which it was sometimes called Chryse [2, 44]. It had been originally settled by Phoenikians, who worked the mines in it, and in the opposite district of the continent [see *Skaptesyle*]; but it was colonized again by the Parians [about B.C. 720], who found in it a Thrakian tribe called Saians, with whom they were continually at war [Thucyd. 4, 104]. This Greek colony became wealthy and powerful and obtained large possessions in the mining district of the opposite shore [Thucyd. 1, 100. Herod. 7, 118].

THEASIDES, c. 85.

A Spartan of influence, son of Leopropes.

THEBANS, the, cc. 108, 118.

Thebes was at this time the most powerful city of Boeotia. It stood on an elevation rising out of the valley of the Asopos, which was a spur of Mt Teumessos. The Thebans were believed to be a Phoenikian colony, led by Kadmos, and their citadel was called the Kadmeia. They had long been at enmity with Athens [5, 77], and, being resisted by Plataea in their claim to be supreme in Boeotia, had continually harassed that town, which accordingly put itself under the protection of Athens.

THERSANDER, c. 52.

A Spartan, ancestor of Argeia wife of Aristodemos [4, 147].

THESSALY, cc. 72, 74, 127.

Thessaly lies between Makedonia on the North, Epeiros on the West, and Phthiotis on the South. It is a great alluvial plain shut in by Mts. watered by one river system, that of the Peneos and its tributaries [7, 129], and was famous for its breed of horses [7, 196]. There was no central government, but there were several leagues of towns, such as that at Larissa, Pherae, Krannon, Pharsalos and others. For certain purposes however it was divided into four great districts,—Thessaliotis, Histiaeotis, Pelasgiotis, Phthiotis,—and there was an officer elected who had some authority over all alike, called a Tagos [Xen. *Hell.* 6, 1, 6—8]; and who had the power of summoning a federal army of all Thessaly. But this was as difficult to do as it was in the case of the Imperial army in the Holy Roman Empire, and Thessaly is seldom recorded to have acted as a united country. Before this period their part in Greek politics consisted almost entirely of constant wars with the Phokians, arising generally from frontier disputes [7, 176; 8, 27—9].

THRAKIA, cc. 33, 95. THRAKIANS, the, cc. 34, 39, 45.

Thrace is the district N. of Makedonia, and bounded on the East by the Euxine. Towards the N.W., the frontier between it and the Celtic tribes was undecided; but Herodotos regards the Danube as dividing it from Skythia [4, 99], and in Roman times the range of Mt Haemos separated it from Moesia. Herodotos mentions 18 distinct Thrakian tribes, and Strabo 22. 'The Thrakians', says Herodotos, 'are 'the most powerful people in the world, except of course the Indians;

'and if they had one head, and would co-operate, I believe that their 'match could not be found anywhere' [5, 3]. They had been subdued by Darios [4, 93], and Megabazos [5, 2], and afterwards served under Xerxes [7, 185].

THRASYLAOS, c. 114.

An Athenian, father of Stesilaos (q. v.).

THYREA, c. 76.

Chief town of Kynuria. It was burnt by the Athenians in B.C. 424 [Thucyd. 4, 57]. Its site is marked by a convent called *Luka*.

TIMO, cc. 134—5.

A priestess of the infernal Goddesses in Paros.

TIRYNS, cc. 76—7, 83.

A strongly fortified town on an isolated hill a few miles S.E. of Argos. Remains of the Cyclopian walls still exist; but it was denuded of its inhabitants, who were removed to Argos, as a punishment for assisting the Greeks against the Persians, contrary to the policy of Argos [B.C. 478. Pausan. 2, 17, 5; 5, 23, 3], and probably from direct hostility to Argos [c. 83].

TISAMENOS, c. 52.

A Spartan, father of Antesion, and grandfather of Argeia wife of Aristodemos [4, 147].

TISANDER, cc. 127—8.

An Athenian, father of Hippokleides.

TISIAS, c. 133.

A native of Paros, father of Lysagoras.

TITORMOS, c. 127.

An Aetolian athlete, renowned for his great strength, who lifted a heavier weight than Milo of Krotona.

TRAPEZUS, c. 127.

A small town in Arkadia, which was afterwards abandoned with some others to form Megalopolis. Some of its inhabitants upon refusing to move were killed, while others migrated to the shores of the Euxine, and there founded another Trapezūs (*Trebizond*) [Paus. 8, 27, 3—4].

TYRRHENIA, c. 22. TYRRHENIANS, the, c. 17.

Etruria was thus called by the Greeks. According to Herodotos it was colonized by Lydians, led by Tyrsenos son of Atys, king of Lydia [1, 94]: and this account was received almost universally by the Romans: hence, for instance, Vergil speaks of the *Lydius* Thybris [*Aen.* 2, 781, cp. 8, 479], though there seems good reason to doubt the fact. The people called themselves Rasenna (Dionys. 1, 30), and at one time extended their power as far north as the Alps, until forced south by the Celts. They early became a powerful naval people, and commanded the Mediterranean until forced from the sea by the growing Karthaginian power; and they were famous for their manufactories, especially for their working of iron.

XANTHIPPOS, cc. 131, 136.

Son of Ariphron, and father of Perikles. He afterwards led the Athenian squadron at Mykale [9, 114], and besieged and took Sestos [9, 114—120].

XERXES, c. 98.

Son of Darios and Atossa, and king of Persia B.C. 485—465.

ZAKYNTHOS, c. 70.

An island 8 miles from the W. coast of the Peloponnese (mod. *Zante*), about 23 miles long. Its chief town was a colony of Achaeans.

ZANKLE, c. 22, 24. ZANKLEANS, the, c. 23.

A Greek colony in Sicily, afterwards called Messene (mod. *Messina*). According to Thucydides [6, 4] it was founded from Cumae in Italy; according to others from Chalkis in Euboea, of which Cumae was a colony. It was after the Ionian revolt occupied by some Saurians [c. 22—4, Thucyd. 6, 4], who in their turn were conquered by Anaxilas, tyrant of Rhegium, who was a member of one of the Messenian families in that town. Accordingly the name of Zankle was changed to Messana; just as in Roman times we find it called Mamertina, when occupied by the Mamertines.

ZEUS *Lakedaemonios*, c. 56.
" *Uranios*, c. 56.
" *Herkeios*, c. 58.

Though Zeus, son of Kronos, is the supreme deity in Greek theology, the father of gods and men, and the supreme controller of all natural phenomena, and subject to nothing but fate, yet he is spoken of under various limiting appellations according to the view in which he is regarded. Thus we find two priests at Sparta one to conduct the worship of Zeus of Heaven, the other of Zeus of Lakedaemon, that is, Zeus regarded as lord of heaven, and the same god regarded in the narrower view as master and protector of Sparta. And he is Zeus Herkeios 'Zeus of the hearth' to the Spartan king, who had an altar dedicated to him in the court of his house as the supreme god of Sparta. So he is called Zeus Hellenius [9, 7] as supreme protector of all Hellas.

ZEUXIDEMOS, c. 71.

Son of Leotychides, who died in his father's lifetime [Pausan. 3, 7, 8].

INDEX TO THE NOTES.

[*The references are by page and line.*]

Printed in the United States
By Bookmasters